D0189996

About Island Press

Island Press is the only nonprofit organization in the United States whose principal purpose is the publication of books on environmental issues and natural resource management. We provide solutions-oriented information to professionals, public officials, business and community leaders, and concerned citizens who are shaping responses to environmental problems.

In 2000, Island Press celebrates its sixteenth anniversary as the leading provider of timely and practical books that take a multidisciplinary approach to critical environmental concerns. Our growing list of titles reflects our commitment to bringing the best of an expanding body of literature to the environmental community throughout North America and the world.

Support for Island Press is provided by The Jenifer Altman Foundation, The Bullitt Foundation, The Mary Flagler Cary Charitable Trust, The Nathan Cummings Foundation, The Geraldine R. Dodge Foundation, The Charles Engelhard Foundation, The Ford Foundation, The Vira I. Heinz Endowment, The W. Alton Jones Foundation, The John D. and Catherine T. MacArthur Foundation, The Andrew W. Mellon Foundation, The Charles Stewart Mott Foundation, The Curtis and Edith Munson Foundation, The National Fish and Wildlife Foundation, The National Science Foundation, The New-Land Foundation, The David and Lucile Packard Foundation, The Pew Charitable Trusts, The Surdna Foundation, The Winslow Foundation, and individual donors.

The Future of U.S. Ocean Policy

Choices for the New Century

The Future of U. S. Ocean Policy

Choices for the New Century

BILIANA CICIN-SAIN
AND ROBERT W. KNECHT

Center for the Study of Marine Policy
Graduate College of Marine Studies
University of Delaware

Island Press
WASHINGTON, D.C. • COVELO, CALIFORNIA

Library of Congress Cataloging-in-Publication Data

Cicin-Sain, Biliana.
The future of U.S. ocean policy : choices for the new century/
Biliana Cicin-Sain and Robert W. Knecht.
p. cm.
Includes bibliographical references and index.
ISBN 1-55963-675-0 (cloth). — ISBN 1-55963-676-9 (paper)
1. Marine resources development—Government policy—United States.
2. Marine resources conservation—Government policy—United States.
I. Knecht, Robert W. II. Title.
GC1020.C53 2000
333.91′ 6415′ 0973—dc21 99-37170
CIP

Printed on recycled, acid-free paper

Printed in Canada
10 9 8 7 6 5 4 3 2 1

Contents

List of Figures and Tables xi
Acknowledgments xiii

One
Introduction: A Time for Reassessment of U.S. Ocean Policy 1
Why This Book? 1
The Evolution of U.S. Ocean Policy 8
Recurrent Tensions in U.S. Ocean Policy 11
The Concept of Ocean Governance 13
The Governance of Ocean Resources and Ocean Space 16
The Intergovernmental Aspects of Ocean Governance 19
Designing a Strengthened Ocean Governance System: Choices for the New Century 25
Organization of the Book 27

Two
From the Founding of the Republic to the 1960s: The Background of American Ocean Policy 31
The Evolution of International and National Ocean Policy 31
The Role of the Ocean in the Historical Development of the United States 35
The Evolution of Resource Management Regimes for Oil and Gas, Fisheries, Wildlife, and Coastal Areas 38
The Politics of Managing Ocean and Coastal Resources: Separate "Iron Triangles" 48
The Beginning of Change: Expanded Governmental Attention to Ocean and Coastal Issues in the 1960s 49

Three
The 1970s: The Enactment of the Innovative Body of Marine Law 53
The Political Dynamics of Policy Initiation and Formulation 53
The National and International Context of the Early 1970s 57
Managing the Nation's Coasts 60
Protecting Marine Mammals 68

Promoting Fisheries: The Adoption of the 200-Mile Limit 77
Adding Environmental Considerations to
Offshore Oil and Gas Development 84
Other Major Laws Enacted During the 1970s 91
Characteristics of the Ocean Laws of the 1970s 95
The Politics of Enactment 100

Four
The 1970s to the 1990s: A Mixed Record of Success in Ocean Policy Implementation 103
The Political and Policy Context of the Period 103
Coastal Management: Successes and Shortcomings. 116
Fisheries Management: Fisheries Depletion Overshadows
Positive Institutional Changes 129
Marine Mammal Protection: Has this Single-Purpose Law
Been the Most Successful? 149
Offshore Oil and Gas Policy: Continued Policy Stalemate
in Most of the Country 168
Controlling Water Pollution: Unresolved Issues 176
Marine Protected Areas: Putting the Program on a
Stronger Footing 184
Taking Stock: Some Successes, Some Failures,
and No Overall Vision 198

Five
The Context of the Late 1990s: Challenges to U.S. Ocean Policy at the Dawn of the New Century 203
The Changed Context of the Late 1990s 203
Challenges to be Faced in the New Century 205
The Overall Ocean and Coastal Wealth of the Nation:
The Need for More Information 208
Navigating the U.S. and World Waters 210
Ocean Resource and Space Utilization Issues 216
Issues in Coastal Management 235

Six

The United States and the World: Regaining Leadership in International Ocean and Coastal Affairs 255

Ratifying the Law of the Sea Convention 256

Implementation of UNCED 262

Suggestions for Regaining U.S. Leadership 276

Seven

Today: Toward a New System of Ocean Governance 279

Basic Problems with the Existing System 279

A Conceptual Framework for Multiple-Use Ocean Governance 284

Needed Ocean Governance Improvements 287

Operationalizing Suggestions for Improvement 296

Some Experiences from Other Nations 305

Building a Coalition for Change 308

Concluding Remarks: Looking to the Future 327

Appendix 329

Notes 337

References 345

Glossary 375

Acronyms 379

Index 383

List of Figures and Tables

Figures

1.1 Map of U.S. Exclusive Economic Zone (EEZ) 2
1.2 Ocean Jurisdictions of U.S. State and Federal Governments 21
4.1 Map of States Participating in the Coastal Zone Management Program 128
4.2 Decision-Making Process for the Development of Fishery Management Plans Under the Magnuson Act 130
4.3 Regional Personnel Breakout, National Marine Fisheries Service 131
4.4 U.S. and Foreign Landings from the U.S. 200-Mile Zone, 1963–1989 144
4.5 Numbers of Stock Groups Classified by Their Status of Utilization for Stocks Under the Purview of National Marine Fisheries Service .. 146
4.6 National OCS Oil and Condensate Production, 1954–1989 170
4.7 National OCS Gas Production 1954–1989 171
4.8 Status of the Federal OCS Leasing Program 173
4.9 Map of National Estuary Program Sites 182
4.10 Map of U.S. Marine Sanctuary Sites 190
4.11 Map of U.S. National Estuarine Research Reserve Sites 197
7.1 Continuum of Policy Integration 292

Tables

1.1 Governmental Functions in the Ocean 15
1.2 Evolution of Government in the U.S. West 16
4.1 Incidence of "Divided" or "Unified" National Government, 1953–2001 106
4.2 State Coastal Zone Management Programs: Year of Federal Approval 118
4.3 Major Amendments to the Coastal Zone Management Act 119
4.4 Composition of the Regional Fishery Management Councils, 1985–1987 133
4.5 Major Amendments to the Fishery Conservation and Management Act 134
4.6 Major Amendments to the 1972 Marine Mammal Protection Act 153
4.7 Estimated Incidental Kill of Dolphins in the Tuna Purse-Seine Fishery in the Eastern Tropical Pacific Ocean, 1972–1995 165
4.8 Subsistence Harvest Levels for Northern Fur Seals in the Pribilof Islands, 1985–1995 166
4.9 California Sea Otter Population Counts by the Fish and Wildlife Service and the California Department of Fish and Game, 1982–1995 167
4.10 Sanctuaries in the U.S. Marine Sanctuaries Program 186
4.11 Estuaries in the National Estuarine Research Reserve System (NERRS) 192
5.1 Level of Economic Activity of Various Ocean and Coastal Sectors 209

5.2 Major Domestic Policy Challenges 247
6.1 United States Participation in Major Recent International Agreements Related to Oceans and Coasts 257
7.1 Features of National Ocean Governance Systems 286
7.2 U.S. National Ocean Policy in the late 1990s: The Political Landscape 311
7.3 A Summary of Needed Changes in National Ocean Policy 325

Acknowledgments

The significant contributions of three research assistants to this book must be acknowledged with special notice. Rosemarie Hinkel, Ampai Harakunarak, and Nigel Bradly contributed importantly to many aspects of the book's preparation: background research, citations, early drafts of some sections, preparation of tables, figures, chronology, glossary, and acronyms, and manuscript corrections and revisions. Their thorough, timely, and cheerful assistance is acknowledged with sincere thanks.

* * *

This book had a long gestation period. We began the work in the early 1980s at a time when the major ocean laws enacted by the United States in the 1970s were in the relatively early stages of implementation and both problems and opportunities were becoming apparent. This was the time, too, when we began our own professional and personal collaboration, and when Bob Knecht made his switch from the federal government to academia.

Writing in the early and mid-1980s, we were able to capture, we hope, through interviews with participants in the passage of a number of the ocean programs we cover in the book, some of the excitement characteristic of the apogee of ocean legislative activity in the 1970s. In the late 1990s, as we confront the complexity of solving our multiple-use ocean and coastal problems, we can lose sight of the fact that over a quarter of a century ago, the United States was truly a pioneer in ocean and coastal policy making.

For support of the work contained in the historical parts of the book (Chapters 1, 2, and 3), we sincerely thank the California Sea Grant College Program (and its director, James J. Sullivan) and the thorough research assistance of then University of California, Santa Barbara, graduate students Laura L. Manning, Phyllis M. Grifman, Lauren Holland, and Robert J. Wilder. Chapter 2 and 3 were written at the Rockefeller Foundation Conference and Study Center in Bellagio, Italy, whose gracious hospitality and inspiring milieu provided the perfect setting for starting a book of this nature.

Professor Jack H. Acher, University of Massachusetts, Boston, worked with us for a time as a collaborator on the book. To him we owe the analysis presented on the passage of the Outer Continental Shelf Lands

Act Amendments in Chapter 3. As well, much of the discussion of the impact of the Reagan administration comes from joint work with Professor Archer that we published, in part, in several co-authored journal articles during this period. Distance and diverging interests prevented us from continuing the collaboration, but we remain very grateful to Professor Archer for his contributions to this book and for his influence on the field of marine policy, expecially his work on the application of the public trust doctrine in the management of America's coasts and oceans.

To our university institution, the University of Delaware, and expecially to the dean of the Graduate College of Marine Studies, Carolyn A. Thoroughgood, we owe a special debt of gratitude for their unfailing support of our work, in the United States and around the world, on integrated coastal and ocean management. Special thanks, too, go to the Delaware Sea Grant Program for its support of a project on multiple-use ocean management, in 1991–1995, on which some of this work is based.

Especially warm thanks go to Catherine C. Johnston, center manager, Center for the Study of Marine Policy, whose efficiency, diligence, and unfailingly helpful manner keeps our research center operating effectively and allows us to write books such as this one. Dosoo Jang, past center chief research assistant, is thanked very sincerely for all his effective work on behalf of the center. Jorge A. Gutierrez's meticulous work on all the tables and figures is gratefully acknowledged, as is the manuscript assistance of Shannon Ferrell, Miriam Balgos, and Susan Bunsick.

Intellectually, we owe a substantial debt to our marine policy colleagues in the Ocean Governance Study Group (OGSG), a group of nearly forty ocean policy scholars in the United States and around the world, which we helped to found in 1991 with the intent of contributing policy analyses on how to make ocean and coastal policy more coherent and cohesive. Some of the ideas presented in the book, especially in Chapter 7, benefitted from intense discussions on national ocean policy at the various workshops and conferences sponsored by the OGSG. While we thank all members of the group, we give special acknowledgment to the members of the Steering Committee: Harry Scheiber, Jon M. Van Dyke, Jack H. Archer, Richard Delaney, David Caron, and M. Casey Jarman.

We also sincerely thank Nancy Foster, assistant administrator for the National Ocean Service, NOAA, for involving us in collaborative efforts on dialogues on national ocean policy during the 1998 Year of the Ocean, which provided a good opportunity for us to interact with the major national ocean and coastal interests and to better understand their perspectives.

We have benefitted from the careful reviews of the book by several individuals—Michael K. Orbach, Marc J. Hershman, Tim Eichenberg, John Twiss, and Jon M. Van Dyke. Their helpful comments were much appreciated and taken into account whenever possible. Any remaining problems or errors in the book are, of course, our sole responsibility.

Our thanks also go to Island Press and its excellent editorial and production staff. Our editor, Dan Sayre, provided just the right combination of firmness and understanding to ensure that the book emerged reasonably on schedule. His editorial comments considerably improved the readibility and flow of the book. On the production side, we acknowledge with sincere thanks the assistance of Christine McGowan in seeing to it that none of the endless pitfulls which typically occur at the end of a book publishing effort such as this were fatal.

Finally, we offer deepfelt thanks to the many ocean policy entrepreneurs in and out of government—in the federal agencies, Congress, ocean user-groups, the states, academia, the environmental community—who have taken chances and used their discretion to help move the United States toward more effective governance of its invaluable oceans and coasts. To them, this book is dedicated.

ONE

Introduction

A Time For Reassessment of U.S. Ocean Policy

The United States has one of the largest and probably the richest 200-mile ocean zone (formally, the Exclusive Economic Zone [EEZ]) of any nation in the world. This "wet America"—about equal in size to the terrestrial United States (see Figure 1.1)—is home to bountiful living and nonliving ocean resources: fisheries, marine mammals, minerals, and other energy resources. It is an area greatly valued by the American people, both for its many uses and for the awe it evokes. Rich fisheries lie off New England, the Pacific Northwest, and Alaska, and in the Gulf of Mexico; large offshore oil and gas deposits exist in the Gulf of Mexico and off California and Alaska; stunningly beautiful beaches line virtually all U.S. shores. And 95 percent of the trade that keeps the nation prosperous is carried on those oceans through ports such as New York-New Jersey, Los Angeles-Long Beach, Houston, and New Orleans. Marine transportation, commercial and recreational fishing, development of offshore oil and gas, swimming and beaching, protecting and viewing marine mammals, military operations, waste disposal, aesthetic enjoyment—these are among the many values that Americans seek to obtain from their ocean.

Why This Book?

It has been thirty-three years since Congress enacted legislation that focused unprecedented attention on the nation's coasts and oceans and led to the establishment of both a vice president–led Marine Sciences Council and the "blue ribbon" Commission on Marine Science, Engineering and Resources. The so-called Stratton Commission examined the U.S. stake in its oceans and coasts in a comprehensive manner, as had never been done before, and in its seminal 1969 report, *Our Nation and the Sea,* provided a blueprint for U.S. action to conserve and manage the bountiful resources and values found there.

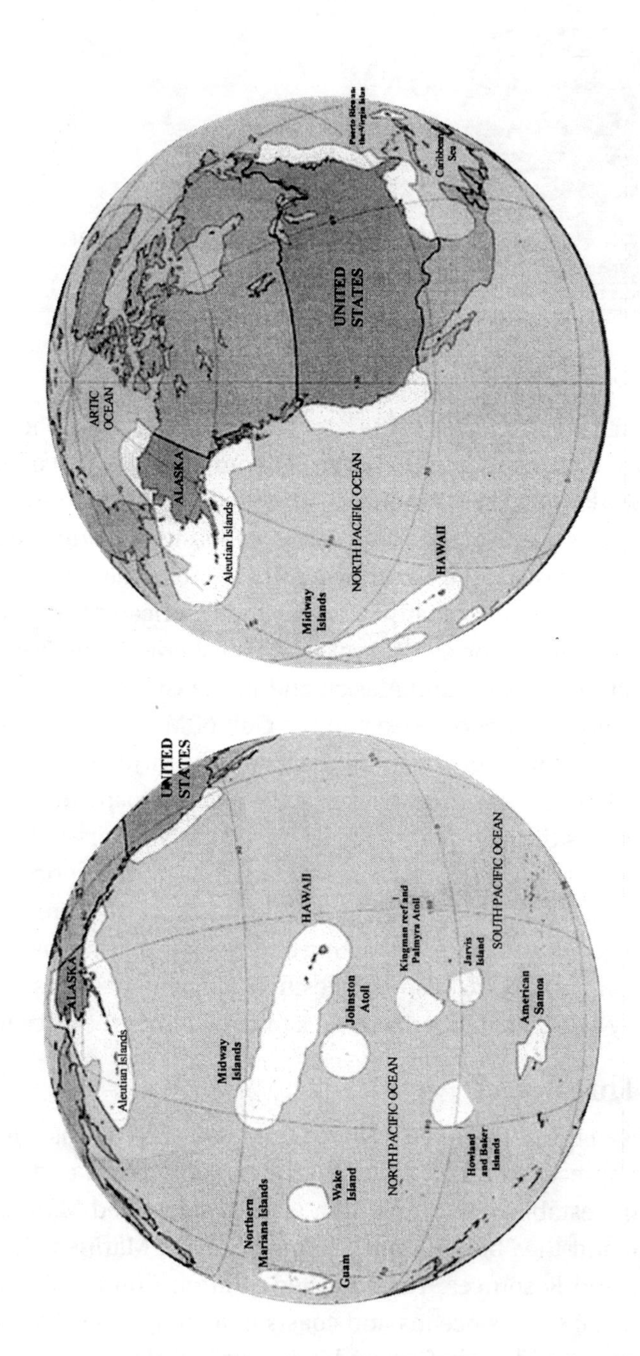

FIGURE 1.1.
Map of U.S. Exclusive Economic Zone (shown in white).
Courtesy of the U.S. Department of State.

No similar comprehensive assessment of U.S. ocean and coastal policy has since been done, and many people now recognize that it is sorely needed to assess the many changes that have taken place since the late 1960s as well as to consider possible future trends and challenges.

The ocean situation in the United States has changed dramatically since the 1969 Stratton report (Knecht, Cicin-Sain, and Foster 1998). The Stratton Commission's good work led directly to the establishment of the nation's ocean agency—the National Oceanic and Atmospheric Administration (NOAA) and to the enactment of innovative coastal zone management legislation. The decade immediately following the Stratton Commission's report saw a rise in environmental consciousness, the emergence of energy use and supply as a major issue, and many new ocean and coastal programs enacted into law—programs dealing with such ocean issues as marine mammals, ports and harbors, water quality, marine sanctuaries, ocean dumping, fisheries, and offshore oil and gas. The subsequent period also saw significant growth in populations in coastal areas, and an attendant rise in conflicts among various users of coastal resources and space. The offshore jurisdiction of the United States was transformed significantly during this time: In 1983, the nation asserted jurisdiction over the 200-mile EEZ, and in 1988 it declared a territorial sea of 12 miles offshore, following the international norms established by the 1982 Law of the Sea Convention. A substantial increase in the interest and capacity of the coastal states and territories to deal with coastal and, increasingly, ocean issues took place during this time, as did the capacity of the educational system in the marine natural sciences and social sciences.

The latter half of the thirty-year period since the Stratton Commission has seen a corresponding burst of activity at the global level. Growing concern, especially in scientific circles, focused on two emerging problems: (1) the prospect that human activities were beginning to change the world's climate and dangerously accelerate the loss of species and biological diversity, and (2) the realization that many societies were living unsustainably and that problems of environment and development were inextricably linked. Concern about these problems led to another seminal event: the United Nations Conference on Environment and Development (the Earth Summit) held in Rio de Janeiro in 1992. As did the activist decade of the 1970s in the United States, the decade of the 1990s has seen the signing of international agreements on climate change and on biodiversity, a comprehensive Law of the Sea Convention finally enter into force, and the development of substantial international programs that deal with integrated coastal management, land-based sources of marine pollution, and the protection and sustainable use of coral reefs. Nations around the world are now experimenting with methods of integrating and harmonizing the multiple uses of their oceans and coasts, and are attempting to operationalize the vision espoused by the oceans chapter of Agenda 21—that the governance of ocean and coastal areas must be "integrated in content and precautionary and anticipatory in ambit." Under Agenda 21, nations cannot solely rely on traditional approaches that govern only one resource or use at a time, but must also consider the effects of one resource or use on other resources, uses, and the environment (UNCED 1992).

The past fifteen years have also seen a fundamental transformation of the international relations regime with important implications for oceans—the end of the Cold War, the collapse of the Soviet Union, economic globalization, the growth of regional economic blocs, and the emergence of the World Trade Organization and of trade and environmental conflicts.

Major Crosscutting Issues Need Addressing in a U.S. Ocean Policy Reassessment

All of these changes, domestic and international, pose both problems and opportunities for the United States and call for a major reassessment of U.S. ocean policy. In particular, there is the need for assessment of crosscutting problems, problems that relate to more than one sector of ocean policy. Specific areas of ocean policy, such as fisheries, have been assessed periodically, but crosscutting issues have typically not been examined since the Stratton Commission report. The crosscutting problems are important and denote deficiencies in the underlying ocean governance regime, deficiencies that must be addressed if the United States is to achieve greater benefits from its offshore zones. Symptoms of such crosscutting problems include the following:

- The United States lacks a strategy for sustainable development of its offshore areas. Even though it declared an EEZ and expanded its territorial sea, it has done little to provide guidance for the governance of these vast ocean areas.
- Conflicts exist among users, among agencies, and between different levels of government over the use of ocean resources and space. Such conflicts have often gone unresolved, incurring significant costs.

- The U.S. approach to ocean governance has largely been through enactment of single-purpose ocean laws, which often neglect not only the effects of one resource or use on other resources or uses and on the environment, but also the cumulative impacts.
- In some cases (such as in offshore oil and gas policy), U.S. policy has oscillated between unmitigated development thrusts and the adoption of wholly conservationist approaches (such as the imposition of moratoria on new development), which prevents the attainment of sustainable development of its ocean.
- Many marine industries in this country are not faring well, in contrast to those in other countries. Examples include loss of shipping to other nations, declines in the fishing industry and the offshore oil and gas industry, and trade deficits in fishery products.
- The growth potential of newer marine and coastal economic activities, such as marine aquaculture and biotechnology, is hampered by the absence of appropriate management frameworks to properly encourage and guide development.
- Although many federal programs deal with the ocean, they tend to be fragmented and lack coherence. Few, if any, mechanisms exist for harmonizing and coordinating the actions of federal ocean agencies, in contrast to the situation in some other nations.
- There are significant problems in intergovernmental relations on ocean issues among federal, state, and local governments, with little real sharing of decision making and of revenues.
- Little attention has been paid to developing policy responses for longer-term issues, such as dealing with the effects of sea-level rise and increased storm frequency resulting from global climate change.
- There has been little nurturing of U.S. capacity for ocean governance—the joining of science and policy—bringing together the natural sciences that help us to understand physical and biological ocean processes; the marine social sciences, which lead to understanding of how humans both affect and are affected by ocean processes and activities; and the managerial capacity that is spread out in many national and state ocean agencies.
- Although the first to enact (in the 1970s) a very elaborate body of law dealing with almost all aspects of the ocean environment, and a major leader in the 1970s in the formulation of an

international constitution for the oceans (the Law of the Sea), the United States is no longer the international leader and pacesetter of ocean policy. It has, at times, been a reluctant participant in some of the major international fora dealing with oceans; whereas other national governments (e.g., Canada, Australia, Korea) have, in recent years, been expending more effort and resources in assessing their ocean and coastal interests and governmental structures and in crafting more integrated national ocean policies.

Our Perspective in Writing This Book

We have come to the writing of this book from the vantage point of more than twenty-five years of writing about, teaching, and participating in national ocean policy, mostly as policy watchers but at times as policy participants. Robert Knecht, in particular, participated directly in national ocean and coastal policy in the period 1972–1980 as the first director of the U.S. coastal management program, as director of new programs in ocean thermal energy conversion and deep seabed minerals, and as a U.S. participant in the Law of the Sea negotiations. Biliana Cicin-Sain served in the federal government as a policy analyst in 1977–1979, working especially on state-federal relations on fisheries issues. In the past two decades, we have served as consultants on ocean and coastal policy to federal agencies such as NOAA, to a number of state governments and ocean/coastal interest groups, to international entities such as the Intergovernmental Oceanographic Commission (which is part of UNESCO), and the World Bank, and have been participants in the international ocean negotiations leading to and following the 1992 Earth Summit.

Our orientation in this book is rooted in political science and public administration. We are concerned with assessing how well the United States is governing its ocean space and resources and with what effects and consequences. In ocean policy, as in any other policy area, we are very aware of the close nexus that exists between "policy" and "politics"—that is, how a particular set of political forces and circumstances is responsible for creating particular policies, and how particular policies can, in turn, change the political dynamics of a particular issue, leading, eventually, to a subsequent change in policy. That is why, in this book, we examine in detail the political forces that were responsible for the enactment of the body of U.S. ocean and coastal law that exists today. If this body of law and policy is to be improved, those same political forces (or some variant of them) must be aligned in support of ocean policy reform.

Our studies of U.S. ocean policy over time have suggested an expression that encapsulates our view of the current policy status: "U.S. ocean policy today is less than the sum of its parts." That is, while the nation has made great strides in certain sectors of ocean policy (e.g., in the control of point sources of marine pollution, the protection of marine mammals, and the establishment of coastal management), overall, the separate parts of the policy don't fit well together. There are instances of conflicting, overlapping, or duplicative policies, and there is no vision of how the various parts may be harmonized, and of how overall guidance and principles may be developed to more effectively govern the offshore domain.

This syndrome of being "less than the sum of its parts" is rooted, in large part, in structural problems in the ocean governance regime, or in insufficient elaboration of the rights and obligations that various parties (federal, state governments, stakeholders) have in coastal and ocean areas. By "regime," we mean the institutional arrangements governing the use of a particular set of resources and/or areas that determine the expectations of the users and their rights and duties as well as the goals that the collective activities seek to achieve. Resource regimes and their behavior have been examined in detail by a number of authors, especially Oran Young (1982). The creation of separate, largely independent management regimes for various resources and uses that are rightfully part of a larger, interconnected system is bound to cause stresses and conflict.

As a result of the body of single-purpose ocean and coastal laws enacted in the 1970s, the ocean governance structure is generally based not on area but rather on the promotion, management, or control of specific ocean resources such as oil and gas or fisheries. The challenge that the nation faces in the twenty-first century is moving from this "first-generation" system of ocean governance—single-use and resource-based—to a "second generation" based on the notion of multiple-use management within designated ocean and coastal areas.

Our reassessment of U.S. ocean policy has benefited from the work of other ocean policy scholars, many of them involved in the Ocean Governance Study Group, who over the years have provided trenchant analyses of the state, in whole or in part, of U.S. ocean policy (e.g., Mangone 1988; Miles 1992; Juda 1996; Juda and Burroughs 1990; Van Dyke 1992; Scheiber 1998; Archer et al. 1994; Hershman 1996; Hershman et al. 1999; Wilder 1998), as well as from recent analyses conducted by the National Research Council (NRC 1997), federal ocean agencies during the 1998 Year of the Ocean (YOTO 1998), the Heinz Center (Heinz Center 1998), and through the National Dialogues carried out by NOAA and other partners (e.g., Knecht, Cicin-Sain, and Foster 1998; Cicin-Sain, Knecht, and Foster 1999) in 1998–1999. An added impetus for completing this work was provided by the congressional consideration, in 1998, of the Oceans Act, which would have created a high-level commission, similar to the Stratton Commission, to create an overall strategy for United States action. While the Oceans Act passed both houses of Congress, it did not become law because differences in the House and Senate versions could not be resolved in a timely fashion (Archer 1999). Nevertheless, consideration of the Oceans Act (which could become law in subsequent Congresses) has rekindled discussion and interest in national ocean policy reassessment among the major interest groups, academics, and the public.

The Intended Audience

This book is intended for all those interested in the oceans and coastal areas of the United States and how public policy is set with regard to their use and conservation. Staff of federal and state environmental and resource agencies with missions that include coasts and oceans will find the book directly relevant to their interests and their programs. Nongovernmental organizations and ocean and coastal user groups (commercial and recreational fishers, recreational boaters, beach-goers, environmental and conservation groups, etc.) will also find, we hope, that the information in the book relates to their work and how they go about it. Staffs and members of Congress or state legislative bodies especially interested in national ocean policy, either broadly or in particular sectors (offshore oil and gas, fisheries, marine mammals, marine protected areas, etc.), and others with roles in the policy formulation process should be assisted in their work by the information and history collected here. Finally, we believe that academics in the field of marine policy will want to consider this book as a possible teaching tool.

The Evolution of U.S. Ocean Policy

In writing this book, we have found it a challenge to include a sufficient discussion of history to make the present situation understandable without overburdening the reader with too much information. And yet, since much of the current situation is a product of recent history, we must dwell on it because it both explains the kind of policy that we have today and, in part, constrains future actions. To explain how we unravel the historical thread of national ocean policy, we must first discuss major periods and recurrent tensions.

Major Periods in the Development of U.S. Ocean Policy

From the founding of the Republic to World War II, a relatively stable system of ocean policy prevailed. The United States favored freedom of navigation over the world's oceans for its naval and commercial vessels, territorial seas of minimum breadth (3 miles), and beneficial exploitation of the ocean's fisheries resources. In contrast, the period from 1945 to the present has been characterized by almost constant change. On the domestic front, this period can be divided roughly into four parts: 1945 to 1970, the 1970s, the 1980s, and the 1990s.

The first period, 1945–1970, was dominated by the struggle between the federal government and the coastal states over jurisdiction of the ocean areas adjacent to the shoreline. By the late 1940s, it had become clear that a substantial amount of oil and gas was contained below the seafloor and that large revenues could be obtained from its extraction. This issue was resolved—temporarily at least—with the enactment, in 1953, of the Submerged Lands Act and the Outer Continental Shelf Lands Act, and by a series of Supreme Court decisions that generally limited the states' jurisdiction and ownership of offshore resources to the 3-mile territorial sea, while establishing federal authority over the outer continental shelf (beyond 3 miles).[1]

The second period, the 1970s, was a time of active ocean lawmaking. Indeed, most of the law and policy that now make up the domestic legal framework for the oceans was put in place during this time. After a period of national studies and reports, and related, in part, to the move to explore "inner space" (the oceans) as a counterpoint to the post-*Sputnik* space program, Congress rather quickly enacted legislation on a wide range of ocean topics. Events of the period added the motives of environmental protection (in the early 1970s) and energy development (in the mid-1970s) to the goal of basic ocean exploration and development. Unfortunately, but understandably, almost all of this ocean legislation was single-purpose in nature, aimed at solving a specific problem (managing fish, protecting marine mammals and endangered species, fostering state coastal management programs, creating marine sanctuaries, developing offshore oil and gas resources, halting ocean dumping of sewage wastes, etc.). Few of the new laws paid more than lip service to the timely identification and effective resolution of possible conflicts with other ocean users.

The 1980s differs from the earlier "active" period in several important respects. Little new ocean legislation was enacted. This period also saw little action on the part of the executive branch, with the exception of the proclamation made by President Reagan in 1983 declaring U.S. control over an EEZ extending 200 miles offshore (U.S. EOP 1983) and the extension of the U.S. territorial sea in 1988 (U.S. Presidential Proclamation 5928, 1989). The focus was instead on implementation, rationalization, and retention of funding for existing programs. The Reagan administration engaged in a determined effort to shift funding obligations and, in some cases, managerial responsibilities from Washington to the states. Efforts were made at both national and state levels to reduce regulatory burdens and to simplify governmental processes. Similarly, the administration attempted to restore market forces as the principal regulator of the economy rather than rely on explicit government intervention. For the oceans, this meant a period of decreased activity—few new programs and less funding for existing programs.

In the 1990s, major changes took place in ocean and coastal policy, particularly at the international level. A major change to the Law of the Sea Convention, in the form of a new agreement revising the deep-seabed mining regime, made the convention acceptable to major industrialized nations such as the United States, which, in 1994, indicated its intention to seek Senate ratification of the treaty. In June 1992, the United Nations Conference on Environment and Development (the Earth Summit) brought forward recommendations for major changes in ocean and coastal management through Chapter 17 of Agenda 21 (the global blueprint on environment and development), and through the Convention on Biological Diversity and the Framework Convention on Climate Change (Cicin-Sain and Knecht 1993).

The 1990s saw a great deal of domestic experimentation in ocean, coastal, and estuarine management, especially at the local level, typically involving increased participation by stakeholders (private sector, environmental interests, academia, state and local governments) in decisions about ocean and coastal resources. The emergence of new management concepts (such as watershed management and ecosystem management) also marked this period. By the end of the 1990s, there emerged a growing consensus about the problems besetting single-sector management of the oceans, and the need to develop more integrated management approaches. Major developments at the end of the decade included initiatives to create a national commission to provide a comprehensive examination of ocean policy and to create a national ocean council (i.e., the Oceans Act of 1998 H.R. 3445 IH, 105th Congress, 2nd Session, March 12, 1998, Commission on Ocean Policy).

Recurrent Tensions in U.S. Ocean Policy[2]

The earlier historical overview of the evolution of American concern with the oceans and of the making of U.S. ocean policy suggests a number of recurrent tensions that have tended to underlie (and continue to underlie) thinking and practice in this area: U.S. interests as a coastal power versus as a maritime nation; internationalism versus unilateralism; federal versus state control over ocean resources; development versus environmental protection; private versus governmental role in resource development.

Sea Power versus Coastal State Power

Since the United States is one of the major naval/maritime nations in the world and also possesses some of the world's richest coastal oceans, there has always been a fundamental tension between asserting the maritime right of nations to roam the sea freely and allowing the coastal states the right to control access to and use of the near-shore ocean. However, starting with the Truman Proclamation of 1945 establishing U.S. control over resources in the continental shelf, and culminating with the Reagan proclamation of the EEZ in 1983, the United States has tended to emphasize its coastal state, rather than its maritime, interests. As we shall see, this trend continues in the 1970s through the 1990s, although throughout this period, the U.S. continues to use its naval forces to assert its interpretations of international law when the actions of other countries adversely affect its freedom of navigation.

Internationalism versus Unilateralism

There has also been tension over whether the United States should lead or follow in the shaping of new international law for the oceans. U.S. foreign policy has often experienced periodic swings between internationalist and isolationist impulses—with the end of the Vietnam War clearly marking a new period of isolationism. Traditionally, ocean policy has emphasized the internationalist dimension. Until 1945, the United States steadfastly supported high seas freedoms and narrow territorial seas and clearly was the leader in creating the new post–World War II international institutions such as the United Nations. Beginning in the

1940s, however, a tendency to act unilaterally vis-à-vis international ocean law has been more dominant. This is certainly as true of the 1945 Truman Proclamation as it is of the 200-mile fishery limit established by the Magnuson Fishery Conservation and Management Act in 1976, much to the opposition of State Department officials. The ultimate manifestation of the "go it alone" philosophy was the U.S. decision not to sign the 1982 Law of the Sea agreement after playing a very important role in shaping the negotiations over a nine-year period. The 1980s also saw the United States use unilateral economic sanctions against nations that acted in ways inconsistent with its environmental policy. In the mid-1990s, the Clinton administration declared its intention to seek ratification of the Law of the Sea treaty, but the pendulum has yet to shift to a more internationally oriented stance.

Federal Jurisdiction versus State Jurisdiction over Ocean Resources

Another recurrent and persistent tension focuses on how much power and authority over ocean resources and users the state and federal governments should wield respectively. State-federal controversies over marine resources reached a high point in the 1950s, with this question occupying a prominent position in the 1952 presidential election. The 1960s, however, saw a definite trend toward an expanded federal role in the management of ocean resources, a trend that was to culminate in the 1970s with the enactment of a number of ocean laws, some of which explicitly wrested management authority away from the states and vested it in the federal government. The 1980s and 1990s, on the other hand, have seen a swing back to states' control, with the states clearly signaling their intention to play a more significant role in the management of resources in federal waters.

Ocean Resources Development versus Environmental Protection

The tension between resource development and environmental protection began to mount at the end of the 1960s, following disasters such as the Santa Barbara oil spill. It clearly dominated much of the debate during the passage of the ocean legislation during the 1970s, with conservation and development groups each winning in separate battles. This tension was manifested at a number of other times during the earlier periods, particularly with regard to the management of fisheries and wildlife resources, and it recurs periodically, with conservationists mounting battles against overfishing and overhunting whenever signs of species depletion become evident. The development-environment tension played a significant role in the ocean controversies of the 1980s. Given the legislative gains made by both environmental and development interests in the 1970s, both camps acquired a variety of tools to block each other's actions well into the 1980s and 1990s, sometimes resulting in policy and programmatic paralysis. The concept of sustainable development, which attempts to reconcile both environment and development goals, entered the picture in the 1990s as a possible avenue for reducing this tension.

Government Role versus Private Role in Resource Development

The extent to which government should take an active role in the actual development of ocean resources and the extent to which it should guide and regulate industry activities is another recurring tension in policy making. In general, and in contrast to the practice of many other nations, the United States has preferred to rely primarily on market forces, the free enterprise system, and the development of nonburdensome legal and regulatory frameworks in its governance of ocean resources. There are perennial calls, however, for the government to take a more active role in the development of such resources as fisheries, oil and gas, or aquaculture. A case in point concerned the debate on the 1978 Outer Continental Shelf Lands Act Amendments when serious proposals were put forward for the federal government itself to actually conduct the offshore oil and gas program through the exploration phase and for the industry role to be restricted to the development and production phase.

To set the stage for the ocean governance challenges of the twenty-first century, it is important to understand the concept of ocean governance, how the needs for that governance have changed over time, and what factors need to be taken into account in designing an ocean governance regime.

The Concept of Ocean Governance

In our view, U.S. national ocean policy in its broadest sense is comprised of two parts: national interests and concerns related to the domestic ocean, and U.S. interests and concerns related to the global ocean. In both cases, there is a need for governance of ocean space and the resources contained therein; hence, ocean governance is a central concept in national ocean policy.

We use the term "ocean governance" to mean the architecture and makeup of the regime used to govern behavior, public and private, relative to an ocean area and the resources and activities contained therein. "Ocean management" means the process by which specific resources or areas are controlled to achieve desired objectives. "Policy" refers to a purposive course of action followed by government or nongovernmental actors in response to some set of perceived problems (Miles 1992). "National ocean policy," therefore, refers to a purposive course of action followed largely by governments in response to a set of perceived problems (or opportunities) related to the oceans and coasts.

In terms of coverage, we apply the term ocean governance to the entire band of ocean adjacent to the United States that is under some form of jurisdiction by this nation: the territorial sea, the EEZ, and that portion of the U.S. continental shelf that, on a geological basis, extends beyond the 200-mile EEZ. The fundamental goal of a system of ocean governance, in our view, is to maximize the long-term benefits to the public from the conservation and use of ocean resources and ocean space.

The Functions of Government in Ocean Areas

To provide these benefits, governments must perform several rather different kinds of functions on behalf of the people they serve. The major functions of government vis-à-vis the ocean are shown in Table 1.1. The federal government is responsible for the first three functions—international relations, national security, and interstate commerce—throughout the territorial sea and the EEZ. While the federal government and the coastal state governments share responsibilities for the last three functions—proprietarial, public trust, and regulatory—the states are generally responsible for them within state waters (0 to 3 miles offshore) and the federal government is responsible beyond that boundary (Knecht 1986).

The Growing Need for Ocean Governance

When, how, and why does the need for governance arise? In the case of primitive societies, the need for government first arose out of the needs for common defense and dispute resolution (Mair 1962). The need to adjudicate disputes authoritatively among individuals or groups that cannot be handled by existing institutions, such as the family, has

TABLE 1.1 Governmental Functions in the Ocean.

FUNCTIONAL AREA	OBJECTIVE
International Relations	Ensure consistency with international agreements of which the United States is a part.
National Security	Maintain the national defense.
Interstate Commerce	Protect free commerce between the states.
Proprietarial	As resource owner, secure maximum earnings for the public.
Regulatory	Protect the public welfare; prevent and mitigate conflicts.
Public Trust	Conserve renewable resources for future generations.

Source: Adapted from Knecht 1986.

traditionally been one of the major reasons for the creation of some form of government. One could speculate that the situation that characterizes the current management of marine areas—where conflicts increasingly cannot be handled by existing means—calls for the emergence of more authoritative conflict management mechanisms.

The evolution of governmental authority in the settlement of the western territories of the United States may also be instructive in the attempt to predict the path that the evolution of ocean governance may take. Put in the context of the early pioneer seeking mineral riches or a prime piece of fertile farmland, the following sequence (abstracted in Table 1.2) roughly describes the evolution of the need for government in a resource-rich area.

In the initial period of exploration and staking of claims, the role of government is limited to a support function of providing maps and keeping records. As the first conflicts occur, a peacekeeping governmental effort (police) is needed to keep order and safeguard property. As uses and users increase and multiple conflicts develop, the need for a more sophisticated form of government arises. As the need for services escalates and conflicts among users heighten, a scheme for general-purpose government that can develop allocation schemes and provide services takes root. Over time, the need to minimize the costs of growth and services, concomitantly, incorporates new planning tools in the governmental process (Cicin-Sain and Knecht 1985).

TABLE 1.2 Evolution of Government in the U.S. West.

PIONEER ACTIVITIES/NEEDS	RESPONSE BY GOVERNMENT
Exploration	Grant from the Crown
Discovery	Maps, resource surveys
Stake claim	Establish registry
First conflicts	Establish law enforcement
Multiple conflicts	Develop allocation, accommodation schemes
Need for services	General-purpose local government
Control growth, costs, congestion, pollution	Develop planning, zoning, environmental regulations

Source: Cicin-Sain and Knecht 1985.

The Governance of Ocean Resources and Ocean Space

This section defines forms of ocean governance and sets forth the kinds of issues related to the utilization of ocean resources and space that decision makers typically need to address.

Forms of Governance

In its most basic form, an ocean governance system (regime) can apply to the conduct of a single activity throughout an entire ocean zone (e.g., fishing in a given state's 3-mile zone), or it can apply to virtually all activities in a more restricted ocean area (e.g., the management regime created by and within a given marine sanctuary). The first type of ocean governance—single-purpose governance—is by far the most common. Except for marine protected areas, such as sanctuaries, very few other broader governance schemes currently exist. To our knowledge, no general-purpose governance regimes (i.e., similar to that used to govern municipalities or counties) have been applied to ocean areas (although portions of cities and counties do extend into the territorial sea).

The current ocean management framework resembles, in many respects, the management of human activities and land resources in suburban areas through a number of overlapping single-purpose special districts. Although these ocean management systems are not formal special districts (formal "units of government" endowed with such powers as the authority to tax and sell bonds), they do, in many ways, operate as special districts, such as water, fire, and park and recreation districts. They clearly have a "governmental character," operate primarily with public funds, tend to be unifunctional, and are set up to serve only certain areas.

As governance tools, special districts have traditionally been decried by some political scientists for their problems of fragmentation, myopia, overlapping jurisdictions, conflicts among units, lack of mechanisms for areawide planning, political invisibility, and low political accountability (Bollens and Schmandt 1965). On the other side of the argument, analysts have posited a preference for the fragmented and polycentric character of the American metropolis, claiming that it operates well through the bargaining and accommodation that takes place among the constituent autonomous government units (Warren 1966).

We agree with the latter viewpoint on the grounds that the essence of the American pluralist federal system is based on multiple points of decision that will always entail some measure of duplication, fragmentation, and overlap. But the polycentric nature of marine management, where no general-purpose government exists and where intersectoral connections are few or nonexistent, does not, in our view, provide a sufficient level of governance for conflict management or other emerging needs.

The Management of Ocean Spatial Use

As suggested in Table 1.2, governance systems evolve to meet new or changing needs. Another key aspect of such systems has to do with the extent to which exclusive uses involving fixed locations and specific geographical areas are involved. Decisions associated with the long-term (twenty to thirty years) commitment of a portion of ocean space to a particular development activity (say, ocean mining or an oil production platform) require a different governance framework than do decisions involving the amount of a certain type of fish that can be taken in a given area. In general, decisions to site one kind of activity in a particular ocean location will necessarily preclude other kinds of activities in that location. These "displacement" effects can be of major importance, for example, when valuable fishing areas are involved or sea lanes used for navigation are affected.

While in the past, the predominant uses of the oceans involved transient spatial use (surface navigation) and transient resource use (fishing), more use of ocean space is now fixed on a given place. At present, there are fixed oil platforms, fixed ocean dumpsites, artificial islands, and military restricted areas and safety zones. We also see areas of the seafloor being used to mount acoustic arrays and other devices to detect and locate underwater objects and to obtain scientific information on temperature, currents, and other oceanographic parameters. It seems likely that all of these types of uses will increase in the future along with the eventual designation and leasing of ocean mining sites.

One characteristic of most uses of fixed ocean space involves some form of connection with the mainland—an oil and gas pipeline, a fleet of ocean dumping barges, electrical power and/or communications cables, a supply line of goods and people. The ocean governance problem, therefore, involves not only the displacement and environmental issues concerned with the use of the offshore site, but also the displacement and environmental impacts of both the shoreland link and the requisite shoreland facilities.

In summary, a governance system dealing with a long-term use of a fixed portion of ocean space will confront these kinds of issues:

- Displacement of other uses by the ocean site itself and any associated safety zone
- Impact on other uses by any transportation/communications link to the shore
- Environmental impacts of the activity and its shoreside link
- Demand for shore facilities and possible displacement of existing users of shore facilities

The Management of Ocean Resource Use

The nature of the management issues raised in connection with ocean resource use depends, of course, on whether nonliving or living marine resources are involved. The types of issues confronted in managing the exploitation of nonliving ocean resources, such as oil and gas or ocean minerals, include the following:

- Efficiency of the mining or extraction processes
- Exercise of due diligence in proceeding with the development
- Effects of the exploitation activity on the surrounding environment and resources
- Demand for shoreside support facilities
- Impacts of any transportation links

- Rents and royalties for extraction of publicly owned resources
- To the extent that a significant commitment of ocean space is involved, the resulting displacements of other ocean uses

Management of the exploitation of living marine resources (fish, shellfish, seaweed, etc.) for consumptive purposes (commercial or recreational) will confront this set of issues:

- Determination of "allowable" harvest levels
- The equitable allocation of portions of the allowable harvest to competing interests
- Determination of permissible techniques to employ in the exploitation activity (fishing gear, kelp-harvesting gear, etc.)
- Protection of areas important to the supporting ecosystem (essential habitat, spawning areas, etc.)
- Impacts of the removal of the resource on other resources in the food web or ecosystem
- To the extent that a significant commitment of ocean space is involved (i.e., in the construction of an artificial reef for sport fishing), the resulting displacements of other ocean uses

Also of interest, of course, is the management of marine water quality, which will involve activities of the following kinds:

- Control of pollutants entering the water column from land-based sources, rivers, or estuaries, and by atmospheric deposition
- Ensuring adequate upland freshwater inflows into bays and estuaries
- Control of ocean dumping activities (solid wastes, chemicals, incineration, etc.)
- Maintenance of effective vessel-traffic separation lanes and other navigational safeguards to minimize risk of accidents and resulting spills

The Intergovernmental Aspects of Ocean Governance

One of the principal complexities in ocean governance is the multiplicity of governmental interests that become involved in ocean governance issues. This is a direct result of both the three-way division of jurisdiction over the coastal ocean—with federal, state, and local governments all having some governance role—and the fragmented, single-purpose approach to ocean management now being used in the United States. These realities bring various agencies at the federal, state, and local government levels into the process at one stage or another. Having a number of governmental participants is not necessarily a bad thing, except that in the existing legal and policy system, state and local representatives can find themselves well informed and concerned about the potential impacts of a proposed ocean project but without the power or authority to significantly influence the decision.

In seeking ways to improve this and similar situations, it is important to understand the various interests of different levels of government as they seek to affect the ocean governance process. The following discussion is influenced by both normative and pragmatic considerations. For example, what kind of a role should the coastal state play in various ocean-use decisions given the nature of the interests involved, and what roles are the states demanding and are they moving into a position where they will have the political power to achieve their wishes?

The Federal Role: Recently Dominant

To the extent that management took place, the coastal states governed ocean resource use for the first 163 years of the nation's history. The 1947 Supreme Court decision in *United States v. California* reversed this situation, temporarily as it turned out, and established the federal government as having jurisdiction over all ocean resources from the tidemark seaward. While the Submerged Lands Act of 1953 returned resource jurisdiction of the 3-mile territorial sea to the coastal states, a companion piece of legislation, the Outer Continental Shelf Lands Act, clearly established federal control over all of the resources of the seabed and continental shelf beyond the territorial sea. Federal control over fisheries resources in the zone 3 to 200 miles offshore was established in 1976 in the Magnuson Fishery Conservation and Management Act. In 1988, the U.S. territorial sea was expanded from 3 to 12 nautical miles by Presidential proclamation. Figure 1.2 depicts the current ocean jurisdictions of the state and federal governments.

Programs that manage or govern the oceans are widely dispersed throughout the federal government and, in general, are poorly coordinated. Major programs exist in the National Oceanic and Atmospheric Administration (NOAA) (which includes the National Marine Fisheries Service, the National Ocean Service, the coastal zone management program, the Sea Grant College Program); the Department of Interior

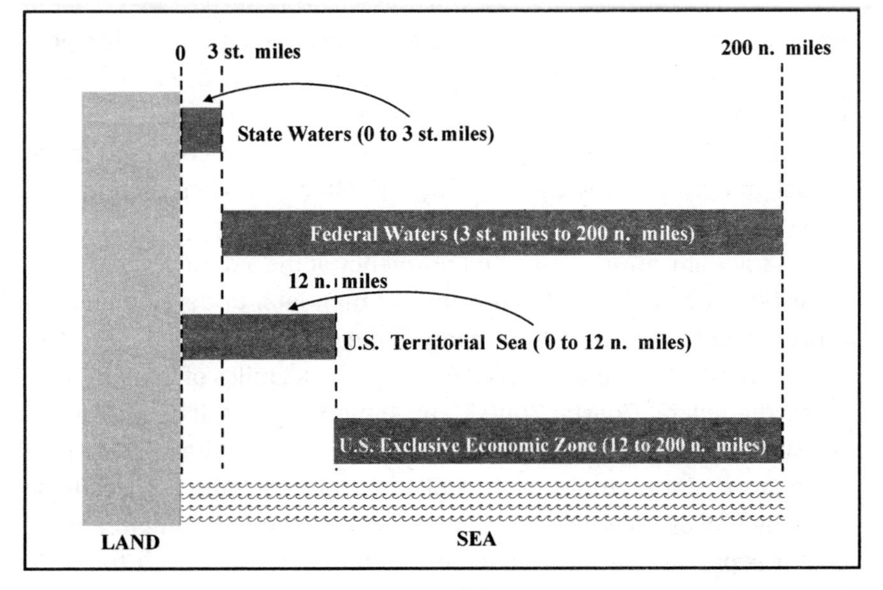

FIGURE 1.2.

Ocean Jurisdictions of U.S. State and Federal Governments.

State Waters: That portion of the U.S. territorial sea under the jurisdiction of the adjacent coastal state (California, Maine, etc.)—3 statute (geographic) miles* in breadth measured from the "baseline," generally the shoreline.

Federal Waters: The remaining portion of the U.S. territorial sea seaward from the 3-mile boundary of state waters out to a distance of 200 nautical miles from the baseline.

U.S. Territorial Sea: A zone of 12 nautical miles in breadth measured from the "baseline" seaward.

U.S. Exclusive Economic Zone (EEZ): A zone of 188 nautical miles in breadth measured from the outer (seaward) boundary of the U.S. territorial sea; thus the outer boundary of the U.S. Exclusive Economic Zone is 200 nautical miles from the baseline.

*However, the seaward jurisdictions of both Texas and Florida extend 3 marine leagues (about 10 statute miles) into the Gulf of Mexico as a result of Supreme Court decisions involving their historic boundaries (see *U.S. v. Louisiana,* 363, US1 [1960] and *U.S. v. Florida,* 363, US121 [1960].

(Minerals Management Service, U.S. Geological Survey); the Environmental Protection Agency; the Defense Department (U.S. Navy and the Army Corps of Engineers); and the Transportation Department (the Coast Guard and the Maritime Administration), to cite the main examples. Since the demise of the Marine Science Council in the early 1970s, there has been no mechanism in the executive branch to establish governmentwide policy for the oceans. Consequently, the ocean programs of the nation tend to be operated as independent entities with little central policy guidance from the administration save that conveyed in the annual budget process operated by the Office of Management and Budget (OMB).

This fractionization means that no agency at the federal level feels responsible (or has the legal authority and mandate) to develop and administer the kind of integrated management program for the broader U.S. ocean (the EEZ) that is called for in the first 3 miles of the territorial sea by the federal Coastal Zone Management Act. Yet, it appears to us that such a program is as urgently needed in much of the EEZ as in the coastal zone. For example, the lack of a federal focal point for ocean-use planning makes it very difficult for federal agencies wishing to initiate new or expanded ocean uses to coordinate their plans with the wide spectrum of existing ocean activities. The lack of a coordinated federal ocean program also causes problems for the coastal states as they attempt to reconcile their coastal management programs with activities occurring in or planned for the adjacent federal waters.

As has been pointed out by Hershman (1987), the federal government, in the name of protecting interstate commerce, also has the authority to manage "spatial" uses in the navigable waters of the Unites States since these uses potentially interfere with free navigation. He also pointed out, as we do, that no federal agency currently has "plenary" responsibility over ocean activities in general, noting that the "public interest" review performed by the U.S. Army Corps of Engineers may be the only example of this kind of general-purpose decision-making mechanism currently in place in the coastal and ocean arena.

The governance of ocean space and resources throughout the EEZ will be of key importance in the years ahead. In addition to promoting interstate commerce, the federal government, of course, must also ensure the security of the nation and must conduct the nation's foreign policy. Any ocean governance scheme, therefore, must take account of these important requirements. In the mid- and late 1970s, the federal government decided that increasing the supply of domestic energy was a matter of national concern and, consequently, endeavored to incorporate this objective into its ocean governance goals. Conserving fisheries and protecting marine mammals and endangered species have also become concerns for the federal government.

As ocean activities increase and resource development accelerates in the EEZ, ocean-use conflicts will continue to grow. Except for the mechanism for resolving conflicts in or near state coastal zones under the "federal consistency" regime of the Coastal Zone Management Act (discussed in Chapters 3 and 4), the only device currently available for the early identification of conflicts is the environmental impact statement (EIS) prepared pursuant to the National Environmental Policy Act. In our judgment, improved ocean coordination is clearly needed at the federal level. Specific options for achieving such coordination are discussed in Chapter 7.

The Coastal State: A Reemerging Participant

Coastal states have extensive interests in the governance of ocean resources and ocean space in the belt of ocean adjacent to their shores. Ocean-use decisions can have important economic, environmental, and social effects on a coastal state. Indeed, coastal states have once again become more involved in ocean and coastal resource management in a number of ways during the last two and a half decades. Most have developed and implemented management programs for their coastal zones; many have become involved in offshore oil planning as they responded to federal (or state) initiatives in this area. And the coastal states have become partners with the federal government and industry in the development of fish management plans.

Referring to the six functions of government listed in Table 1.1, coastal states have strong interests in how these governmental roles are performed in federal waters since major segments of their coastal populations can be affected. Here are four specific concerns of the coastal states in the management of "federal" ocean resources:

1. Coastal states are "home" to most users of these resources; hence coastal states and their citizens have major economic interest in the conservation and rational development of the federal ocean zone and its resources.

2. Coastal states have important "public trust" responsibilities and economic interests within their own jurisdictions to protect; hence, they have to be concerned with "spillovers" and other adverse environmental effects emanating from poorly regulated uses in adjacent federal waters that could damage state resources.

3. New federally encouraged ocean activities compete with other ocean activities of economic interest to the coastal states and their communities for access to valuable and limited coastal space for necessary shoreside facilities (ports, processing plants, etc.), often at the expense of the traditional users.

4. Coastal states are "owners" (proprietors) of the ocean resources lying immediately adjacent to the federal ocean zone and hence are concerned with "drainage" of state oil and gas fields and other actions that could reduce the value of state resources.

Currently, the coastal states have major governance responsibility only over the first 3 miles of ocean lying immediately adjacent to their shorelines. Despite the fact that they have important interests in uses and resources beyond this 3-mile limit, their ability to influence ocean decisions of the federal government in federal waters is very limited.

A coastal state's position on ocean governance issues will usually reflect the interests and concerns of its coastal population. People inhabiting the coastal zone are apt to see the ocean and its resources in more immediate and more tangible terms than does the inland population. Not surprisingly, these interests are likely to involve (1) economic consideration such as earning a living, (2) having a high-quality environment, and (3) having an influence over government decisions that affect the coastal population. Except perhaps in states dominated by a large offshore oil industry, such as Louisiana, this will mean that the coastal state, in its ocean governance role, will tend to be more concerned than the federal government about the well-being of fishing, both commercial and recreational; about the quality of the environment and the coastal tourism business dependent on it; and about coastal recreation and clean water and beaches. These ocean governance goals will be just as significant and meaningful to many coastal states and their communities as, for example, the goal of energy self-sufficiency might be for the federal government. In our judgment, therefore, an ocean governance scheme that permits the coastal states to give force and effect to such concerns is more likely to be stable and enduring than a system that does not routinely accommodate such interests.

Designing a Strengthened Ocean Governance System: Choices for the New Century

The review just presented and the early chapters in this book describe the way in which the present ocean governance regime in the United States has been constructed. The current set of arrangements evolved over an extended period of time, generally one at a time, in response to a particular need or problem. Certainly, no grand design was followed; the system merely grew as and when needed, each piece being added to the last with little consideration of the evolving whole. A growth spurt occurred during the 1970s when the bulk of the legislation making up the current regime was enacted. And this first-generation scheme, ad hoc as it is, served the nation reasonably well while the pressures to use marine resources were relatively light and the conflicts between uses were rare. Only recently, when the use of marine resources began to increase and the interactions between uses became problematic, have the weaknesses of the current scheme become obvious.

Creating a strengthened second-generation[3] scheme for the governance of U.S. ocean resources and U.S. ocean and coastal interests in general is going to require some hard choices early on in the next century. The tension between the conservation of ocean resources and their development and use will have to be accommodated; and it is hoped that the evolving concept of sustainability will be helpful in this regard. Appropriate apportionment of governance responsibilities between the national level and the state and local levels of government will need to be addressed explicitly, especially with regard to activities in the recently expanded portion of the territorial sea. Also, given the scale of many ocean activities, it is likely that at least some aspects of ocean governance will be centered at the regional (multistate) level: The Gulf of Mexico and New England are cases in point.

Another significant challenge in fashioning a second-generation ocean governance framework involves the achievement of a proper balance between domestic concerns and international obligations under the Law of the Sea Convention and other international ocean agreements; for example, balancing U.S. interests in the management of the resources of its EEZ with the U.S. obligation not to infringe on the rights of other nations to use its EEZ waters for navigation and other non-resource-related purposes. Also the need to ensure that the best science undergirds the policies and processes of a strengthened governance system will continue to grow in importance as the intensity of use grows and the economic stakes multiply. But neither can a new system ignore ethical considerations. We believe that resource management measures will increasingly be scrutinized for their capacity to exercise effective stewardship over renewable resources and for their sensitivity to the needs of future generations.

Some of the More Difficult Choices

As a prelude to more detailed discussion of these issues in Chapter 7, we list a few of the more difficult choices below:

1. The extent to which a coherent and authoritative national ocean and coastal policy is articulated similar, for example, to that issued by the Commonwealth of Australia in 1998.[4]

2. The content of such a national ocean and coastal policy statement relative to overarching national goals such as sustainable use of marine and coastal resources and priority for protection of productive ecosystems over extraction of nonrenewable resources.

3. The content of a national ocean policy statement relative to a number of pressing sectoral issues such as:

   - Will the United States take a firm stand with regard to rebuilding its fishery stocks?
   - Will protection of marine mammals shift toward a more balanced approach?
   - Will the nation begin to attack the non-point-source pollution problem and other land-based sources of marine pollution in an effective manner?
   - Does the United States need a national port policy to guide selective port and channel deepening programs?
   - What kinds of public policy reform are most likely to break the stalemate in the development of new offshore oil and gas resources?
   - Should the coastal states begin to play a stronger role in the recently expanded U.S. territorial sea?

Choices will also have to be made with regard to some key institutional questions. For example, can the United States hope to have a coherent and dynamic national policy for its coasts and oceans without some sort of institutional mechanism at the national level to better coordinate the activities and programs of the numerous federal agencies active in the ocean and coastal arena?

These are some of the choices facing the nation in the years ahead. We hope that this book will provide some help in making these choices. (See Chapter 7 for a more detailed discussion of these issues.)

Organization of the Book

Our discussion of the evolution of U.S. ocean policy, the current challenges it faces, and the alternatives for improving the underlying ocean governance regime is organized largely in a chronological fashion, emphasizing in particular the post-1960s period. We address, in some detail, the origin, evolution, and performance over time of programs designed to manage major ocean and coastal resources and uses of ocean and coastal space—programs in coastal management, fisheries management, marine mammal protection, oil and gas development, clean water, and marine and estuarine protected areas. In a less detailed fashion, we also address current challenges in other areas of national ocean policy: national security, marine transportation, marine aquaculture, coastal tourism, and marine biotechnology.

The reader should note that there are a number of important ocean policy topics that we are unable to address due to time and resource constraints: for example, U.S. maritime boundaries with its continental neighbors and in the Pacific Islands,[5] ocean exploration,[6] the development and status of the ocean sciences,[7] marine education (in both the natural and the social sciences).[8]

We begin with a brief historical background on the evolution of U.S. concerns with the oceans. Chapter 2, "From the Founding of the Republic to the 1960s," attempts to telescope several hundred years of U.S. ocean policy making into a succinct account. The goal is to show the origins of some of the ocean-issue areas that are still very much at the top of the national agenda.

Chapter 3, "The 1970s: The Enactment of the Innovative Body of Marine Law," focuses on a period of extraordinary congressional activity vis-à-vis the oceans that saw the establishment of the dozen ocean and coastal laws that comprise the current system of ocean governance. We explore, in some detail, the nexus between the politics of the time and the resulting policies—how many single-focus interest groups fought separate battles in Congress to achieve protection or promotion of their preferred part of the marine environment—from fisheries to marine mammals, to oil and gas development. Using, in part, original data sources, such as interviews with participants in this policy enactment process, we unravel the dynamics of enactment of new programs in coastal management, marine mammal management, fisheries management, oil and gas development, clean water, and marine and estuarine protected areas.

In Chapter 4, "The 1970s to the 1990s: A Mixed Record of Success in Ocean Policy Implementation," we tackle the very difficult question of how the ocean policies enacted in the 1970s were implemented. We also examine the nexus between policy and politics, focusing particular attention on the impact of Reagan administration policies on the ocean policies and programs enacted a few years earlier. To the extent allowed by the available data, we trace the implementation and achievements (or lack thereof) of the major ocean policies enacted in the 1970s: the 1972 Coastal Zone Management Act, the 1976 Fishery Conservation and Management Act, the 1972 Marine Mammal Protection Act, the 1978 Outer Continental Shelf Lands Act Amendments, the 1972 Clean Water Act, and the 1972 Marine Protection, Research, and Sanctuaries Act. Our examination of these major laws shows how U.S. national ocean policy can succeed in a particular part but fail in the "sum of the parts," that is, in the integration of these parts into a coherent whole.

Chapter 5, "The Context of the Late 1990s: Challenges to U.S. Ocean Policy at the Dawn of the New Century," focuses on today's changed context and on challenges that U.S. ocean policy will face in the next several years: navigating the U.S. and world waters (marine transportation and ports, U.S. defense interests); protecting and maintaining coastal infrastructure (especially to take advantage of growing coastal tourism); improving ocean resource management programs (in fisheries, marine mammal protection, oil and gas development, marine protected areas); developing appropriate policy frameworks for emergent uses (such as marine aquaculture, marine biotechnology, and marine tourism); and developing long-term policies for issues such as sea-level rise.

The challenge of regaining U.S. leadership on international ocean issues in the twenty-first century are laid out in Chapter 6, "The United States and the World." The chapter focuses, in particular, on three important global agreements where the nation is failing to have appropriate and timely input and influence—in the implementation of the 1982 Law of the Sea Convention, in the Convention on Biological Diversity, and in the Framework Convention on Climate Change.

In Chapter 7, "Toward a New System of Ocean Governance," we review major problems in the current system of ocean governance, outline the major features of a multiple-use ocean governance regime, discuss a series of possible governance improvements and reforms, and provide specific suggestions for operationalizing such improvements, particularly in relation to the creation of a national strategy for the U.S. EEZ, the creation of an interagency mechanism for coordinating national ocean policy, and the crafting of effective partnerships with the coastal states. We present a characterization of the political landscape in the ocean and coastal area at the end of the decade along with a discussion concerning the kind of coalition building that will likely be necessary for the improvement of U.S. national ocean policy.

TWO
From the Founding of the Republic to the 1960s
The Background of American Ocean Policy

To understand the current state of U.S. ocean policy and its likely directions in the future, we must first understand its antecedents. We must understand, as well, the major political forces that have shaped past and present ocean developments in the United States. This chapter unravels a number of different historical threads that set the stage for the major changes that ocean policy underwent in the 1970s. In a broad overview, we first sketch the evolution of international and national ocean policy. Next, we explore the role that the ocean has played in the historical development of the nation. We then discuss the evolution, up until the late 1960s, of management regimes for using and protecting important ocean and coastal resources—oil and gas, fisheries, wildlife resources, and the coastal zone—as well as the political dynamics that have characterized the management of these resources. Subsequently, we analyze the efforts made in the late 1960s to begin crafting a more concerted approach to national ocean policy.

The Evolution of International and National Ocean Policy

A relatively stable system of ocean law and policy prevailed from the founding of the American republic to the end of World War II, both internationally and within the United States. Internationally, the dominant paradigm governing the world's oceans was the notion of "freedom of the seas," pioneered by Hugo Grotius, a Dutch jurist, in 1608 (Van Deman Magoffin 1916). Grotius's famed treatise, *Mare Liberum,* persuasively argued that because the continents of the world were separated by the sea into a number of distinct land areas that could not develop without intercourse with the others, there is a "natural law" to the effect that the oceans were meant to be perpetually open for free trade and communications between nations (Grotius 1633). Moreover, because it was practically impossible to occupy, divide, or apportion the fluid and mobile sea, Grotius argued that the ocean could not be considered "property" and owned as such (Friedheim 1979).

The Freedom of the Seas Doctrine

The freedom of the seas doctrine was developed as a reaction to efforts made in the 1400s to divide up the oceans of the world. To settle disputes between the maritime powers of Spain and Portugal, in 1493, Pope Alexander VI decreed in a papal bull that all lands discovered or that would be discovered 100 leagues west of a line drawn between the Azores and the Cape Verde Islands belonged to Spain, while all lands east of that line (extending all the way around the globe) belonged to Portugal. Soon Spain was claiming exclusive navigation rights to the western Atlantic, Gulf of Mexico, and the Pacific, while Portugal was doing the same with regard to the south Atlantic and Indian Oceans (Mangone 1988). As the Dutch expanded their trade with the East Indies in the late 1500s and early 1600s, they ran into the exclusive navigation claims of the Portuguese and often became engaged in combat with Portuguese vessels. It is said that the Dutch East India Company, in defense of its trading routes, sponsored the international law work by Grotius, who worked as counsel for the company (Mangone 1988).

As coastal nations came to support the concept of freedom of the seas, they also saw the need to control the band of sea immediately adjacent to their shorelines. Without such a protective zone, armed ships could sail menacingly close to a nation's shore and interfere with its commerce and security. Thus arose the legal concept of the territorial sea. It quickly became generally accepted that nations could establish control over the area immediately adjacent to their coastlines. Within this zone, coastal states could exert their police powers, set customs, and control fishing. It was understood, however, that coastal nations were not to interfere with the "innocent passage" of foreign vessels in their territorial seas, an indication of the paramount importance that nations attached to the freedom-of-the-seas principle.

There was much discussion in international law circles as to how wide this belt should be. The prevailing notion at the time was that a coastal state might exercise jurisdiction over a marginal belt of sea in the same way it did over its land territory: namely, to the extent its forces allowed. Thus, in the late 1700s, 3 miles became the norm for the width of the territorial sea corresponding to the range of the most lethal weapon then available to coastal states: the cannon.

For the New World, the freedom-of-the-seas doctrine was of critical importance in resupplying the vulnerable English colonies at Jamestown and Plymouth. As the American colonies became the United States, the new nation embraced the existing tenets of international ocean law. As its shipping and naval interests grew, the United States favored the concept of narrow territorial seas and encouraged other nations to do the same. Thus, as early as 1793, Thomas Jefferson, then secretary of state, set 3 miles as the width of the U.S. territorial sea in a letter to the government of Britian and France, at that time involved in a war just off the American shores (U.S. Congress 1817). In the same letter, however, Jefferson spoke in favor of reserving the country's rights to "as broad a margin of protected navigation as any nation whatever" (quoted in Hollick 1981, p. 19). For most of the next 150 years, the policy of claiming a 3-mile territorial sea while supporting almost total freedom of the high seas served U.S. interests well.

Thus, before World War II, coastal nations possessed relatively narrow territorial seas. National activity beyond territorial waters was limited primarily to navigation, coastal fishing, and, in a few cases, to distant-water fishing and cable laying. The involvement of the United States in the oceans during this period was also limited. Besides shipping and naval maneuvers, fishing in coastal waters, and some distant-water whaling, a few shallow-water oil wells off the shores of California and Louisiana represented the bulk of the nation's ocean-oriented efforts.

The Enclosure Movement

In contrast to the earlier stability, the period following 1945 has seen almost constant activity, both internationally and nationally, with respect to ocean law and policy. The major new development was the "enclosure" by coastal nations of ocean space adjacent to their coastlines (Friedheim 1979), culminating in the 1980s with the worldwide acceptance of national control over the resources contained within 200-mile ocean zones.

It all started in 1945 with a unilateral move by the United States to assert jurisdiction over the resources of the continental shelf, following the realization that significant oil and gas deposits lay offshore. On September 28, 1945, President Truman issued a proclamation declaring that ". . . the United States regards the natural resources of the subsoil and seabed of the continental shelf beneath the high seas but contiguous to the coasts of the United States as appertaining to the United States, subject to its jurisdiction and control . . . The character as high seas of the waters above the continental shelf and the right to their free and unimpeded navigation are in no way thus affected" (U.S. Presidential Proclamation 2667 of 28 September 1945).[1]

The U.S. move in 1945 to claim jurisdiction over the resources of its continental shelves precipitated extended-jurisdiction claims by some Latin American countries in the late 1940s and the 1950s (Szekeley 1979). By 1955, several nations were claiming extended jurisdiction beyond theretofore accepted narrow territorial seas—some of them extending authority as much as 200 miles offshore. These extensive claims of national jurisdiction over ocean space led to pressures for an international conference to restore order and coherence to an increasingly chaotic situation.

The first Law of the Sea conference held under the auspices of the United Nations in Geneva in 1958 drafted four conventions dealing with the high seas, the territorial sea and contiguous zone, fisheries, and the continental shelf. The 1958 conference, however, reached no agreement on a major contentious issue—the width of the territorial sea. A 1960 Law of the Sea conference, called specifically to deal with this matter, also failed to reach agreement, and the expansive claims of coastal states continued (Hollick 1981, Chapter 5). Finally, after the United States, the Soviet Union, and other maritime nations acceded to the demands of developing countries to include consideration of an international regime for the resources of the deep seabed, agreement was reached to hold a third conference in 1973.

In 1982, after nine years of effort, that third conference did agree on provisions giving legitimacy to 200-mile exclusive economic zones, while protecting most navigational freedoms and establishing 12 miles as the maximum breadth of territorial seas. The 320 articles contained in the agreement addressed virtually all ocean issues and, with the possible exception of the regime for the deep seabed, established international norms for ocean activities for the future.[2] Even nations that declined to be parties to the convention in 1982 (most prominently, the United States, because of its objections to the seabed mining provisions),[3] considered much of the text of the treaty as the best manifestation of current international ocean law and practice. (Changes in the 1990s in the U.S. position regarding the Law of the Sea are explained briefly later in this chapter and more fully in Chapter 6.)

The Role of the Ocean in the Historical Development of the United States[4]

Prior to the advent of aviation, the sea provided the only means of access to islands or continents surrounded by water. Thus the sea has been of great importance to nations that started out as colonies of a mother country on another continent.

Transportation and Commerce

In the case of the United States, while the very earliest Americans may have migrated over the Bering Strait, the majority of the population arrived by sea, largely between the 1600s and the mid–twentieth century. Wave after wave of Europeans and Asians arrived with the aim of making a better life. In addition, hundreds of thousands of Africans were forcibly brought by sea to this country and sold as slaves. The ports that received the new Americans—New York, Philadelphia, Boston, Baltimore, Charleston, San Francisco—became diverse and prosperous cities and the principal connections between the growing young country and the Old World.

The sea was a critically important lifeline in the early 1600s to the small and struggling colonies that clung precariously to the edge of the continent. Failure of a ship to return with necessary supplies could (and did) spell disaster. The sea continued to be important to the colonies as they grew and developed, with the oceans becoming the essential transportation link in the increasing commerce of the region. The exchange of agricultural products such as cotton and tobacco for goods manufactured in Europe and Great Britain became a vital element in the growing prosperity of the New World. The colonies also soon established a flourishing and lucrative trade between the sugar islands of the Caribbean and the slave-trading regions of Africa. Shipping was by far the easiest way to move goods between the expanding settlements along the East Coast, and by the mid–nineteenth century, clipper ships had united the east and west coasts of the United States by means of the long and arduous trip around the South American continent. Later on, maritime trade routes were established between Asia and the United States, and the West Coast ports also began to grow.

Defense of the Nation

Periodically, the sea took on another dimension, as long-standing tensions between the nations of Europe extended into the waters of the New World and American seaports became home to the armed naval ships of

Great Britain, France, and other nations. The Revolutionary War was followed by a series of small and not-so-small conflicts, with naval forces playing the predominant role. In the Civil War, naval forces played decisive roles at several key points in the conflict, including the critically important blockade of Confederate ports. The U.S. Navy gradually grew in strength and importance, becoming by the end of World War II the largest and most powerful navy in the world. Since that time, U.S. seapower has taken on more specialized roles (e.g., the operation of the underwater ballistic missile fleet), with the conventional surface ships of the navy no longer seen as necessarily the best or only way to project U.S. military strength to various parts of the world. Yet, ensuring the mobility of U.S. naval forces continues to be an important consideration in policy making.

The Ocean as Food Source

From the beginning, Americans looked to the sea for food. This nation was richly endowed with world-class fishing grounds off both the Northeast coast and the Pacific Northwest and in the Gulf of Mexico. Long before the settlement of the colonies in the seventeenth century, scores of European vessels transported many tons of fish from American waters to Europe. By the end of the seventeenth century, Massachusetts was exporting about 11,000 tons of dried codfish a year (Mangone 1988, p. 25). As the number of fishermen increased and techniques improved, fish catches steadily rose over time. In 1938, the value of fish landed by U.S. fishermen topped that of any other nation (Mangone 1988, p. 129). Perhaps the most ambitious "fishing" was that undertaken by whalers out of such ports as New Bedford and Nantucket in Massachusetts. Their expeditions would extend over many thousands of miles and last for many months.

Turning Inward

The nineteenth century saw the United States turn its attention inward as it sought to explore and settle the interior and western parts of the country. After the Civil War, the northeastern portion of the nation resumed its rapid industrialization. During this period, railroads and highways were built, and farming, grazing, and mineral exploration increased greatly. The nation was preoccupied with an assessment of its resources and the beginning of their exploitation and use. No longer was the ocean regarded as critical to the day-to-day survival of the nation.

For many Americans living in the nation's heartland, the ocean was of little direct concern.

The period from 1910 to 1945 was one of considerable turbulence and unrest in the United States. The Great Depression of the 1930s followed a brief interlude of relative tranquility after World War I. The depression, in turn, was broken only by the economic boom brought on by World War II. While these great wars vividly demonstrated the effect of superior naval forces, it became clear by the end of World War II that nuclear weapons and an effective air force would reduce the relative importance of the navy in the future.

World War II and the Postwar Prosperity

World War II had a significant impact on the way Americans thought about and viewed the ocean. As Orbach (1982b) eloquently described it:

> The War had gotten hundreds of thousands of Americans out on and across the oceans, many for the first and last time. They felt the sway of the troopship's deck; they rejoiced at the sight of sunrise at sea, and of landfall; they lounged, and fought, under the fabled palms of the South Seas. They finally had something with which to connect Melville, Conrad and London. Regardless of the conditions under which they encountered these experiences, they were *oceanic* experiences, felt and remembered by the better part of an entire generation of American males. (44)

With the cessation of hostilities in 1945, a great pent-up economic demand was released in the United States. More and better housing was needed; the transportation system needed an overhaul and updating; factories and farms required modernization and expansion; the infrastructure of the nation's cities needed improvement; new water plants, sewer plants, and airports were needed. These pressures led to a building spree of unprecedented proportions.

Virtually everything in the United States grew between 1945 and 1970—its population, the size of its cities, personal income, spending (private and public), taxes, the number of automobiles and second homes, the amount of leisure time. These changes, of course, had their impacts. Given that two-thirds of the population was living in the coastal areas of the country, a good share of the postwar development occurred along or near the nation's shorelines. In fact, the nation's ocean beaches had become one of the two most favored vacation destinations of Americans (the mountains were the other). Wetlands and marshes were filled for marinas, second homes, and resort hotels. New industrial operations, power-generating facilities, and sewage plants began to increasingly contaminate the nation's rivers and estuaries. Important fish and wildlife habitats were unwittingly being destroyed.

By the mid-1960s, the nation was becoming aware that important parts of the natural environment were in trouble. Rachel Carson's book *Silent Spring* (1962) and Jacques Cousteau's television shows sounded the early alarms. Citizens began calling for government action to protect the nation's ocean, coastal, and estuarine environments, and a number of studies were launched by the federal government.

The Evolution of Resource Management Regimes for Oil and Gas, Fisheries, Wildlife, and Coastal Areas

Before examining the series of events in the late 1960s that would be catalytic in the enactment of significant ocean legislation in the early 1970s, it is important to review the evolution of management practices regarding major ocean and coastal resources. How were these resources being managed in the late 1960s, by whom, and with what results?

The management of these offshore resources was traditionally thought to be the purview of state governments. Between the late 1930s and the 1970s, however, this situation was to undergo significant change, with the federal government taking on an increasingly larger role. Prominent among the factors responsible for the growth in the federal role were the growing economic importance of oil and gas resources coupled with a new technological ability to recover offshore resources, a perception that foreigners were seriously overfishing waters near the United States, and the alarming depletion of marine mammal populations worldwide.

Offshore Oil and Gas Development: Always Controversial[5]

The federal government first asserted a significant management role in the case of offshore oil development. Prior to the 1940s, both the federal government and the coastal states assumed that the states owned any oil and gas deposits in the submerged "tidelands" adjacent to their coasts, as a remnant of the jurisdiction that the original thirteen colonies had over the marginal sea (Miller 1984). Indeed, in California, where offshore oil production began in 1896, the state had been formally leasing nearshore tracts for oil development as early as the 1920s (Cicin-Sain 1986).

Growing Intergovernmental Conflict With the realization that significant oil and gas resources existed offshore and the growing economic importance of these resources, the federal government began to challenge the states' claim to resource ownership through administrative action in 1937. In a 1945 landmark suit brought to the Supreme Court, *United States v. California* 332 U.S. 19 (1947), the federal government argued that the territorial sea concept of the marginal sea did not arise until after the Revolution; hence there were no existing property rights for the states to succeed to at the time of independence (Miller 1984). In 1947, the Court surprised many observers by handing down a decision saying that the federal government, in fact, required "paramount rights" in coastal waters in order to fulfill its foreign policy, national security, and interstate commerce responsibilities to the nation as a whole and that it was impossible to split off the resource questions from these broader issues (*United States v. California* 332 U.S. 19 [1947]).

Understandably, the coastal states and ultimately the Congress reacted sharply to the Supreme Court decision and, in 1952, ownership of the submerged lands[6] became a prominent issue in the presidential election, with Eisenhower supporting the case of the states. The compromise finally attained through congressional action in 1953 was a strict division of authority between federal and state governments, with the Submerged Lands Act (SLA) of 1953 giving the states title to the submerged lands and resources of the territorial sea, including "the right and power to manage, lease, develop and use" them, while the Outer Continental Shelf Lands Act (OCSLA) of 1953 established that the "subsoil and seabed of the outer Continental Shelf appertain to the United States and are subject to its jurisdiction, control, and power of disposition."

As it became clear that substantial amounts of oil and gas existed in the Gulf of Mexico, probably off the shores of California and Alaska, and perhaps off the East Coast, the coastal states, individually and in groups, sued the federal government for extended boundaries. All of these efforts failed, except for those of Florida (Gulf coast only) and Texas. The Supreme Court ruled that in these two cases the pre-admission boundaries of about 9 nautical miles had existed at the time of those states' admission into the Union (*United States v. Florida et al.*, 363 U.S. 121 [1960] and *United States v. Louisiana et al.*, 363 U.S. 1 [1960]).

Federal Leasing of Offshore Areas The OCSLA was a very succinct law that basically gave the federal government (through the secretary of the interior) the mandate to lease offshore lands under U.S. control to the highest bidder in the petroleum industry with very few conditions and strings attached. The act generally "followed the Truman proclamation of 1945 in affirming the rights of the United States to the natural resources of the continental shelf, while stating that the character of the superjacent waters with regard to fishing and navigation remained unchanged" (Mangone 1988, p. 188).

Under the provisions of the OCSLA, the federal government began leasing federal offshore tracts for the first time in 1954 in the Gulf of Mexico off the coast of Louisiana. A second lease sale involving federal waters off Texas soon followed in the same year. Federal waters off the coasts of Florida, California, and Oregon/Washington were leased in 1959, 1963, and 1964, respectively. Lease sales in waters off Alaska and in the Atlantic outer continental shelf (OCS) were held in 1976 (U.S. Department of Interior, Minerals Management Service, 1986, p. 4). By the end of 1985, there had been eighty-seven lease sales in the federal outer continental shelf, resulting in the issuance of leases for over 40 million acres of federal offshore lands (U.S. Department of Interior, Minerals Management Service 1987, p. 13).

The acceleration of leasing of federal OCS lands that had taken place in the 1960s suffered a significant setback in 1969 when a major oil spill took place in the Santa Barbara Channel off the southeastern coast of California. On January 29, 1969, Union Oil's Platform A in the Santa Barbara Channel blew out and an enormous amount of oil (estimates ran as high as 3 million gallons) spewed forth and blanketed an area of 660 square miles, affecting over 150 miles of coastline, and killing many birds and other marine life (Nash, Mann, and Olsen 1972, p. 22).

The Santa Barbara oil spill proved to be a catalytic event in the rise of the environmental movement, a movement that, in the 1970s, was to fight to place significant restrictions on how offshore oil planning and development were conducted. Images of the extensive damage to beautiful beach areas, the plight of seabirds covered with oil, the dead marine life—widely transmitted through the extensive media coverage of the spill—contributed to the growing perception that resource development could pose serious threats to the environment.

Fisheries Resources: Foreign Threats

Prior to the 1970s, fisheries in the United States were either not managed or only partially managed by the states or international fishery entities. The role of the federal government during this time was essentially a nonregulatory one—limited to the gathering of fishery statistics, the provision of services to fishers (such as loans for vessel construction), and the pursuit of international fishery agreements.

State Management Laissez-faire management of fisheries, both in the United States and in Europe, characterized much of the ninteenth century. Up to 1850 or so, the fisheries of the oceans and large inland lakes were generally considered inexhaustible (Nielsen 1976). Regulation of fisheries, moreover, was thought to be unnecessary, impractical, and threatening to the personal liberty of fishers.

By the time the nineteenth century drew to a close, however, there were clear signs that some fisheries were in trouble as a result of overfishing, improper fishing, or destruction of spawning streams by industrial development. The popular press and fishery professionals made calls for increased regulation by the state fish commissions, which were present in most of the states (Nielsen 1976). Some forms of fishing regulation followed, but these were primarily aimed at reducing the problem by institutionalizing inefficiencies in the fishing enterprise through various restrictions on fishing methods and gear. As Wenk (1972) elaborates, "Limitations on bait, fishing gear, boat sizes, duration of season, as well as on species harvested and catch limits, were developed state by state, ostensibly as conservation measures. Closer examination revealed that the great majority of 'scientifically based' regulations promulgated on the pretext of conserving stocks were in fact politically expedient devices to allocate the resource among competing fishermen, simply to reduce political pressures and resolve economic conflicts" (p. 305). "The inevitable result has been rules which increase cost, are awkward to administer, and are cumbersome to enforce" (U.S. COMSER 1969, p. 95).

Fishery managers were often unable to prevent the decline of specific fisheries. For example, in California (which led the nation in both the volume and value of fisheries landed at the end of World War I), fishery managers were unable to prevent the decline of the highly profitable California sardine industry. As McEvoy (1983) explains:

> The roots of late twentieth-century fishery management problems in California lay deep in the history of the industry as it underwent its technological revolution during the World War I era. At that time fishers and entrepreneurs, emboldened by opportunity and the apparent boundlessness of the ocean's resources, tore at the fisheries with little regard to the fact that they had histories of their own and required

management. Government trustees of the people's fishery resources, unprepared for the scale and complexity of the task that industry thrust upon them and hamstrung by their own lack of control over the economic forces that drove the industry, could do little to temper the rapacity of the onslaught. (pp. 520–521)

The Beginning of a Federal Role The federal government first became involved in fisheries in 1871 when the Office of the Commissioner of Fish and Fisheries was established to study problems of overfishing. The commission—transferred to the new Department of Commerce and Labor and renamed the Bureau of Fisheries in 1903—got its first real management mandate with the enactment of legislation controlling fur seal fisheries, salmon fisheries, and other fisheries in all areas subject to U.S. jurisdiction off the territory of Alaska (Greenberg and Shapiro 1982, p. 646). The bureau worked vigorously to protect salmon stocks and fur seal populations in Alaska by means of both domestic regulations and international agreements. But it was not until 1915 that the federal government began to exert some leverage over the states and their lagging programs in fisheries conservation. Under the Appropriations Act of 1915, the Congress required the states to have adequate laws for the protection of fish in order to be eligible to receive the benefits of the Bureau of Fisheries species propagation program.

International Negotiations During this time, the Bureau of Fisheries became increasingly preoccupied with international fishing issues. Two major fishery conflicts had languished for nearly a hundred years. The first, with Great Britain, involved fishing in the waters off what are now the Atlantic provinces of Canada. After extended controversy, several agreements were ultimately made allowing expanded access to Canadian fishing grounds by American fishers in exchange for allowing Canadian fish to enter the United States without duty (Mangone 1988, pp. 45–46). The second dispute, also involving Great Britain on behalf of Canada, pertained to the taking of fur seals in the Bering Sea. The U.S. acquisition of Alaska from Russia in 1867 brought about wholesale destruction of the seal herds. In 1869, the U.S. Congress enacted the first conservation laws applying to fur seals, but the quotas were set too high to reverse the decline in stocks. Also, after 1879, vessels from Japan, Russia, Canada, and the United States began the indiscriminate taking of seals on the high seas, substantially worsening the problem. Agreements drafted in 1894 and 1897 proved inadequate and failed to gain the approval of all four nations. Only in 1911 was a fur seal convention finally agreed to that eliminated pelagic sealing (sealing done on the high seas) and created a four-nation, conservation-oriented cooperative effort to address the problem (Mangone 1988, pp. 46–48).

The U.S. government's involvement in international fisheries problems continued and ultimately resulted in international agreements with Canada in the 1920s and 1930s to create (and successively strengthen) the first international fishing commission to manage a high-seas fishery—the Pacific Ocean halibut fishery off British Columbia and the states of Washington and Alaska. Agreements were also reached with Canada to manage Fraser River sockeye salmon in 1937, after forty-five years of discussion and debate and after regulation by Washington State and British Columbia had proved ineffective (Mangone 1988, pp. 153–154). The federal government also participated in international discussions in 1918, 1935, and 1937 to prevent the further decline in whale populations worldwide, with a major convention—the International Convention for the Regulation of Whaling—being concluded in Washington, D.C., in 1946 (Mangone 1988, p. 163). In 1949, the U.S. government signed its first multinational fisheries agreement covering the northwest Atlantic—the International Convention for the Northwest Atlantic Fisheries (ICNAF) (Mangone 1988, p. 157). International fisheries commissions were also established for tunas—the Inter-American Tropical Tuna Commission in 1949 and the International Convention for the Conservation of Atlantic Tuna in 1967.

Interstate Commissions On the domestic front, the federal government made efforts to encourage the states to better manage the fisheries adjacent to their shores. One significant move in this direction was the creation of three interstate marine fisheries commissions by means of interstate compacts approved by Congress between 1942 and 1949 (the Atlantic States Marine Fisheries Commission in 1942, with fifteen participating states; the Pacific Marine Fisheries Commission in 1947, with five states; and the Gulf States Marine Fisheries Commission in 1949, also with five states).

Created to promote and better utilize marine fisheries through the development of a joint program of fisheries promotion and protection and the prevention of the physical waste of fisheries from any cause, the commissions, however, had no regulatory or management authority except that granted by the member states. Consequently, they operated mainly as fora for inquiry, debate, and advice. The lack of power of the commissions to require individual states to adopt particular management measures considerably hampered their ability to enact and implement meaningful interstate management plans.

Growing Threats from Foreign Fishing In 1956, the United States placed second only to Japan in world fishery catch. A little over a decade later, in 1968, the U.S. position had dropped to sixth place (Wenk 1972, p. 303).

A major factor accounting for the decline in the American catch was the growing efficiency of the foreign fleets that came to harvest fish off America's shores. The former Soviet Union, in particular, had dramatically modernized its fishing fleet through a capital investment of over $4 billion, following World War II (Wenk 1972, p. 303). Beginning in the mid-1950s, a new type of fishing vessel—the factory trawler—began to cross the Atlantic to fish in America's waters in the North Atlantic. These "floating hotels" were huge, worked around the clock in almost all weather conditions, and processed and froze all the fish they caught (Brooks 1984, pp. 52–53). Their harvesting power was awesome, as explained by William Warner using a hypothetical land-based analogy:

> First, assume a vast continental forest, free for the cutting or only ineffectively guarded. Then try to imagine a mobile and completely self-contained timber-cutting machine that could smash through the roughest trails of the forest, cut down the trees, mill them, and deliver consumer-ready lumber in half the time of normal logging and milling operations. This was exactly what factory trawlers did—this was exactly their effect on fish—in the forests of the deep. It could not long go unnoticed. (quoted in Brooks 1984, p. 53)

To improve the position of the U.S. fishing industry, the federal government slowly began to adopt a series of measures designed to encourage the development of the industry. The 1954 Saltonstall-Kennedy Act provided that 30 percent of the custom duties collected on imported fish be allocated to research and marketing to promote the use of domestic fish. Seeking to assist the commercial fishing industry, in 1960, Congress approved a federal fishing vessel construction subsidy program that provided up to one-third (later one-half) of the cost of new fishing boats if they were to be used in fisheries threatened by foreign imports of fish (Mangone 1988, p. 139). Federal assistance continued in 1964 with the passage of the Commercial Fisheries and Research Development Act (88-309), which authorized a program of grants to the states (ultimately at a level of $36 million per year) to study and improve commercial fisheries; and in 1965, with the passage of the first Anadromous Fish Conservation Act (89-304), also a cooperative grants program with the states.

On the regulatory side, up to the mid-sixties, the U.S. government had stoutly maintained that exclusive fisheries rights of the world's coastal nations did not extend beyond the 3-mile territorial sea. Indeed, in 1964, the United States enacted legislation that provided for reimbursement for fines paid by owners of any U.S. fishing boats seized by foreign governments for "unlawful" fishing within their extended jurisdiction zones (Mangone 1988, p. 138). But in 1966, after numerous extensions of jurisdictions by other nations and with increased foreign fishing immediately adjacent to U.S. shores, the State Department shifted its position to support a 3- to 12-mile-wide exclusive fishery zone for the United States, and legislation creating such a zone was enacted in 1966 (Mangone 1988, p. 140).

Fisheries at the End of the 1960s: Troubled Prospects By the end of the 1960s, it had become apparent that American fisheries and the fishing industry were threatened: Stocks were overfished, and both stocks and fishers were seriously threatened by the superior catching power of foreign fleets. The industry itself had not been modernized and largely remained a small-scale, family-centered operation. The industry was characterized by small vessels, overcapitalization, and overfishing. There was a lack, too, of vertical economic integration of the industry—a discontinuity between fish catching and fish processing. State management of fisheries was largely ineffective, and states were often unable to resist industry pressures for providing continued access in cases where the status of the stocks warranted more conservationist approaches. In many cases, state measures to regulate fisheries, although ostensibly designed for conservation purposes, were, in fact, handicaps to technological innovation and were designed to protect the status quo (Wenk 1972, p. 304).

Other problems were evident as well. At the international level, it was becoming apparent that the multilateral treaties and organizations to which the United States was a party were not working out as effective management tools. Harvest quota levels established through bargaining among diverse national actors often exceeded maximum sustainable yields; enforcement, which was vested in each member country with respect to its own vessels, often turned out to be nonexistent. At the sub-national level, little was being done to manage the many fish species that swam across state boundaries. As an influential analysis put it, "Although fish migrate freely across State lines, [no] single instance of systematic programs being prepared jointly by two or more States for the management or development of their fisheries resources [could be identified] . . . [while] three interstate commissions exist . . . none has regulatory powers or adequate staff . . ." (U.S. COMSER 1969, pp. 95–96). By the end of the 1960s, pressures for change began to mount.

Wildlife Hunting and Fishing as an "American Right"

In feudal England, until the mid–nineteenth century, wildlife law was an instrument for the landed class to use to maintain and enforce its privileges (Lund 1980, p. 8). They did this through a "qualification system" that allowed only certain classes (feudal lords and later the landed gentry) the privilege of hunting and fishing, the ability to possess certain weapons, and the right to eat certain animals (Lund 1980, p. 8; Bean 1983, pp. 10–11). This class-based system, coupled with management measures such as closed seasons, restrictions on gear, and protected areas (the same type of management measure used today), were effective, although discriminatory, in achieving conservation goals by limiting entry to the use of wildlife resources (Lund 1980, p. 11).

The United States adopted the same type of management methods, with the exception of the qualification system, which, because of its undemocratic nature, conflicted with American ideals. In contrast to the English system, game hunting in the United States was seen not as a sporting matter but as an essential source of food and clothing (Lund 1980, p. 19). Free access to hunting and fishing and the right to bear arms were inextricably linked in early American history. For example, even privately owned lands (as long as they were undeveloped), were open for all people to fish and hunt, regardless of the will of the landowner (Lund 1980, p. 24). This put great pressure on wildlife resources, and by the end of the ninteenth century, many species were being decimated. A major problem was the relative ease with which a few people could decimate a fish population by casting a net over the neck of a watercourse and taking all the fish that they caught (p. 27).

Lund (1980) has described early eighteenth-century examples of controls imposed on taking of fish and game: "[d]eer near Christian settlements, fowl close to towns, fish in convenient fishing holes, rivers and river systems: all were managed by protective rules" (pp. 28, 29). Closed seasons were also introduced; for example, oysters sessile in their beds at maturity, and deer in winter or breeding seasons. Lund notes, however, that enforcement of early rules was difficult and ineffective, and many species continued to decline (p. 31).

In addition to overfishing and overhunting, other pressures were combining to further decimate fish and wildlife populations. Chief among these were the destruction of habitat for agricultural land or towns, the damming of rivers, and sedimentation of waterways associated with this destruction (Lund 1980, p. 106). Legislative answers to these problems did not always come within the traditional body of wildlife law that had hitherto controlled takings and imposed various restrictions on hunting and fishing. Rather, the growing body of environmental laws from the middle of this century onward has provided, either explicitly or implicitly, for the value of species, and the need for habitat protection. Great damage to species, however, had already been done by this time; in particular, the destruction of forest habitat by residential and commercial construction.

Up until the 1970s, wildlife resources were managed largely by the states through an ad hoc and species-by-species approach. A "cozy" relationship often existed between state wildlife managers and the "hook and bullet boys" (fishers and hunters). By the end of the 1960s, new animal protection groups began mobilizing to establish animal protective legislation at the federal level, thus circumventing these established political relationships at the state level.

Coastal Areas: The Need to Control Rampant Development

Use of the nation's coastal areas was not regulated by either the federal or the state governments until the mid-1960s, except through state and local laws generally dealing with land-use planning and zoning. The unprecedented building spree that had characterized the postwar prosperity had, by the 1960s, intensified pressures on the use of coastal areas and taken a heavy toll on the nation's estuaries and wetlands, unwittingly destroying valuable fish and wildlife habitats.

In the mid-1960s, coastal states began to take initiatives establishing measures to manage the multiple uses of the coastal zone. As early as 1963, the California legislature created an interim program to begin management of San Francisco Bay. Alarmed at the rate at which San Francisco Bay was being filled for development, citizens formed the nation's first "Save the Bay" group in 1960 and lobbied for state action (Hildreth and Johnson 1985). The Bay Conservation and Development

Commission (BCDC), made permanent by state legislation in 1965, became the nation's first coastal management program and served as a model for the statewide coastal program put in place in California by popular initiative in 1972.

Several other coastal states also began programs in the early 1960s. Massachusetts initiated a regulatory program to protect its coastal wetlands in 1963 and expanded the program in 1965. The first efforts at actually applying the zoning technique to the shoreline began in 1969 along the Great Lakes in Michigan and Wisconsin. Prompted by proposals for additional industrial development in Delaware Bay and Narragansett Bay, the states of Delaware and Rhode Island enacted coastal legislation in 1971 (Peterson 1999; Olsen and Lee 1991). Concerned with inappropriate coastal development, Oregon and Washington also created state coastal laws in 1971 (Hildreth and Johnson 1985).

By the late 1960s, this state initiative had moved to the national level, where a variety of interest groups and prestigious study commissions would recommend the enactment of a national coastal zone management program. The politics of the enactment of this program are discussed in Chapter 3.

The Politics of Managing Ocean and Coastal Resources: Separate "Iron Triangles"

Prior to the 1970s, the political subsystem for governing the use of living marine and coastal resources (fisheries and wildlife) could be characterized as largely state-dominated, with the federal government playing a limited, primarily service-oriented role. Relationships among entities in this political system represented a relatively closed system of interactions among a narrow set of interests, linking state wildlife management agencies with federal agencies such as the National Marine Fisheries Service and the U.S. Fish and Wildlife Service, with the relevant congressional committees, and with interest groups such as fishing, hunting, and wildlife management groups (Cicin-Sain 1982a).

These "subgovernments" or "iron triangles" (denoting a continuous, stable, and predictable set of interactions among a small number of participants found in the legislature, executive branch, and interest groups) as political scientists call them (e.g., Lowi 1969; Palley 1994), operated relatively independently from one another. That is, a particular subgovernment network existed for commercial fisheries, another for recreational fisheries, yet another for marine mammal management, and so on. Decision processes within these entities were largely insulated from external political forces, with individual networks enjoying political autonomy and control over their substantive areas of concern.

In the management of oil and gas, similar trends of an independent subgovernment could also be found among the relevant congressional committees, the Department of Interior, and the oil industry. In particular, a close relationship was said to exist between the Department of the Interior and the "majors" (the major petroleum companies such as Standard Oil of New Jersey, Mobil, Texaco, Standard of California, and Gulf Oil). As one scholar put it, the secretary of the interior acted "as the chief administrative lobbyist within the federal system for the de facto government of oil" (Engler 1961, p. 60). In the area of coastal management, given that this was a new area of governmental activity, no subgovernment or "iron triangle" existed during this time period.

With the legislation enacted in the 1970s, these subgovernment relationships were to undergo considerable change and become more open, increasingly complex, and less predictable, significantly complicating the political dynamics of national ocean policy (see Chapter 3).

The Beginning of Change: Expanded Governmental Attention to Ocean and Coastal Issues in the 1960s[7]

The 1960s saw an expansion of governmental attention to the oceans, a major impetus for which came from scientific quarters. The largest and most ambitious worldwide effort to study the earth and its behavior had taken place in 1957–1958, during the International Geophysical Year (IGY). The IGY involved more than seventy nations and the largest network of observing stations ever assembled. The crowning event of the IGY was to have been the launching by the United States of the world's first orbiting satellite, planned for midsummer 1957. Problems in thc satellite program, then under the direction of the navy, caused the launch to be delayed. It is an understatement to say that U.S. officials were dismayed when the former USSR launched its own satellite, *Sputnik,* into space on October 4, 1957, well before the U.S. launch. To make matters worse, the Soviet Union launched a second satellite into orbit on November 3, 1957, putting the first living organism (a dog) into orbit. The first U.S. satellite, *Explorer I,* was finally launched on January 31, 1958, but at 30 pounds, it appeared a minor effort compared to the 1,120-pound *Sputnik.*

These events kindled extensive soul-searching in the United States, and much of the blame for the delayed space program fell on the educational system, particularly on science and mathematics instruction. Proposals for all kinds of support for American sciences and engineering emanated from many quarters, and a number of them were quickly adopted. As Wenk (1972) put it, "[W]ith one countdown, science was launched from the quiet and seclusion of the laboratory into the orbit of national policy" (p. 41).

The IGY and the *Sputnik* launch, combined with the election of John F. Kennedy as president in 1960, created a significant window of opportunity for federal funding of ocean science and engineering. Policy entrepreneurs armed with a variety of proposals were prepared to capitalize on the opening. The National Academy of Sciences had issued a major report on the ocean sciences in 1959, presenting a blueprint for a significant increase in federal support for ocean sciences. In the educational realm, Athelstan Spillhaus in 1963 advanced the concept of Sea Grant as an ocean analogue to the successful Land Grant College Program, an idea that was ultimately enacted into legislation in 1966 (Miloy 1983).

The Stratton Commission and the Marine Sciences Council

The major piece of ocean legislation propelled into law by these forces was the Marine Resources and Engineering Development Act of 1966. That legislation moved beyond ocean science into issues of the organization of the national ocean program and the improved coordination of federal ocean activities. The principal devices called for in the legislation were a Marine Sciences Council and a Commission on Marine Sciences, Engineering, and Resources (COMSER). The Marine Council began meeting in mid-1966 under the vigorous leadership of Vice President Humphrey and Executive Secretary Edward Wenk Jr. The five annual reports of the Marine Council (through 1973) showed a range of activities from the creation of the Sea Grant College Program in 1966 to the launching of the International Decade of Ocean Exploration in 1968. A companion blue-ribbon commission, COMSER, came into being in early 1967 with the appointment of a distinguished panel of experts from within and outside of government, chaired by Julius Stratton, former president of the Massachusetts Institute of Technology.

COMSER, which ultimately became known as the Stratton Commission, issued in January 1969 a comprehensive and forward-looking report entitled *Our Nation and the Sea* (U.S. COMSER 1969), reviewing the status of most areas of American ocean policy. Several important themes emerged from the Stratton report. First, it called for a centralization of the federal government's ocean effort if the full benefits of the nation's marine and coastal resources were to be realized. Concomitantly, the report called for the creation of a civilian ocean and atmosphere agency to undertake the full range of actions needed to realize the effective use of the sea. Second, the report stated the urgent need for a concerted effort to begin planning and managing the nation's coastal zones. It advocated more research and recommended a federal-state program in coastal zone management. Finally, the report highlighted the need for a much-expanded program in ocean science, technology, and engineering, at both the national and global levels.

The work of the Stratton Commission would have a very substantial impact on the national ocean program in the years ahead. Within a year, its recommendations calling for a new federal oceans agency were being implemented. In 1970, President Nixon issued Reorganization Order Number 4 creating the National Oceanic and Atmospheric Administration (NOAA) within the Department of Commerce. Although not as inclusive as had been recommended by the Stratton Commission (it did not include the Coast Guard or the Maritime Administration), the reorganization did bring together many of the key ocean programs theretofore scattered throughout various federal departments. Moved into the new administration were the Environmental Sciences Services Administration (ESSA, already in the Department of Commerce), the Bureau of Commercial Fisheries (from the Interior Department), and the Sea Grant Program (from the National Science Foundation). The work of the Stratton Commission also provided a compelling argument for a national coastal zone management program and would lead directly to legislation creating such a program.

Thus, by the end of the sixties, elements of a national ocean policy began to emerge for the first time. At the same time, a national consensus—that substantially more ought to be done by the federal government to capitalize on the great benefits that the oceans were seen to hold—was beginning to develop. Committee members and staff in both houses of Congress were excited by the prospects. The years between 1966 and 1969 also represented a peak in high-level executive branch attention to national ocean policy after World War II. President Lyndon B. Johnson, for example, discussed the importance of the oceans in his 1968 State of the Union address and delivered a major ocean policy speech on the occasion of the dedication of a new oceanographic vessel in 1967 (Abel 1981).

Toward the end of the 1960s, the outlines of a new problem stream began to be visible. New political groups began to mobilize and focus attention on the degraded state of the nation's natural environment. The origins of this problem stream and the way in which recognition of these problems affected national ocean policy are the subject of Chapter 3.

THREE

The 1970s

The Enactment of the Innovative Body of Marine Law

A plethora of new regulatory and management-oriented ocean laws were enacted by Congress during the 1970s. What factors were responsible for this outpouring? Where did the initiative for these new policies originate? How did these policy changes gain a place on the always crowded national agenda? What forces contributed to the final shape of these measures?

In seeking the answer to these questions, we analyze the domestic and international forces that led to the enactment of the innovative body of ocean law in the 1970s, focusing in particular on the political interactions among Congress, the executive branch, and interest groups in the shaping of these laws. To illustrate the nature of politics during this period, we discuss the passage of several major ocean and coastal laws—the Coastal Zone Management Act (CZMA), the Marine Mammal Protection Act (MMPA), the Magnuson Fishery Conservation and Management Act (MFCMA), and the Outer Continental Shelf Lands Act Amendments (OCSLAA). We also discuss, in less detailed fashion, the enactment of the Clean Water Act, the Marine Protection, Research, and Sanctuaries Act, and the Endangered Species Act. We describe the major features of each act, emphasizing its novel or innovative aspects and discuss the political dynamics involved in congressional enactment. We place particular emphasis on the features of each act that would later prove problematic in the implementation stage.

The Political Dynamics of Policy Initiation and Formulation

Policy initiation represents the first step in the policy process, whereby the large number of problems that government *could* address is reduced to the much smaller number of problems that government *will* address. The study of policy initiation, in effect, is the study of change. Significant obstacles to change exist at individual, organizational, and societal levels, and these must be overcome before problems can be placed on the public agenda (Brewer and deLeon 1983, Chapter 2). Policy formulation, the next step in the policy cycle, concerns the process of deliberation and bargaining that occurs after sufficient momentum has been built up to induce decision makers to "do something" about an issue. A narrowing of debate takes place among key actors found in the congressional committees, in the relevant administrative agencies, and in the affected interest groups. Out of these political interactions and consideration of the merits of alternative policy options, a policy decision ultimately emerges and a particular piece of legislation is enacted.

Studying the origins of policy and the congressional dynamics that accompany the enactment process is imperative for understanding what happens, and why, later in the implementation stage. Legislative enactment often contains multiple (and sometimes contradictory) goals and objectives—the result of bargaining and compromise reached among many diverse interests. To more fully understand how and why statutory goals and objectives are translated and interpreted by a variety of bureaucratic and interest-group actors in implementation, one must first understand the combination of forces that are responsible for the passage of an act in Congress.

Where does the impetus for a particular policy come from? From inside or outside the government? Or from a combination of both? What general contextual forces, domestic or international, set the stage for the emergence of an issue? Which groups "pick up" the issue and make it their own and then persuade others of the "rightness" of their cause? Synthetic analyses by political scientists, focused on such policy areas as poverty, child abuse, consumer protection, and the environment, have identified different patterns of policy initiation, which provide a comparative vantage point from which to examine the origins of the body of ocean law enacted in the 1970s (Downs 1972; Cobb and Elder 1975; Nelson 1978; Lineberry 1980; Jones 1984).

Summarizing a number of studies, Lineberry (1980), for example, provides four explanations for why issues are placed on the public agenda:

1. The presence of a crisis or a sudden worsening in some state of affairs (e.g., the Great Depression of the 1930s, which gave rise to the spate of New Deal legislation).

2. A slow but perceptible change in objective conditions (generally, worsening conditions).

3. The presence of "policy entrepreneurs," either in government or out, who pursue and promote the issue, often inspired by a fervent belief in the "rightness" of their cause (Ralph Nader's role in the promotion of auto safety regulation is often cited as a case in point).

4. The catalyzing role of the media—in letting people know about crises, about deteriorating conditions, or publicizing the activities of policy entrepreneurs.

In a study of the emergence of protective legislation against child abuse, Nelson (1978) highlights the importance of a number of related factors:

1. The presence of studies that, over time, thoroughly documented the deplorable conditions of child abuse and attracted considerable media attention.

2. The importance of "issue naming"—the label of "child abuse," in this case, evoking as it did uniformly negative images, made this problem an issue against which no one could oppose taking action.

3. The availability of "slack resources" in Congress—the presence of talent and resources in a relevant congressional committee that was "searching for an issue" (p. 35).

While many analysts tend to think of policy initiation in linear terms (i.e., conditions worsen or crisis occurs, studies highlight the seriousness of a problem, entrepreneurs bring the attention of the problem to relevant decision makers, and so on), John Kingdon (1995) postulates that major changes in policy are determined by developments in three separate streams—"problem," "politics," and "policy"—which operate largely independently of one another, according to their own rules and dynamics (pp. 86–88). Major policy changes occur only when there is a confluence of the three streams.

In the problem stream, problems are brought to the attention of policy makers by systematic indicators (such as governmental indices, budgets, and studies), by focusing events such as crises and disasters, or by feedback from the operation of current programs. The generation of proposals in the policy stream, resembles, Kingdon (1995) posits, a process of natural selection in biology:

> Much as molecules floated around in what biologists call the "primeval soup" before life came into being, so ideas float around in these communities. Many ideas are possible, much as many molecules would be possible. Ideas become prominent and fade. There is a long process of "softening up": ideas are floated, bills introduced, speeches made; proposals are drafted, then amended in response to reaction and floated again. Ideas confront one another (much as molecules bumped into one another) and combine with one another in various ways. The "soup" changes not only through the appearance of wholly new elements, but even more by the recombination of previously existing elements. While many ideas float around in this policy primeval soup, the ones that last, as in a natural selection system, meet some criteria. Some ideas survive and prosper; some proposals are taken more seriously than others. (pp. 116–117)

The political stream, in turn, is influenced by such factors as swings in national mood, national elections, party control of the executive and legislative branches of government, composition of key congressional committees, the degree of interest and commitment of congressional leaders to particular policy areas, and interest group pressure campaigns.

At critical times, the separate streams of problems, policies, and politics come together. "Solutions become joined to problems, and both of them are joined to favorable political forces" (Kingdon 1995, p. 194). "Windows of opportunity" open for brief periods when significant problems are recognized more or less at the same time as the development of a propitious political climate occurs. Policy proposals, even remotely relevant to the perceived problem, which happen to be ready at this time have a good opportunity to gain a place on the national agenda. Policy entrepreneurs, in and out of government, play a key role in joining solutions to problems and in merging both problems and solutions with politics (Kingdon 1995, Chapter 8).

The National and International Context of the Early 1970s

The early 1970s, a time of relative calm and optimism, in spite of the continuing war in Vietnam, proved to be a dynamic period in global ocean and environmental affairs. The 1969–1972 period, especially, was an interlude during which the relative prosperity of most industrialized nations coupled with the optimistic outlook of newly independent developing nations, permitted a shifting of attention, temporarily at least, away from the Cold War tensions between East and West and from bread-and-butter economic issues to such questions as the rescue of rare and endangered species, the protection of the environment, and the cleaning up of the excesses of highly industrialized and consumer-oriented societies. This climate of optimism was particularly evident in the United States, where preoccupation with the turmoil in Southeast Asia was beginning to subside and the economic strains to be caused by the oil embargoes of 1973 and 1974 had yet to become manifest.

For the first time since the world family had been enlarged with nearly a hundred recently independent nations, several important international conferences were to focus on environmental concerns. In 1972, most of the nations of the world gathered in Stockholm for the first United Nations Conference on the Human Environment. One of the outcomes of this conference was the creation of the U.N. Environment Programme, now based in Nairobi, Kenya. Other international conservation programs were also launched during this period, including the Convention on International Trade in Endangered Species of Wildlife, Fauna, and Flora, which, when adopted in 1973, called for the prohibition of trade in specimens of species vulnerable to extinction (Bean 1983, pp. 272–275).

Another international development that was soon to have a direct bearing on ocean management in the United States also got underway during this period: Under the aegis of the United Nations, discussions began on the need for revising international law as it applied to the oceans—the law of the sea. As discussed in Chapter 2, President Truman's proclamation in 1945 of U.S. jurisdiction over the resources of the continental shelves adjacent to the United States had started a wholesale movement by coastal nations around the world to enclose as national territory increasing areas of ocean bordering their shores. Maritime nations such as the United States and the former Soviet Union urged a U.N. conference on this issue. Developing nations, on the other hand, wanted discussions to begin on the subject of the sharing of the wealth to be derived from the minerals of the deep seabed of the ocean. Information had become available in the mid-1960s suggesting that millions of tons of nodules containing valuable amounts of scarce metals such as cobalt, nickel, and manganese were scattered on the seafloors of the deep oceans. Developing nations wanted to see an international authority created to manage the exploitation of these minerals—resources they saw as the common heritage of mankind.[1]

In 1970, agreement was reached on a broad, inclusive agenda for such a conference, and plans were developed to open the substantive negotiations in Caracas, Venezuela, in April 1974. The Third United Nations Law of the Sea Conference (UNCLOS III) was to extend over eight years, finally concluding a draft convention in April 1982 in New York City. Even though, in the end, the United States and several other industrialized nations were not able to agree on certain of its provisions (i.e., those dealing with seabed mining),[2] the conference itself served, during the decade of the 1970s, to focus the attention of many nations of the world on ocean problems and opportunities.

Growing Environmental Concern

In the United States, the late 1960s and early 1970s was a period of strongly heightened interest in the environment. The huge oil spill off Santa Barbara, California, in January 1969 attracted widespread attention and concern. While the first Earth Day (April 22, 1970) may have signaled the official start of this movement, earlier books such as Rachel Carson's *Silent Spring* (1962), played a very important role in heightening public concern about environmental degradation. It was as if the nation, and particularly its young people, were beginning to tire of the long and divisive debate over the war in Southeast Asia and were ready to turn their energies to more positive and rewarding issues closer to home.

Public awareness of environmental issues rose dramatically during this period. For example, in 1965 only 69 percent of the public considered air pollution a somewhat serious or very serious problem; whereas in the 1970s, 85 percent of the public regarded pollution as a serious issue (Whitaker 1976, pp. 23–24).

Growing environmental awareness on the part of the general public was fueled by the mobilization of ecologically oriented environmental groups (such as Friends of the Earth, Defenders of Wildlife, and the

Society for Animal Protective Legislation) that added new concepts of ecology to older notions of conservation held by more established traditional wildlife management groups such as the National Wildlife Federation (Cicin-Sain 1982a).

In a separate but related development, as discussed in Chapter 2, the 1960s had seen an awakening of a substantive interest in the oceans and their potential benefit. In a kind of counterpoint to the space exploits of the late 1950s and 1960s, programs were being designed to explore "inner space," as the ocean depths were then being called. Congress, together with the National Academy of Sciences and groups in the executive branch, instituted a series of studies dealing with the problems and opportunities bound up in coastal and ocean resources. The best-known of these—the Stratton Commission's report *Our Nation and the Sea*, issued in January 1969 (COMSER 1969)—presented a very optimistic assessment of the ocean resources potentially available to the United States. A counterpart to the innovative and effective academic research program of the last century—the land grant college system—was established in 1966 in the form of a system of "sea grant" colleges (Abel 1981; Miloy 1983). By the early 1970s, the first fruits of a focused program of marine resources–oriented research were becoming evident.

A Resurgent Congress—But a Weakened Presidency

The 1970s also saw some striking changes in the functioning of the Congress and the presidency and in the relations between these two institutions. This was a period of congressional resurgence that witnessed a highly active and reform-minded Congress. During this time, Congress underwent a number of significant changes in its organization. Perhaps most striking were (1) the loss of traditional power for full committee chairmen; (2) growth in the number of congressional subcommittees (e.g., from 257 in 1970 to 299 in 1975); and (3) great growth in congressional staff (e.g., from about 5,000 to 7,000 for House members), marking a significant increase in the professional competence and expertise available to lawmakers (Patterson 1978).

Key members of Congress with an interest in the oceans—such as Senators Stevens (R-Alaska), Pell (D-Rhode Island), Hollings (D-South Carolina), and Magnuson (D-Washington); and Representatives Dingell (D-Michigan), Murphy (D-New York), Breaux (D-Louisiana), and Studds (D-Massachusetts)—lent strong, substantive leadership to the ocean programs movement, with the assistance of highly committed and competent staff in such committees as the House Committee on Merchant Marine and Fisheries and the Senate Committee on Commerce, Science, and Transportation.

In contrast to the resurgence in Congress, this period saw a significantly weakened presidency—the result of the perceived failures of the "imperial presidency" as manifested in the conduct of American involvement in the Vietnam War, in the Watergate scandal, and in the perceived failure of the Great Society programs (Greenstein 1978). Moreover, during the 1960s and 1970s, the rate of turnover in the presidency was unprecedented in modern American history. Between 1933 and 1961, three presidents held office for the equivalent of seven four-year terms, a period of twenty-eight years; whereas from 1961 to 1981, five different presidents held office, averaging just one four-year term each.

The weakened presidency, coupled with the rapid turnover in the office, meant that, in the 1970s, Congress became the prime moving force in domestic policy. This was particularly true in the case of ocean policy, where new issues related to ocean resources were embraced by congressional entrepreneurs casting about for "uncolonized" issue areas. During the seventies, the presidency provided little substantive direction for ocean policy. With the exception of federal reorganization efforts affecting ocean matters, the three presidents during this period (Nixon, Ford, and Carter) displayed little substantive concern with ocean affairs, each predominately preoccupied with other areas of policy—Nixon and Ford most prominently with foreign affairs and with the domestic devolution of federal functions to state and local governments, and Carter with the wholesale reorganization of the federal government and with the forging of comprehensive energy and urban policies.

In summary, the growing international and national interest in the oceans and the environment, the mobilization of new environmental interest groups, the resurgence of Congress and, in particular, the proliferation and specialization of congressional subcommittees, and the weakened presidency all contributed to the spate of new ocean legislation in the 1970s.

Managing the Nation's Coasts[3]

While enactment of the Coastal Zone Management Act of 1972 (*CZMA* 1972, Sections 1451 et. seq.) involved at least a half-dozen years of study and debate, the idea of giving special planning and management attention to the coastal zone was a fundamentally new one in the United States. The effort to manage San Francisco Bay in the early 1960s was perhaps the first program of this type (Odell 1972). Within a few years, coastal resource protection programs were also developing in several other states, usually in direct response to perceived threats of one type or another (Heath 1971). But these early state and local coastal management efforts tended not to be comprehensive, suffered from lack of funding, and could not, under the aegis of state legislation, affect federal coastal activities.

Legislative Origins of the Coastal Zone Management Act

The new federal legislation authorized the establishment of a federal program to encourage and assist the states in developing and implementing programs to manage their coastal areas, within zones defined in the act to include the 3-mile territorial sea adjacent to state shorelines and an unspecified width of shoreland, "the uses of which have a direct and significant impact on the coastal waters," depending on the particular geographic situation in the state (*CZMA* 1972, Section 1453 [1]). Even though the program was fully voluntary and included no penalties for failure to participate, the legislation represented the federal government's most aggressive step yet regarding the management of land and water uses at state and local levels.

The origins of this legislation can be traced to at least four important sources. First and probably most influential were the recommendations contained in the final report of the Stratton Commission (U.S. COMSER 1969), which, as a part of its call for greater public benefits from coastal and ocean resources, explicitly urged federal coastal zone management legislation to assist the states in managing such resources and to require federal programs and activities to conform to state coastal policies (U.S. COMSER 1969). The Marine Science, Engineering and Resources Council,[4] working during the same period with an interagency panel of federal agencies (the Committee on Multiple Uses of the Coastal Zone), examined coastal zone management (CZM) issues and needs and, with the assistance of the Department of the Interior, began drafting legislative approaches, some of which found their way into the CZMA. Also of major importance were the findings of two closely related studies—the National Estuarine Pollution Study and the National Estuary Study, both of which were published in 1970. These studies, authorized by Public Law 89-753 and Public Law 70-454, and conducted by the Interior Department, presented a clear picture of the extent of pollution in the nation's estuaries and the urgent need for remedial action. Finally, from an entirely different quarter, the idea of comprehensive land-use planning was beginning to gain some acceptance. Two studies, the report of the Public Land Law Review Commission (U.S. Congress, Senate 1967) and the American Law Institute's Model Land Development Code (Reilly et al. 1973), called attention to this need and provided suggestions for specific approaches.

Forces Active during Enactment

Enactment of the CZM legislation in 1972 occurred at the time when the environmental movement had already become a strong and effective force in U.S. politics. Older conservation groups such as the Audubon Society and the Sierra Club, joined by newer organizations such as the Friends of the Earth and the Natural Resources Defense Council, collectively had a major voice in shaping many environmental bills enacted between 1969 and 1975. The active participation of these groups ensured that the legislation incorporated the highest degree of environmental protection that was politically feasible.

Another contextual factor was the contemporaneous movement toward a stronger state role in land-use planning. The National Governors Conference passed strong land-use planning resolutions at its 1969 and 1970 meetings, and some of the recommendations of the American Land Institute Model Land Code were beginning to find their way into state legislation (e.g., Vermont and Florida) (Reilly et al. 1973).

Finally, some of the individual coastal states themselves had already become effective forces on behalf of coastal protection by the time that the federal legislation was being considered for final action. At least five ocean states (California, Oregon, Washington, Rhode Island, and Delaware) and three Great Lakes states (Michigan, Wisconsin, and Minnesota), had by then embarked on coastal or shoreline programs. These states, some with coastal management programs already underway, became valued participants in the process of debating and shaping the national legislation. Being closer to the real needs for coastal management and more aware of the obstacles and constraints posed, they had a significant impact on the final content of the legislation (U.S. Congress, Senate 1976).

Coastal Planning versus Land-Use Planning While many external forces are active during the legislative process, often the most influential, in retrospect, turn out to be those interests that actually drafted early versions of the legislation. Relatively few groups or individuals have the expertise, the time, or the inclination to produce full drafts of complex legislation; most find that they can adequately protect and/or advance their interests through reviewing the drafts of others and by suggesting specific changes in language or concepts. In the case of the CZM legislation, early, relatively comprehensive drafts were produced by the Interior Department (as a part of its work with the interagency Panel on Multiple Uses of the Coastal Zone), the newly formed Coastal States Organization, and the staffs of the Senate Committee on Commerce, Science, and Transportation and the House Committee on Merchant Marine and Fisheries.

It is important to note that two members of the House Merchant Marine and Fisheries Committee (Alton Lennon [D-North Carolina], chairman of the oceanography subcommittee, and Charles Mosher [D-Ohio], ranking Republican member of that subcommittee) were advisory members of the Stratton Commission and participated in the formulation of the recommendation for CZM legislation. Similarly, Senator Warren G. Magnuson (D-Washington), chairman of the Senate Committee on Commerce, Science, and Transportation, and Senator Norris Cotton (R-New Hampshire) were also advisory members of the commission (U.S. COMSER 1969). As is customary in these instances, the congressmen and the senators were supported in their work on the commission by members of their committee staffs, generally the same individuals on whom the task of drafting legislation of this type would fall. It is not surprising, then, that drafts of the legislation introduced by Magnuson and Lennon in 1969 closely followed the Stratton Commission's recommendations released only a few months earlier.

Following the formation of the Coastal States Organization (CSO) by the National Governors Conference in late 1969, that group produced a draft of CZM legislation in early 1970, which was provided to both the House and the Senate committees. The CSO draft, while also embracing many of the Stratton Commission's recommendations, included some new land-use planning elements taken from the American Land Institute Model Land Development Code.

At this juncture, a major split occurred in the legislative process. The Nixon administration, an initial proponent of CZM legislation, reversed its position and in February 1971 endorsed, instead, comprehensive national land-use legislation dealing with all areas of critical environmental concern, not simply the coastal zone. Furthermore, the administration wanted the new land-use program to be housed in the

Department of the Interior and not in the Commerce Department, which since 1970 had been home to the new National Oceanic and Atmospheric Administration, the agency recommended by the Stratton Commission to direct the national CZM program. Thus, the two most contentious issues surrounding the passage of the CZM act came into focus: Should CZM be part of a national land-use program or be independent, and which department should manage the program—Interior or Commerce?

The concept of national land-use planning, however, had not yet reached political acceptability. Unexpectedly strong and persistent opposition to the land-use legislation came from a wide range of development-oriented interests (real estate, energy producers, commercial and industrial concerns), and from wary local governments and reluctant agricultural interests. As these interests slowed the movement toward a national land-use bill, the momentum for CZM legislation resumed.

The question of which cabinet department, Commerce or Interior, would manage the national CZM program, however, was settled in favor of Commerce and NOAA only at the very last moment by the House-Senate Conference Committee in the process of reconciling differences between House and Senate versions of the bill (U.S. Congress, Senate 1976, p. 7). Interestingly, the House-Senate compromise that gave the program to Commerce and NOAA was based on the supposition that a soon-to-be-passed land-use bill would give the Interior Department responsibility for that program and, at the same time, a measure of control over land use in the landward portions of the coastal zone (U.S. Congress, Senate 1976, p. 6). Shortly after the enactment of the CZM legislation, however, Congress defeated further attempts to pass the national land-use legislation, and the movement for national land-use planning subsided significantly.

On October 27, 1972, President Nixon signed the CZM legislation into law after the measure recommended by the Conference Committee had been overwhelmingly adopted by both the Senate (68 to 0) and the House (376 to 6). Final House figures mask a spirited last-minute attempt to restore the Interior Department as lead agency of the program. In President Nixon's signing message, he made clear his administration's preference for a national land-use program housed in Interior, stating that he had signed the measure "as a first step toward a more comprehensive program" (U.S Congress, Senate 1976, p. 459).

Indeed, the Nixon administration declined to request appropriations to begin implementation of the newly authorized CZM Act after enactment, presumably awaiting the passage of land-use legislation. In the fall of 1973, however, key senators held several administration energy bills "hostage" until the administration agreed to request funding for the CZM program.

An Intergovernmental Approach

Major features of the 1972 CZM legislation can be summarized as follows:

1. Recognizing that conflicts existed among competing uses in the coastal zone (often conservation versus development), as well as among the overlapping authorities of the three levels of government, the act designated one level of government—the state—as the appropriate one to take the lead in developing plans to manage land and water uses in the coastal zone.

2. The program was a voluntary one, with no sanctions imposed against states choosing not to prepare CZM programs under the act.

3. Federal grants-in-aid were to be made available for a limited period to state governments (and through them, to local governments) for developing CZM programs, and after federal approval of state programs, grants were to be available to the states for implementing and administering such programs.

4. Federal agencies were called upon to assist states in the CZM process and, after program approval, to act consistently with the policies contained in their coastal plans.

5. The criteria describing the requirements for federal approval of state programs all dealt essentially with process and not with substantive standards.

Federal financial assistance to encourage states to undertake planning endeavors of various types was, of course, not new. In this sense, the CZM Act took its place alongside programs in urban planning (administered by the Department of Housing and Urban Development), water quality planning (administered by the Environmental Protection

Agency), and the oldest forms of planning assistance—water resources planning (administered by the Department of Interior). Under these programs, state and local governments were given federal grants to develop plans, which were ultimately expected to lead to increased state and local government capacity in these three program areas. But the CZM Act appeared to go further than any of these in giving the states a voice in the actions of federal agencies affecting their coastal zones once the state CZM program had been approved at the national level. The real strength of the novel "federal consistency" provisions established by Section 307 of the act was to be determined later as states attempted to use this power to gain greater control in the federal offshore oil and gas leasing program.

Taken together, the various provisions of Section 307 required, in effect, that *federal activities, federal development projects, federal licenses and permits, and federal assistance action* be consistent with approved state CZM programs. While escape clauses existed for each of the above categories of federal actions, these were judged to be relatively narrow and limited. Sections 307 (C)(1) and (3) employ the term "affecting the coastal zone"; thus federal actions occurring *outside* the coastal zone but affecting it also fall within the scope of the federal consistency requirements, adding an element of "extraterritoriality" to a state's power under the CZM Act. This aspect was to become very important in the late 1970s and early 1980s as states began to differ with the federal government over offshore oil leasing policies in federal waters.

The CZM Act embodied an unusual form of federalism. The underlying concept of the act—the driving force, as it were—was contained in the following quid pro quo: If states adopted CZM programs that were judged to meet the national interest and other tests found in the act, and if the states were willing to enforce their programs in relation to local and state interests (public and private), federal agencies would be bound by the coastal policies contained in those programs as well. Although states would appear, therefore, to be in a strong position, proponents of this approach to CZM soon realized that local interests, if not properly accommodated, could frustrate state CZM efforts in the state legislature, and federal agencies, jealous of their prerogatives, could complicate if not fully prevent federal approval of a state program. Hence, the realities of the situation required accommodation and balancing of federal, state, and local interests (Archer 1989).

Interestingly, the act did not require states to achieve specified substantive goals in the CZM programs, but instead required that state programs address the coastal problems and opportunities present in the state by means of a set of enforceable policies aimed at improving management within the coastal zone. This orientation is evident in the statement of congressional findings contained in the act:

> The key to more effective protection and use of the land and water resources of the coastal zone is to encourage the states to exercise their full authority over the lands and waters in the coastal zone, by assisting the states, in cooperation with federal and local governments and other vitally affected interests, in developing land and water use programs for the coastal zone, including unified policies, criteria, standards, methods, and processes for dealing with land and water use decisions of more than local significance. (*CZMA* 1972, Section 1451[i])

Tensions in the Law

Often in the process of framing legislation, conflicting goals or policies are incorporated into one bill; the give and take of the enactment process often plants the seeds of the most substantial problems that will be faced during the implementation stage. Compromises and accommodations made hastily in order to secure the few key votes needed for passage often present major problems later to those faced with implementation. Of course, there are many occasions when vague and even inconsistent language is intentionally inserted into a measure—when committee members cannot agree on more definitive wording or when the drafters of the legislation want the implementers of the legislation to have wider latitude in their interpretation of the measure. Problems can thus arise later during implementation either because issues were dealt with vaguely or inconsistently during the legislative process, or because they were not addressed at all, intentionally or not.

The broad, general language of the CZM Act resulted in building in at least three major tensions during the legislative process that were to play important roles in the implementation stage. The first tension was between land and water. Was CZM to be a program directed principally toward the management of the land lying directly on the water's edge, typically very valuable for many purposes and much sought after by a variety of users? Or was it to attempt to handle both land and water, recognizing that different techniques, different interests, and different uses occurred in each? The second fundamental tension concerned whether the act would focus on conservation or on management of resources. Were CZM programs largely to be protective in their orientation, aimed principally at conservation and restoration of the fragile but exceptionally valuable natural areas (wetlands, estuaries, beaches, and dunes) found in the coastal zone, or were they to be management-oriented, striving for a reasonable balance between conservation and development and seeking to accommodate facilities best located in the coastal zone? The third tension was intergovernmental in nature. Was the program to be a top-down effort, with federal guidelines and policies strongly shaping the state and local CZM plan development process, or was the program to be built from the bottom up, fashioned mainly at the local and state levels, with the federal role largely one of providing financial and technical assistance (Archer 1989)?

In addition, the legislation contained the roots of several other problems. First, the requirement that state plans "provide for adequate consideration of the national interest involved in the siting of facilities necessary to meet requirements which are often other than local in nature" as a prerequisite for their approval by the federal government was largely undefined in the legislation and ultimately was to result in considerable litigation and delay. Second, the question as to whether the federal consistency provisions of the act applied to lease sales conducted as a part of the federal offshore leasing program also was unclear (Archer and Bondareff 1988). Eventually, this issue was to be decided by the U.S. Supreme Court. Finally, nothing in the 1972 legislation indicated how long the state could expect federal financial assistance for coastal zone management to continue, and as a result, expectations differed greatly.

Protecting Marine Mammals

The Marine Mammal Protection Act (P.L. 92-522, hereinafter MMPA) was enacted in 1972 in response to rising worldwide concern about the status of marine mammals and detrimental impacts by humans on these species (*MMPA* 1972). Explicitly aimed at protecting marine mammal populations, the MMPA represented a key ocean policy action by establishing that high priority be given to protecting these living creatures of the marine environment. The Endangered Species Act (P.L. 93-205, hereinafter ESA), extended its protective mantle to all threatened and endangered species, whether on land or in the ocean—thus providing an additional level of protection for marine creatures found to be threatened or endangered (*ESA* 1973). Taken together, the MMPA and the ESA have proven quite effective in restricting and, in some cases, preventing the exploitation of other ocean resources (such as offshore oil and gas) in instances that might adversely affect these creatures.

The Marine Mammal Protection Act[5]

Humanity's fascination with marine mammals is long-standing. Mammals ourselves, we regard them as unique embodiments of our species' attachment to and fascination with the sea. They can be likened to humans because of their attributes and behavior; for example, the intelligence and communicative skills of porpoises, or the nurturing behavior of sea otter females toward their young. In some respects, their capabilities and behavior can be viewed as transcending those of human beings. The mammoth whales that traverse great distances through different regions of this water planet, embodying mystery and adventure, strength and grace, are among the most fascinating creatures on earth.

By the late 1960s a number of scientific studies and reports, and the actions of newly formed wildlife protection groups, had focused worldwide attention on the threats of extinction or depletion faced by marine mammal populations as a result of human activities. Three factors stand out as catalytic in the rise of the marine mammal protection movement. The first was the growing concern with the decline in whale populations around the world and the inability of the International Whaling Commission to reduce the alarming depletion of whales caused by overharvesting. A second factor was public reaction to the harvesting of baby harp seals in Canada (in the Gulf of St. Lawrence and the northwest Atlantic) and of Pribilof fur seals (in the Aleutian Islands in Alaska) through clubbing (furriers preferred clubbing seals to death instead of shooting them in order to preserve their pelts intact).[6] In a massive mobilization and media campaign, the Committee for Humane Legislation—a large animals rights and animal welfare organization—focused public attention and criticism on the harvesting of both species of seals (Manning 1990). The third catalyst was increased public awareness of the vast numbers of porpoises being drowned incidental to commercial fishing. As many as 200,000 to 400,000 porpoises were being drowned each year in giant purse-seine tuna nets (U.S. Congress 1972).

Key Features of the Act

Congressional discussions of the need for marine mammal protective legislation underscored the numerous hazards to which marine mammals were being exposed—for example, degradation of the ocean environment (through ocean dumping, pesticides and heavy metal contamination); people's increasing take of the fish stocks upon which the mammals depend; and accidental death due to vessel traffic. The cumulative effect of these hazards and the inadequacy of existing knowledge of marine mammals in general led Congress to conclude that protective action regarding these animals was required. As one congressman commented, "As far as could be done, we have endeavored to build such a conservative bias into the legislation" (U.S. Congress, House 1971).

At its core, the MMPA is a protective act. Protection was to be attained through a moratorium on the take[7] of marine mammals (a "complete cessation of the taking of marine mammals and a complete ban on the importation into the United States of marine mammals and marine mammal products" Section 1362 [7], (16 U.S.C. § 1362 [7]), a long-term research effort to better understand marine mammals and their environment, and a set of prohibitions and penalties for violation of the act's provisions.

A number of exceptions to the moratorium, under permitting conditions, were allowed—most prominently, those for scientific research and display and for take incidental to commercial fishing operations (particularly relevant to the tuna fishery, where porpoises were habitually caught and injured in purse-seining operations). Exemptions to the moratorium were allowed for the subsistence and craft-making use of Alaskan natives, and for the Pribilof Islands fur seal harvest (*MMPA* 1972, Section 1371). Stiff penalties were imposed on violations of the taking provisions of the act (Section 1375).

In addition, the MMPA established a highly detailed formal procedure whereby the moratorium could be waived and taking and importation permitted. As Bean (1983) notes, prior to any waiver the secretary must determine that the affected species or population be within the range of optimum sustainable population and that the taking authorized by that waiver does not reduce it below that level. As of the date of Bean's book (1983), only one moratorium decision under the full administrative process required by the act had been issued (p. 299).

Beyond the immediate (and tangible) protection offered by the moratorium, there was also a strong concern in the act with the enhancement of marine mammals and the ecosystems of which they are a part. The act pioneered an ecosystemic view of the marine environment by declaring the following:

> Marine mammals have proven themselves to be resources of great international significance, esthetic and recreational as well as economic, and it is the sense of the Congress that they should be protected and encouraged to develop to the greatest extent feasible commensurate with sound policies of resource management and that the primary objective of their management should be to maintain the health and stability of the marine ecosystem. Whenever consistent with this primary objective, it should be the goal to obtain an optimum sustainable population keeping in mind the optimum carrying capacity of the habitat. (*MMPA* 1972, Section 1361 [6])

This language marked a departure from the species-by-species harvest orientation that had heretofore been dominant in marine mammal management. "Optimum sustainable population (OSP)" replaced "maximum sustainable yield"[8] as a management objective, and a holistic approach to the ecosystem was stressed. Unfortunately, implementation of these innovative concepts was hampered by the tautological manner in which "optimum sustainable population" and "optimum carrying capacity (OCC)" were originally defined in the legislation—OCC was used to define OSP, and OSP to define OCC.[9]

Implementation of the act was split between two federal departments. The Department of Commerce (through the National Marine Fisheries Service [NMFS] in NOAA) was given authority over cetaceans (whales, dolphins, and porpoises) and all members, except walruses, of the order Pinnipedia (seals and sea lions). The Department of the Interior (through the United States Fish and Wildlife Service) was given authority over all other marine mammals (sea otters, polar bears, walruses, manatees, and dugongs (MMPA 1972, Section 1362 [11] [B]).

In addition, the act established a three-person Marine Mammal Commission and a Committee of Scientific Advisors to oversee and conduct research studies and the coordination of management activities (*MMPA* 1972, Sections 1401–1403). The commission is an advisory body with no regulatory or rule-making authority under the MMPA, and is charged with the following duties: (1) reviewing and studying existing domestic and international laws concerning marine mammals; (2) continuing monitoring of the condition of marine mammals, research programs, methods of management, and so on; (3) conducting studies it deems necessary in conjunction with conservation and protection; and (4) making policy recommendations to the various agencies charged with research, regulation, and enforcement related to marine mammals.

The MMPA represented a significant departure from previous management practice, wresting, as it did, management control away from the states. Coastal states, such as California and Alaska, had long been managing and adequately protecting (or so they thought) their marine mammal populations. Their loss of management authority would produce significant conflicts in the implementation process (such as in the case of sea otter management in California [see e.g., Cicin-Sain 1982c]). While shifting management authority to the federal level, however, the act provided for the opportunity for the states to request a return of management authority. Under Section 109 (b)(g), under state petition, if the secretary (of Interior or Commerce, whichever was applicable) determined that state laws and regulations were consistent with the act, management authority could be returned to the state. However, although Alaska petitioned for return of management authority in the mid-1970s, no management authority has been returned to any state.

An Internecine Fight

Several bills protecting marine mammals were debated in Congress from March 1971, when Senator Fred Harris (D-Oklahoma) and Representative David Pryor (D-Arkansas) first introduced the Ocean Mammal Protection Act, to October 1972 when the Marine Mammal Protection Act was signed into law by President Nixon. During this time, a major congressional debate centered not on the desirability of enacting protective legislation, but rather on the different degrees of protection that should be afforded to these marine creatures.[10]

Engineered by the animal protection groups organized in the late 1960s and early 1970s, such as the Committee for Humane Legislation, the Pryor measure in the House (H.R. 6558) called for stringent protection by imposing an outright prohibition against the taking or importation of marine mammals or marine mammal products. This stringent approach, which viewed marine mammals as a "moral resource"[11] to be preserved (Grandy 1981) was supported by a coalition of recently organized environmental groups, such as the Fund for Animals, Friends of the Earth, the Humane Society of the United States, Committee for Humane Legislation, Society for Animal Protective Legislation, International Fund for Animal Welfare, Defenders of Wildlife, and World

Federation for the Protection of Animals. The newer groups promoted concepts relating to the moral and aesthetic value of wildlife resources and to the rights of animals in preference to prevailing theories about wildlife management that theretofore had emphasized use coupled with conservation (Cicin-Sain 1982a).

Throughout debates on the MMPA, the newer environmental groups skillfully used the media to dramatize their cause—the plight of animals worldwide. They were described by an inside observer as "leading a citizens' lobby in an attack on a Congress not yet used to this kind of pressure" (Robinson 1981). In fact, during the controversial raid on Cambodia at the height of the Vietnam War, congressional mail was dominated by letters protesting the use of beagles in scientific experiments and calling for the protection of marine mammals (Robinson 1981).

The debate on the management practices proposed in the MMPA was an internecine struggle between factions of the environmental movement. More established wildlife groups, which had traditionally emphasized management of wildlife for both use and conservation, mobilized against the strong protectionist approach embodied in the House bill. A coalition of twelve national conservation organizations (including the World Wildlife Fund, the Wildlife Society, the North American Wildlife Foundation, the Citizens Committee on Natural Resources, the Wildlife Management Institute, the International Association of Game, Fish and Conservation commissioners, and the Audubon Society) mobilized their contacts in both the Congress and the Nixon administration to defeat the strictly protectionist measure.

Among those favoring a management-oriented act were the Departments of State, Interior, and Commerce, the Council for Environmental Quality, the fur and tuna industries, and a significant portion of (although not the entire) scientific community. The furrier interests (represented primarily by the Fouke Company, based in South Carolina and one of the nation's largest fur traders), sought to prevent language in the MMPA that, in their view, would put their industry out of business, and thus fought to achieve the weakest bill possible (Poser 1981). The state interests, represented by the International Association of Fish and Game Commissioners, were primarily concerned with the potential implications of the bill on marine mammal management programs, such as Alaska's. While the state representatives recognized the need for some form of federal control over offshore species (such as whales and porpoises), because the states had neither the jurisdiction nor the resources to handle such species, their major lobbying effort was focused on incorporating procedures and definitions in the legislation that could be implemented through familiar management practices, and on retaining for the state as much traditional jurisdiction as possible (Lenzini 1981). They were to lose on both accounts, as the MMPA ultimately incorporated novel concepts of management (e.g., optimum sustainable populations and optimum carrying capacity) and generally preempted state management of marine mammals.

The management-oriented groups introduced a substitute measure (H.R. 10420), sponsored by Congressman Glenn Anderson (D-California), which provided protection while allowing for significant exemptions for Native Americans and for take incidental to commercial fishing operations. This measure was drafted by Frank Potter (staff member for Congressman John Dingell) (D-Michigan), Lee Talbott (Council on Environmental Quality), and Carlton Ray (Johns Hopkins University). According to Potter, this drafting team believed that a protectionism emphasis should be built into a management framework, and saw a need to draft a bill that would coexist with the Fur Seal Treaty, thus continuing the ban on pelagic hunting, while providing a new perspective on the management on marine mammals (Potter 1981). They drafted the bill on behalf of Congressman Dingell, who gave his approval and allowed Congressman Anderson to sponsor it. Combining protection of marine mammals with exemptions for incidental take and other uses of marine mammals, the bill corresponded well with the interests of Anderson's constituents in California—the home of tuna fishermen as well as of leaders of the marine mammal protection movement (Potter 1981).

Hearings by the Subcommittee on Fisheries and Wildlife Conservation of the House Committee on Merchant Marine and Fisheries in September 1971 debated both options, and the committee unanimously reported out the management-oriented bill, the Anderson-Pelly initiative, incorporating amendments allowing take for Native Americans' subsistence and arts and crafts purposes and take incidental to commercial fishing operations. A weak version of this bill ("weak" by the standards of the protectionists) was defeated in December 1971 by the protectionism groups in another dramatic instance of use of media resources. On the eve of the vote, editorials blasting the management-oriented measure and urging defeat or postponement appeared in the *New York Times*, *Washington Post*, and *Washington-Star News*.

In early 1972, a bill affording stronger protection passed the House, and the conservation battle moved on to the Senate. After considerable debate, a relatively weak Senate bill emerged, managed by Senator Hollings (D-South Carolina) and Senator Ted Stevens (R-Alaska). This was apparently due to the fact that the protectionist groups focused their efforts on obtaining an alternative, much stronger bill, thus losing ground in the bargaining that took place on the Senate bill that eventually emerged (U.S. Congress, Senate 1972). In the midst of the Senate deliberation process, in another display of media prowess, Alice Harrington, head of Friends of Animals and Committee for Humane Legislation, published an article entitled "The People of the Sea Are Faced with a New War," accusing Senators Hollings and Stevens of being beholden to commercial interests (the South Carolina–based Fouke Company in the case of Hollings; for Stevens, the Pribilof fur seal hunt).

In view of the later conflicts in which the MMPA would clash seriously with fishery management practices at both federal and state levels, it is surprising that only one fishing group presented testimony: the lobby representing the tuna fishery—a group significantly affected by the proposal because of its purse-seining operations in which large numbers of porpoises were being drowned. The group's testimony, moreover, emphasized the steps that the fishers themselves had taken to improve fishing gear to allow the porpoises to escape unharmed from nets.

The absence of other marine-sector interests such as fisheries and offshore oil in the enactment of the MMPA underscores the very separate nature of the political processes involved in each marine sector at the beginning of the decade, and helps to explain, in part, the serious conflicts that were ultimately to occur among different ocean users.

The final version of the act, reconciled in a conference committee in July 1972 and signed by President Nixon in October 1972, although not offering the maximum degree of protection sought by the protectionists, nevertheless represented a significant and unprecedented step in the conservation of marine mammals, calling for a moratorium on the taking of marine mammals and the importation of marine mammal products and introducing a number of broad and far-reaching concepts of management.

The impetus for the MMPA came from and was sustained by the more radical protectionist wing of the wildlife conservation movement that provided the external pressure needed to get the act seriously considered. Without such unrelenting pressure, marine mammal protection probably would not have been enacted. It was, however, concerned members of the Congress and federal officials who were responsible for crafting the more pragmatic act that ultimately emerged from the process. Some critics of the act decried the fact that the law was drafted hastily and enacted quickly because of its popularity with the public. Internal inconsistencies in the law as well as vagueness in some of its central concepts (e.g., OSP), they maintained, would create significant difficulties in interpretation (Parsons 1981).

Issues for Implementation

Several contentious issues surrounded the passage of the MMPA and led to conflict in the implementation stage (Baysinger 1981). First, controversies surrounding incidental take related to tuna fishing and continuation of the fur seal hunt (issues initially won by the pro-management groups) would remain alive because of continuing opposition by the protectionist groups. Second, the Congress elected to maintain the existing division of authority between the Departments of Commerce and the Interior. NOAA, in the Department of Commerce, already had responsibility for research and management of cetaceans (whales, dolphins, porpoises) and seals, while Interior had authority over walruses, sea otters, polar bears, and manatees. Although the Congress indicated its dissatisfaction with this jurisdictional split, it retained the status quo because it expected the creation of a Department of Natural Resources, which would merge the two programs (Potter 1981). If such a department were not created "within the reasonably near future," Congress planned to reexamine the question and to consider consolidating the marine mammal program (Coggins 1975).

Third, the ambiguity in central concepts, such as "optimum sustainable population" and "depletion" of a species was to create interpretation problems in the implementation stage. The concept of OSP was not well understood at the time of passage, was ill-defined in the legislation and somewhat circular. It is interesting to note that OSP was left intentionally vague, according to one of the framers of the concept (Potter 1981). OSP provided a new perspective that would have to be defined by rethinking old management concepts during the actual implementation process. This approach was taken apparently because the proponents of OSP had more trust in the sympathies of the "environmental mafia" found in the federal agencies that would implement the concept than in the willingness of Congress to adopt the OSP, had its implications been fully spelled out.

Finally, the ecosystems view embodied in the MMPA, while representing a significant advance over the previous species-by-species approach, was itself a limited concept, because it did not, in our view, adequately address important human needs (food, energy) dependent upon ocean resources (fish, oil and gas) that may be in conflict with protecting marine mammals. This omission was to lead, in some cases, to proponents of the MMPA promoting pristine images of marine ecosystems in which the present weight of human presence is ignored (for example, in the case of the sea otter in California, where protectionists have argued that sea otter populations should be restored to levels prevailing in the 1800s, before the colonization of California and before the establishment of profitable fisheries along California's coast).

Promoting Fisheries: The Adoption of the 200-Mile Limit

The adoption of the Magnuson Fishery Conservation and Management Act of 1976 (Magnuson Act, P.L. 94-265, hereinafter MFCMA),[12] after several years of debate on earlier versions of a 200-mile limit, was the result of the confluence of several international and domestic contextual factors and the concerted action of policy entrepreneurs in Congress who, together with fishing and environmental interest groups, crafted the language of the act and shepherded it through the congressional process.

Among the most important contextual factors leading to the passage of the act were (1) the growing concern with the worldwide overexploitation of fish stocks; (2) the decline of U.S. fisheries as a result of a dramatic increase in harvesting by foreign fleets; and (3) the lack of progress, by 1975, of the Law of the Sea negotiations. By the early 1970s, through a number of studies and reports, it had become clear that overexploitation of fishery stocks was proceeding at an alarming rate; according to scientists' reports, approximately 25 percent of U.S. stocks were overfished or threatened with overfishing (U.S. Congress, House 1975).

In the United States, fishing by technologically superior foreign fleets, such as those of the former Soviet Union and Poland, had resulted in the serious depletion of at least ten major commercial stocks, causing hardship for American fishers as well as raising the specter of ecological disaster.[13] Coupled with these developments, the lack of progress at the 1975 session of the Law of the Sea negotiations in Caracas lent support to those who thought it imperative to "go it alone" in order to prevent further overexploitation, instead of awaiting the doubtful results of the ongoing negotiations.

A New Era of Fishery Management

The Magnuson Act established U.S. "exclusive management authority" over all fish, anadromous species,[14] and continental shelf living marine resources (with the exception of highly migratory species[15] found within the newly created fishery conservation zone (FCZ),[16] spanning from the 3-mile limit of the territorial sea to 200 miles offshore. The act created a national program for the conservation and management of the fishery resources of the United States "to prevent over-fishing, to rebuild over-fished stocks, to ensure conservation, and to realize the full potential of the Nation's fishery resources," including the development of underutilized fisheries (*MFCMA* 1976, Section 1801 [a] [6]; [b] [6]). The Magnuson Act thus marked the beginning of a new era of fishery management in the United States. It signaled, in effect, that new management systems had to be created to protect and enhance the economic and social benefits derived from commercial and recreational fishing.

The act introduced four major, novel features in fishery management. First, it established federal management jurisdiction over the FCZ, an area in which fisheries largely had been unmanaged, or only partially managed by international authorities, or managed indirectly by state governments. This action brought under the control of the federal government an area comparable in size to the land mass of the continental United States (over 2 million square nautical miles, within which are found 15 to 20 percent of the world's traditionally harvested marine fishery resources) (U.S. Department of Commerce 1981a). Under Section 306 of the act, the states retained control over fishing activities in the territorial sea with one major exception: State authority could be preempted by the federal government in cases where state action (or lack thereof) adversely affected the carrying out of the fishery management plan for a fishery that was predominantly found in the fishery conservation zone (*MFCMA* 1976, Section 1856 [b] [B]).

Second, the act set new national standards for managing fisheries and called for the preparation of fishery management plans (*MFCMA* 1976, Section 1851). Among the most important of the seven national standards promulgated was the goal of attaining the "optimum yield (OY)" from each fishery, defined as "the amount of fish which will provide the greatest overall benefit to the Nation, with particular reference to food production and recreational opportunities and which is prescribed as such on the basis of the maximum sustainable yield from such fishery, as modified by any relevant economic, social or ecological factor" (Section 1802 [18] [A] [B]). The emphasis on social and economic benefits in the standard of optimum yield represented a significant move away from prevailing management theories that previously had relied primarily on biological criteria for management, that is, the concept of maximum sustainable yield. Among the other national standards promulgated, fisheries had to be managed as a unit throughout their range, using the best scientific information available and promoting efficiency in the utilization of fishery resources without, however, discriminating among residents of various states and without having economic allocation as the sole purpose of management.

Third, and probably the major innovation contained in the act, was the establishment of a complex system of management, relying on decision making at two levels: preparation of fishery management plans by newly created regional councils, and their ultimate approval or disapproval by the Secretary of Commerce via the national office of NOAA's National Marine Fisheries Service. Eight regional fishery management councils were established (New England, Mid-Atlantic, South Atlantic, Caribbean, Gulf, Pacific, North Pacific, and Western Pacific),[17] each with responsibility for developing plans for the fisheries within its jurisdiction. Membership on the councils included, as voting members, the principal state officials responsible for fishery management in each of the states involved, the regional director of the National Marine Fisheries Service for the geographical area concerned, and a number of "knowledgeable" "public" members, their number varying from council to council, who were nominated by state governors and appointed by the Secretary of Commerce and who generally represent the fishing industry and the public.[18] Among nonvoting members were found a number of federal agency representatives from the Fish and Wildlife Service, the Coast Guard, and the Department of State, and the executive director of the Marine Fisheries Commission of the area involved.[19]

The Magnuson Act can be viewed as an agent of change in the type of federalism prevalent in the fishery management policy arena. As discussed earlier in Chapter 2, prior to the Magnuson Act, fishery management largely was conducted within the purview of state action, with the federal government playing mainly a support services role and with occasional interstate cooperation in the form of the marine fisheries commissions. By establishing federal management authority and creating regional councils that incorporated state, federal, and interest group participation in the decision-making process, as well as dividing policymaking power between the regional council level and the national Department of Commerce level, the Magnuson Act crafted an unusual system of intergovernmental policy making, one without ready analogue in other policy arenas (Rogalski 1980).

Fourth, with regard to foreign fishing, the MFCMA eliminated all fishing by foreign nations, except under the terms of new governing international fishery agreements (GIFAs) to be negotiated between the United States and foreign nations. Under the GIFAs, foreign nations could apply for permits to harvest "that portion of the optimum yield of [a] fishery that will not be harvested by vessels of the United States"—the "total allowable level of foreign fishing" or TALFF (*MFCMA* 1976, Section 1801 [d] [2] [A]). While major authority for negotiating the GIFAs rested with the Secretaries of Commerce and State, the Congress reserved for itself a significant oversight role—a "veto" power, in effect, as GIFAs had to be transmitted to Congress for a period of sixty calendar days, during which time Congress had the opportunity to prohibit the GIFA from entering into force (Section 1823). Once GIFAs were approved, foreign nations could apply for and receive, in exchange for fees, fishing permits for harvesting under strictly defined conditions (Section 1824).

A final important aspect of the foreign fishing provisions established under the Magnuson Act concerned Section 205, Import Prohibitions, which continued the traditional U.S. position of protecting its high-seas tuna fisheries by excluding highly migratory species from national zones of extended jurisdiction (*MFCMA* 1976, Section 1825). Under the provisions of this section, an import embargo could be imposed on the seafood exports of nations that deny fishing access to U.S. vessels, seize such vessels, fail to negotiate in good faith to conclude an international fishery agreement, or fail to comply with existing fishery agreements.

While we stress the novel aspects of each of the acts under consideration, it is important to point out what each act omits or does not fully address. In the case of the fisheries law, several important omissions should be noted. For example, while development of fishery resources is called for in the act (*MFCMA* 1976, Section 1801 [b] [6]), no specific measures to put this goal into effect are prescribed. This omission led to the enactment of the 1980 amendments to the act encouraging the formation of joint fishing ventures. Similarly, while noting that amendments to the Magnuson Act may be necessary in the event a Law of the Sea Treaty enters in force for the United States, the act omitted any discussion of the inconsistencies between the U.S. approach to fishery management and the model of extended jurisdiction advanced in the Law of the Sea convention,[20] and continued the U.S. traditional policy of excluding highly migratory species from national extended jurisdiction zones—a stance clearly in contrast to international practice.

The Magnuson Act was also largely silent on issues of state-federal interaction on fisheries that span multiple jurisdictions—state-federal areas or interstate areas. Although Section 306 treated questions of division of authority between the states and the federal government regarding the territorial sea and the FCZ, it was ambiguous concerning the precise conditions under which the federal government may preempt state authority. Similarly, the act generally ignored questions related to coordination with programs to manage or protect other marine resources or uses, such as the protection of marine mammals. The act included only a brief, technical amendment to the Marine Mammal Protection Act, and made no reference to means whereby the newly enacted national policy of protecting and promoting the nation's fishery resources could ultimately be reconciled with the national policy previously established in the MMPA of giving priority to the protection of marine mammals.

How the Bill Was Crafted

The effort to enact the 200-mile bill was led by a number of policy entrepreneurs in Congress, most notably Congressmen Studds, Leggett, and Murphy on the House side and Senators Magnuson, Stephens, and Kennedy on the Senate side. Among congressional staffers most active in the drafting of the legislation, according to the personal interviews we conducted on the passage of the Magnuson Act,[21] were Bud Walsh and Steve Perlis from the Senate Committee on Commerce, Science and Transportation, and Ned Everett, Dick Sharood, Rich Norlin, and George Mannina from the House Committee on Merchant Marine and Fisheries and the members' staffs. According to a congressional staff member who was instrumental in writing the original version of H.R. 200, Studds's staff teamed up with Senator Magnuson's staff, who were drafting a Senate version of the bill, and the respective versions of the bill were introduced at the same time. At first, the bill contained provisions only for "kicking the foreigners out"; it was only after several iterations that the notion of a domestic management regime was also incorporated (Norlin 1977).

These efforts were strongly supported by environmental and fishing groups, such as the National Federation of Fishermen (NFF), a group representing coastal fishers[22] who organized in 1973 to push for enactment of a 200-mile limit (Norlin 1977). Among coastal fishers, New

England groups, in particular, favored enactment in order to curtail the extensive harvesting by foreign fleets off the shores of New England. Other groups favoring the legislation included the National Coalition for Marine Conservation (one of the major national sports fishing groups), the American League of Anglers, fishing tackle manufacturers, and the Sierra Club.

Opposition to the legislation came primarily from the administration—the Departments of State, Treasury, Defense, and the Joint Chiefs of Staff—and from the powerful and well-organized lobbies representing the tuna and shrimp fisheries, which traditionally had depended on fishing within the 200-mile zones of Latin American nations and hence feared retaliatory action from other nations.

The major arguments for and against the enactment of the 200-mile limit revolved around the costs and benefits of awaiting the results of the Law of the Sea negotiations as opposed to proceeding with unilateral action. State Department spokespersons vehemently argued that unilateral action would adversely affect the LOS negotiations; as Ambassador Moore put it, "... if passed and enacted into law ... [a 200-mile limit bill] ... would ... breach solemn ocean treaty obligations of the United States, severely harm United States oceans and foreign relations interests, and mark a step backward in mankind's efforts to achieve a stable order for the world's oceans" (quoted in McKernan 1976). State Department representatives, through an interagency committee set up to deal with LOS matters, issued memoranda to the other agencies to coordinate their opposition (Robinson 1977). State Department staff were reported to have "gone around NOAA/NMFS issuing orders and telling people to keep quiet and not to support the bill" (Robinson 1977).

Similarly, the tuna and shrimp lobbies, who were primarily concerned with the probable retaliation of other nations to a unilateral U.S. declaration of extended jurisdiction, successfully blocked the 200-mile-limit initiative for several years, stalling the proceedings by having the bill referred to other congressional committees in addition to the House Committee on Merchant Marine and Fisheries (Utz 1977). On the proponent side, the policy entrepreneurs in Congress worked very closely together with representatives of coastal fishers, primarily the National Federation of Fishermen, in drafting the bill and in generating support for it. A symbiotic relationship, in effect, developed between the interest groups and the House and Senate staffs. Much of the bill was written by this small group in the evenings during the 1975 LOS meeting in Caracas (Norlin

1977). The central concept of "optimum yield" is reported to have been coined by several members of this group over drinks in a Caracas bar (Sloan 1977).

While it is clear that lobbyists for coastal fishers had significant input in the drafting of the act, it is less clear whether support for the measure on the part of ordinary fishers was well informed. The fact that fishers regarded it as simply a "get rid of the foreigners" law is suggested by the shock expressed by many of them at the domestic management aspects of the MFCMA, once the law had been enacted.[23]

Other parts of the fishing industry, such as the processors represented in the National Fisheries Institute (NFI), at first opposed the bill because the organization represented distant-water as well as coastal packers, and the international interests of the former group could have been hurt by unilateral action. After the disappointing 1975 LOS session, however, NFI switched its position to unanimous support and began actively to lobby on behalf of the measure (Weddig 1977).

Outside this relatively small group of supporters, strong support for the measure was generated in New England among environmental groups and others. As one participant put it, "it was a weird coalition in the House—commie haters who wanted to keep the Russians out, environmentalists . . . ; everyone in New England was working like crazy to get the bill passed . . . ; a number of liberals were taken in by the Administration's rhetoric and opposed the bill in fear of retaliation" (Norlin 1977).

The act faced major difficulties at several stages: getting extended jurisdiction identified as major legislation, securing other congressional co-sponsors (the bill ultimately was co-sponsored by 175 members), and scheduling hearings. It is reported that Congressman Dingell, then Chairman of the House Committee on Merchant Marine and Fisheries (1973–1974) and closely allied to the tuna industry, successfully inhibited for a time the holding of the requisite hearings (Norlin 1977). Similarly, the shrimp and tuna lobbyists, working through other contacts in the House and Senate, succeeded in having the bill referred to additional committees, such as Foreign Relations and Armed Services Committees. After two years of such delaying tactics, the proponents of the 200-mile limit introduced the bill on the first day of the next session and got it marked as H.R. 200, a fortuitous symbolic label for fostering issue recognition (Utz 1977). The bill was passed on March 30, 1976, and signed into law by President Ford on April 13, 1976 (U.S. GSA 1979).

Adding Environmental Considerations to Offshore Oil and Gas Development

Two years after the passage of the Magnuson Act, and after several years of deliberation, the Congress enacted the Outer Continental Shelf Lands Act Amendments of 1978 (P.L. 95-372, hereinafter referred to as OCSLAA 1978). This legislation substantially modified and extended the original Outer Continental Shelf Lands Act of 1953 that established the legal and policy framework within which the federal government made areas of the seabed of the U.S. continental shelf beyond 3 miles available for leasing by industry for oil and gas development. The 1953 act, a companion measure to the 1953 Submerged Lands Act (43 U.S.C. 1301 et. seq.), which gave the coastal states ownership of seabed resources within 3 miles of their shorelines, was a very general piece of legislation dealing principally with the procedures of leasing, royalties, and diligence (U.S. Congress, House 1977). Environmental and safety matters and involvement of other affected interests had not been of concern at that time.

The situation began to change significantly in the late 1960s. The large oil-well blowout that occurred in federal waters off Santa Barbara, California, in January 1969, triggered a wave of public concern over safety standards, the effects of offshore drilling on the marine environment, and government offshore leasing policies in general. President Nixon's announcement in 1971 that his administration intended to increase substantially the amount of acreage leased on the outer continental shelf (OCS) first alerted environmentalists and others to what was perceived as a major threat to the marine environment and its living resources. Subsequently, when Nixon announced a threefold increase in 1972 (to 3 million acres) and a tenfold increase in 1974 (to 10 million acres) in offshore leasing targets, state and local officials also became alarmed (U.S. Congress, House 1977). Especially concerned were officials from "frontier areas," new areas scheduled to be opened up to offshore drilling off New England, the middle Atlantic states, central and northern California, and Alaska. Reports of the effects of the large North Sea oil buildup on small communities in Scotland added to the apprehension felt by these U.S. coastal areas (Baldwin and Baldwin 1975). The oil embargoes of 1973 and 1974, of course, strengthened the resolve of the Nixon and Ford administrations to increase offshore production and led to heightened concern among environmental interests, state and local governments, and fishing and coastal tourism groups. Litigation involving offshore oil development became a regular occurrence (Eichenberg and Archer 1987).

A Congressional Initiative

Congress had two major goals in amending the 1953 act—to expedite the development of offshore resources and, while doing so, to protect the marine and coastal environment. The ongoing energy crisis and the widespread public alarm at blowouts and tanker spills had compelled a general reexamination of both the national energy policy and the program to develop offshore oil and gas resources. Dependence on foreign oil had risen to almost 50 percent of the total U.S. consumption by the mid-1970s, of which 36 percent came from the Middle East (U.S. Congress, House 1977). It was widely believed that this level of dependence on energy resources from volatile and politically unstable sources made the U.S. economy extremely vulnerable to oil embargoes, reduced what was then a favorable balance of trade, and adversely affected national security interests. It was only natural that Congress would turn its attention to developing the energy resources of the U.S. outer continental shelf to reduce this dangerous dependence on foreign oil. OCS resource estimates encouraged the view that energy independence could be achieved by accelerating OCS production (U.S. Congress, House 1977, pp. 65–74). Yet many members of Congress saw increased energy production from the OCS principally as means to bridge the gap from dependence on U.S. and world oil and gas resources to a future fueled by alternative sources of energy (solar, geothermal, nuclear, coal, gasification, oil shale, etc.) (Murphy and Belsky 1980).

Upon examination, the 1953 act looked woefully inadequate to its critics in the Congress for the task of bringing about a dramatic increase in offshore energy production in the 1970s. Coastal states, environmentalists, and the public, in turn, focused attention on the act's lack of environmental planning, standards, and safeguards. Interior's preference for large, up-front bonus payments in addition to royalties on production from OCS leases was strongly criticized, because this practice was thought to exclude small companies and to reduce competition. Administration of the leasing program was viewed as "essentially a closed process involving the Secretary of the Interior and the oil industry" (U.S. Congress, House 1977, p. 54). The 1953 act made no provision for mitigating the impacts of OCS development upon state and local communities, which feared that they would bear the burden of any expansion in offshore energy production, or upon competing users of ocean space and resources, such as fishers. The act offered neither assistance to state and local governments in planning for, nor compensation to fishers for economic loss resulting from OCS development. Finally, a frequently expressed criticism of the 1953 act was that it "essentially [was] a carte blanche delegation of authority to the Secretary of Interior" (U.S. Congress, House 1977).

The Congress intended the new OCSLAA to address these perceived inadequacies of the 1953 act. To limit the broad discretion allowed the secretary under the earlier act to manage OCS development, the OCSLAA required the development of a five-year leasing program, describing the size, timing, and location of proposed lease sales and prohibiting the lease of any area not included in the plan. Alternative bidding methods (in addition to bonus bids and fixed royalties) were authorized and required on an experimental basis. Environmental studies to assess the impacts of OCS projects on the human, marine, and coastal environments were mandated. OCS development was divided into four distinct stages: preparation of a five-year lease plan, lease sales, exploration, and development. Separate plans and, in most cases, environmental impact statements (EISs) for exploration and development were required. Before conducting lease sales and approving OCS exploration and development plans and projects, the secretary had to consult with governors of affected states and accept their recommendations if the secretary found them to be in the national interest. Funds were created for oil-spill contingencies and to compensate fishers for damages and loss of fishing gear due to OCS activities.

In a functional sense, using the categories defined by Murphy and Belsky (1980), the OCSLAA dealt with six major topics: expedited development; planning and information; competition; safe operations; citizen, state, and local input; and financial assistance to affected states.

Expedited Development The legislation required better information to be made available to affected interests and permitted greater input into the decision-making process, thereby, it was hoped, reducing uncertainty, delay, and litigation. It also provided for simplified court procedures, again with a view toward reducing dilatory tactics. Federal agency responsibilities, especially with regard to the leasing program itself, were clarified, with the objective of reducing uncertainty and delay (*OCSLAA* 1978, Sections 1344, 1349).

Planning and Information The Department of the Interior was required to produce and disseminate a comprehensive five-year leasing program, taking account of a range of national interests beyond solely national energy needs. Furthermore, Interior was required to begin an OCS information program that would make available to state and local governments information needed to adequately plan for and accommodate the effects of the OCS development (*OCSLAA* 1978, Sections 1344–1345).

Competition The legislation authorized and required the use of a number of alternative bidding schemes, in addition to up-front bonus bidding and fixed royalties, as a way of encouraging small companies to participate in the OCS program and promoting increased competition (*OCSLAA* 1978, Section 1337).

Safe Operations The act required, for the first time, that all aspects of the OCS program explicitly consider the environmental consequences of proposed actions. At least six months of environmental studies were required before leasing in new areas, and a full environmental impact statement had to be prepared for at least one development and production plan for offshore fields in a frontier area. All safety regulations had to include the "best available and safest technologies economically feasible." Finally, new oil-spill and fishing contingency funds were created to ensure that damages caused by oil spills or interference with fishing activities (damage or loss of gear only) were compensated (*OCSLAA* 1978, Sections 1346–1348).

Citizen, State, and Local Input Provisions of the legislation sought to involve affected interests—states, and through states, local governments, other ocean users, and the public—more directly in the OCS decision-making process. Explicit consultations with governors of affected states were required, and public participation was allowed when appropriate and possible. It should be noted, however, that the new provisions did not go so far as to allow coastal states to veto a proposed OCS action, and while the role of the states under the federal consistency provisions of the CZMA with regard to exploration and development plans was made more explicit, the time allowed for state review was reduced from six months to three (*OCSLAA* 1978, Sections 1344–1345).

Financial Assistance to Affected States The legislation further refined the Coastal Energy Impact Program (CEIP), enacted as a part of the amendments to the Coastal Zone Management Act in 1976 (*CZMA*, Section 1456a), and authorized a new set of OCS grants to assist states in carrying out their responsibilities under the legislation (*OCSLAA* 1978, Section 1351). The legislation, while increasing and liberalizing financial assistance to states and local governments and creating a new categorical grant program, did not, as some had hoped, contain provisions for a direct sharing of federal OCS revenues with the coastal states.

A Clash of Interests

Legislative efforts to revise the 1953 OCS Act began in 1974, when the Senate Committee on Interior and Insular Affairs reported out a comprehensive OCS reform law that was then promptly passed by the Senate (U.S. Congress, House 1977). This action had been preceded by extensive hearings conducted by that Senate committee each year during the period 1971–1974. The National Ocean Policy Study (NOPS), established by the Senate Committee on Commerce, Science, and Transportation, also was very actively investigating OCS activities during this period, as were subcommittees of the House Committees on Merchant Marine and Fisheries, Interior and Insular Affairs, and Judiciary. House action on the 1974 Senate bill did not follow due to jurisdictional conflicts between these committees and lack of time (U.S. Congress, House 1977, pp. 95–100).

The Senate, after adding some additional features, again passed an OCS bill in 1975, but once more the House did not act. Instead, congressional proponents of OCS reform legislation, realizing that disputes among the three primary committees claiming jurisdiction could stall the legislation indefinitely, requested an unusual action of the House leadership—the creation of a new, select committee for the sole purpose of handling an OCS bill. In April 1975, Speaker O'Neill agreed and established the Ad Hoc Select Committee on the Outer Continental Shelf. Made up of members and staffs from the three committees and chaired by the chairman of the Committee on Merchant Marine and Fisheries, the Select Committee held an extensive series of hearings, made field trips to the North Sea oil development area, examined the situation in Alaska, and by May 1976 was able to report out a comprehensive measure. After many hours of debate and after eighty amendments were dealt with, a reform OCS bill finally passed the House of Representatives in June 1976.

House and Senate conferees had little trouble agreeing on a version of the bill that largely followed the House-passed measure, but strong opposition by the large oil companies and the Ford administration succeeded in getting the bill recommitted to committee just three days before the end of the legislative session, thus effectively killing the bill for that session. Support from the newly elected Carter administration, however, fundamentally changed the situation, and during the second session of the 95th Congress, the much studied and debated measure finally passed both houses of Congress and, on September 18, 1978, was signed into law by President Carter.

As can be surmised from this description, the outside forces surrounding the passage of the legislation were divided rather clearly into two opposing camps. The environmental and conservation groups, joined by national associations representing state and local governments and fishing interests, supported passage of the legislation. The oil companies, their trade associations, and other business groups (e.g., the U.S. Chamber of Commerce) generally opposed its passage. Prior to January 1977 and the beginning of the Carter administration, the Department of the Interior strongly opposed any effort aimed at "reforming" the OCS Act, claiming that it was working satisfactorily,[24] while the Department of Commerce with its coastal zone management, fisheries and environmental responsibilities, quietly assisted the legislative efforts by responding positively to congressional requests for "technical assistance." After the Carter inauguration and upon the appointment of Cecil D. Andrus as secretary of the interior, the new administration also urged OCS legislative reform and more open support and assistance from Commerce became possible.[25]

In contrast to some legislative enactment, where the major forces shaping the program were outside of government, in this case the major force pushing for reform was inside—the members and staff of the House Ad Hoc Select Committee. Although admittedly building on the earlier work by House and Senate groups (the Senate NOPS activities, the hearings of the Senate and House Interior and Insular Affairs Committees and those of the House Committee on Merchant Marine and Fisheries), the Ad Hoc Committee was staffed competently with enthusiastic and well-informed individuals, had established a deadline by which to produce a bill, and was able to work in an atmosphere devoid of often time-wasting and acrimonious bickering over committee jurisdiction. In addition, resources were made available to hold a full slate of field hearings and to collect and analyze the necessary information. This degree of concentration, in itself unusual, when coupled with the priority attached to the legislation by the House leadership, ensured that legislation would emerge even in spite of exceptionally strong opposition by powerful and well-financed interests.

The Ad Hoc Select Committee was advised and supported by a relatively wide range of outside interests that wished, for one reason or another, to see offshore oil and gas policy reform. These included national environmental and conservation organizations such as the Natural Resources Defense Council, the Sierra Club, and the Environmental Policy Institute; groups representing state and local governments (National Governors Conference, National League of Cities, National Association of Counties) and individual states; and organizations representing fishing interests such as the National Federation of Fishermen. The opposition came from the large oil companies and their trade associations such as the American Petroleum Institute, the Western Oil and Gas Association, and the National Ocean Industries Association (U.S. Congress, House 1977).

Thus, the "closed process" in which the secretary of the interior and the industry had managed OCS oil and gas leasing, exploration, and development since the mid-1950s—akin to an "iron triangle": stable, exclusive, and opposed to any change that might expand the number of actors—was subjected to a barrage of criticism from a "network" comprised of coastal states and local governments, environmental organizations, Congress, and others. As noted earlier, the most effective proponents of change were members and staff of the committees of Congress directly concerned with offshore energy development. The fact that the industry and Interior lacked strong congressional support was the clearest sign that changes in the national policy respecting OCS energy development would occur. Missing one leg of the iron triangle meant that the structure was inherently unstable and unable to resist the forces pushing for change.

The groups in opposition had two principal fears: (1) that the legislation would strengthen the hand of those opposing offshore oil development, thereby opening the way for further litigation and delay, and (2) that the legislation might put the federal government in the oil business by requiring that the Department of the Interior conduct exploratory drilling operations itself in proposed lease areas in order to know more precisely how much oil and gas was likely to be found on the public land to be leased. Federally conducted exploration could, of course, substantially reduce the prospects of an oil company acquiring what later turned out to be a very rich field for a relatively small bonus bid. Ultimately, when the oil and gas interests saw that some legislation was inevitable, they sought, in the final stages of the debate, to limit the scope and impact of any new bill. It is interesting to note that some small oil companies supported the legislation in the hope that it would increase competition by allowing them to move into offshore areas (U.S. Congress, House 1977, p. 117).

In considering the heated history of congressional enactment of the OCSLAA, however, it would be a mistake to assume that the industry and Interior were routed by the 1978 amendments. The proponents of a rapid expansion in developing OCS oil and gas resources achieved a remarkable success; in the face of widespread concern about the environmental effects of OCS oil development, the new law enshrined as a national priority a quick and dramatic increase in OCS energy production.[26] And although the amendments set out a detailed OCS development process, required Interior and the industry to accept environmental safeguards, and provided for consultation with the states before final decisions are made, many of these changes had been anticipated by the agency before Congress codified them in the 1978 amendments.[27] It is unlikely, of course, that Interior would have modified its OCS leasing, exploration, and development process except for the pressure exerted upon it by the Congress and its allies. And certainly, codifying these new procedures removed them to a degree from the realm of agency discretion—statutory requirements must be met. But how they are met is an interesting and different question, which we discuss in later chapters. Nevertheless, acceptance of these changes, however reluctant, by agency officials and industry hands would still leave them in effective control of the OCS oil and gas development program and preserve intact a large measure of the agency's discretion to run the program as it chose, sometimes with little apparent regard for the views of governors, local governments, environmental groups, and other federal agencies.

Other Major Laws Enacted During the 1970s

Among other laws relevant to oceans and coasts enacted during this period, three stand out: the 1972 Federal Water Pollution Control Act (popularly known as the Clean Water Act); the Marine Research, Protection, and Sanctuaries Act of 1972, and the Endangered Species Act of 1973.

The 1972 Clean Water Act

By the early 1970s, the state of America's water quality had reached crisis proportions. In 1971, a task force launched by Ralph Nader produced a detailed, although largely anecdotal, report, *Water Wasteland,* detailing the serious state of water quality in the United States. Also in 1971, the *Second Annual Report* of the President's Council on Environmental Quality (CEQ) confirmed many of the findings of the Nader report (Adler et al. 1993).

Up to 1972, efforts to address water pollution occurred at both the federal and the state levels. The Water Pollution Control Act of 1948 provided the first federal funds for state water pollution control programs, with federal subsidies further increasing following passage of the Federal Water Pollution Control Act of 1956 (c. 518, 70 *Stat.* 498). These two acts did not, however, provide enforceable mandates or standards in association with the federal funding that was being allocated. The first attempt to provide standards on a consistent, national level came in 1965 with the creation of the federal Water Pollution Control Administration, which was given the mandate of requiring states to develop water-quality standards for interstate waters.

The continued evolution of environmental attitudes in the United States created a demand for enforceable standards and regulations for water quality, thus providing a clear vision for the 1972 Clean Water Act (Adler et al. 1993). The objective of the Clean Water Act of 1972 (P.L. 92-500, hereinafter CWA 1972) was "*to restore and maintain the chemical, physical and biological integrity of the Nation's waters*" (Section 101 [a]). The Environmental Protection Agency was designated as the lead federal agency in charge of implementing the CWA. The CWA is, of course, the principal legislation governing the quality of the nation's coastal waters and, hence, is directly relevant to national ocean policy.

Adler et al. (1993) note that the language of the objective of the CWA is important in two respects. First, by defining the target for water quality in terms of ecosystem integrity, Congress intended to return water quality to a clean state, rather than merely aiming to stop pollution. Second, by directing that *maintaining* water quality is a principal objective, it is clearly intended to actively protect waters that had, to date, escaped the impact of pollution.

In addition to the overarching objective, there were also several subsidiary and interim goals of the CWA. The first of these was known as "zero discharge," and stated that it "*is a national goal that the discharge of pollutants into the navigable waters be eliminated by 1985*"

(Section 101 [a] [1]). Next was the "fishable and swimmable waters" goal, which stated that "*it is the national goal that wherever attainable, an interim goal of water quality which provides for the protection and propagation of fish, shellfish, and wildlife, and provides for recreation in and on the water be achieved by July 1, 1983*" (Section 101 [a] [2]). The final goal was "no toxics in toxic amounts," and stated "*it is the national policy that the discharge of toxic pollutants in toxic amounts be prohibited*" (Section 101 [a] [3]).

Other notable features of the 1972 CWA were that pollution control was to be achieved by reducing pollutants, rather than by diluting them in receiving waters—this had the effect of changing the burden from the regulators to the dischargers; and all point sources of pollution were required to obtain permits, with tough new requirements attached. States were required to develop water-quality standards for in-state waters, to identify all waters not meeting these standards, to calculate the additional pollution reductions needed to achieve the standard, and to incorporate this into their permitting system.

The Marine Research, Protection, and Sanctuaries Act of 1972

Today's national marine sanctuaries system can be traced back to eleven bills introduced in the late 1960s to establish sanctuaries in specific areas off the coasts of California, Massachusetts, and New Hampshire. These bills were introduced in response to public concern arising from the dumping of nerve gas and oil wastes off the Florida coast, and from the Santa Barbara oil spill of 1969 (Tarnas 1988).

There were four common concepts central to these bills. First, a mineral exploitation moratorium was called for in all areas proposed as sanctuaries. Also, these protected areas were to be free from pollution and used as laboratories for scientific research. Third, they were to be preserved for their recreation, conservation, ecological, or aesthetic values. The last concept emerged in 1971 following hearings on the various bills—that sanctuaries should be multiple-use areas in which management regulations were aimed at allowing activities compatible with protecting the recognized values of an area (Tarnas 1988).

The Marine Protection, Research, and Sanctuaries Act (P.L. 92-532, hereinafter MPRSA 1972) was passed in 1972 with threefold aims: Title I of the act regulated the dumping of pollutants into marine waters; Title II directed that research be done on the effects of ocean dumping; and Title III established the national marine sanctuaries program with the purpose of conservation of resources. There were however, differences of opinion within Congress as to whether the intention of the new Title III should be preservation oriented (i.e., to exclude humans and human activity) or a multiple-use system that provided for compatible and balanced uses of marine sanctuaries (Tarnas 1988). Tarnas reports that the members of the House Merchant Marine and Fisheries Committee believed the program's emphasis to be preservation, while Representatives Keith and Pelly were of the view that Title III was more for multiple use as long as it was consistent, and balanced, with the primary purpose of preservation (p. 277).

The controversies and confusion over the role of the MPRSA continued for several years until, in 1984, Congress amended the legislation to expand and alter the purposes and policies of the act. The intent of the act was stated more clearly to allow for comprehensive, multiple-use management of special marine areas, to be achieved by protecting identified areas through control of the allowable uses, and by excluding incompatible uses of the sites (Tarnas 1988).

The Endangered Species Act of 1973

Building upon earlier 1966 and 1969 legislation and a 1973 international agreement, the Endangered Species Act (ESA) (P.L. 92–205), was enacted a year after the MMPA and extended protection to all threatened and endangered species, terrestrial and marine alike—adding, in effect, an additional layer of protection for threatened and endangered marine mammals such as whales, sea otters, and walruses. The law authorized the listing of threatened and endangered species and the identification and protection of habitats critical to their survival. It also called for the formulation and implementation of recovery plans for endangered species. Section 7 of the legislation required federal agencies to obtain approval from the U.S. Fish and Wildlife Service or the National Marine Fisheries Service before taking any action that could affect a listed species or its habitat.

The act recognized that endangered species of wildlife and plants "are of aesthetic, ecological, educational, historical and scientific value to the Nation and its people" (16 U.S.C. Section 1531 [a] [3]) and declared as its purpose providing "a means whereby the ecosystem upon which [they] depend may be conserved" (Section 1531 [b]). Two different degrees of vulnerability were recognized: "endangered" and "threatened." Endangered was defined to mean "any species which is in danger of extinction throughout all or a significant portion of its range (Section 1532 [c]). Threatened is defined as "any species which is likely to become an endangered species within the foreseeable future throughout all or a significant portion of its range" (Section 1532 [20]). The act contained the authority to list as threatened or endangered individual populations of a given species even though the species was healthy in other locations (Bean 1983).

The third element of the ESA was "critical habitat," which the act, in its original 1973 form, neither defined nor contained provisions for its designation, yet federal agencies were given the duty not to modify or destroy such habitat. As defined later (in amendments made in 1978 following administrative action by the U.S. Fish and Wildlife Service and the National Marine Fisheries Service), critical habitat is an area or areas "essential to the conservation of the species." The act in its original form contained exceptions for certain specific cases of economic hardship, for Alaskan natives involved in the production of authentic native articles of handicraft and clothing, for wildlife already possessed at the time the act was passed, and for scientific or propagation purposes (Bean 1983).

Taken together, the MMPA and the ESA put into place an interlocking regime involving policies and procedures to achieve long-term protection (and where needed, recovery) of marine mammal populations and other threatened or endangered species.

Characteristics of the Ocean Laws of the 1970s

Five major themes characterize the marine laws enacted by Congress in the 1970s: (1) a significant increase in the scope of governmental activity; (2) centralization of functions and enhancement of the federal role; (3) a sectoral orientation, or a use-by-use approach; (4) the lack of explicit priorities among ocean uses and the absence of mechanisms for resolving conflicts among them; and (5) the introduction of innovative management concepts and structures.

An Expanded Government Role

In contrast to the privatizing trend that most domestic policy would undergo in the 1980s, the 1970s witnessed a clear assertion of the appropriateness of a strong governmental role in the management of ocean resources and uses. The laws enacted in the 1970s represent significant increases in the scope of governmental activity—by expanding existing governmental roles or by crafting essentially new functions. The Magnuson Act, as noted earlier, created an entirely new zone to be managed under U.S. control through a highly distinctive system of governance combining regional and national elements. As discussed in Chapter 2, fisheries in this area had previously been unmanaged or only partially managed by the states and by international organizations. The Coastal Zone Management Act, in effect, added a new function that had not been performed previously: planning for the use and protection of the coastal zone through federal assistance to the states. The Marine Mammal Protection Act significantly increased the scope of governmental protection of marine mammals, which had previously been only partially protected by the states or by international treaties. In response to energy shortages in the early seventies, the Outer Continental Shelf Lands Act Amendments, at the end of the decade, called for more aggressive management of U.S. outer continental shelf (OCS) resources through a more elaborate system of planning, increased industry competition, and tightened environmental controls.

In this respect, then, the 1970s saw the pendulum swing toward emphasizing the "coastal state" rather than the "sea power" interests of the United States, and the crafting of a domestic law framework for managing the most important resources and uses of the ocean and coastal areas. Unilateral rather than international action, too, was emphasized during this period, particularly with regard to the enactment of the Magnuson Act, a move that was at best premature relative to the ongoing Law of the Sea deliberations. In other instances, however, "international" orientations continued to be important, particularly in the MMPA, whose roots were closely related to the internationally based wildlife protection movement.

Enhancement of the Federal Role

The increased scope of governmental activity vis-à-vis the oceans that occurred in the 1970s involved primarily an enhancement of the federal role, and, with very few exceptions, a steady centralization of functions at the national level. The pendulum, in this case, clearly swung to the federal side of the state-federal continuum. The Magnuson Act established a new regional management structure, under federal control, to manage the new 200-mile zone. The Marine Mammal Protection Act wrested control away from the states and established federal supremacy over the protection of marine mammals. Although oriented toward the states, the Coastal Zone Management Act nevertheless established a strong federal presence in the coastal zone by offering tangible incentives to the states to implement planning goals established at the federal level. While the

OCSLAA strengthened the role of the states in the oil and gas leasing process, the basic thrust of these amendments was to reaffirm and more carefully define the federal supremacy over OCS resources previously established by the Outer Continental Shelf Lands Act of 1953.

Although the main effect of these marine laws was to centralize regulatory functions at the federal level through a single-purpose approach, most of these acts also contained provisions delineating explicit roles for the states, as well as provisions that could be used to affect the operation of other marine sectors. The "consistency" provision of the CZMA, as discussed earlier, gave states leverage over federal activities within or affecting their coastal zones by requiring that certain types of federal actions be conducted in a manner consistent with approved state coastal plans. While the MMPA preempted state management of marine mammals and gave the federal government protective control over these species, it also provided opportunities for states to regain authority taken from them under certain conditions. While the Magnuson Act gave the federal government ultimate control over fishery resources, it also established a complex system of regional decision making for fisheries management in which the states had extensive input. Moreover, while federal preemption of state fishery activities in state waters was possible under Section 306 (b) (1) of the act, this was permitted only under certain, rather stringent, conditions. Through Section 19, the OCSLAA provided new opportunities for state review of the OCS leasing process. Taken together and viewed in retrospect, these provisions, while they created the appearance of a cooperative approach to marine resources management and raised expectations, frequently led to disappointment, conflict, and litigation since they did not institutionalize truly equitable partnerships between the federal government and the coastal states.

A Sectoral Approach

With the possible exception of the CZMA and the Marine Protection, Research, and Sanctuaries Act (P.L. 92-532, *MPRSA* 1972), the body of ocean law enacted in the 1970s was sectorally oriented and single-purpose in nature, reflecting a use-by-use approach. This is not only true of the acts highlighted in this chapter (the MMPA, the Magnuson Act, and the OCSLAA) but also of other ocean legislation enacted during this period—such as the Deepwater Port Act and the Clean Water Act. All represent single-purpose approaches to specific ocean uses or activities. In this approach, each marine resource or use (e.g., commercial fisheries, marine mammals, oil and gas operations) is managed under separate statutory authority by different agencies with narrowly defined missions.

This segmented approach is in marked contrast to the ocean system itself, which is characterized by a high level of interaction among resources and processes—each resource dependent on others in a complex system. Fisheries and marine mammals, to cite only one example, are naturally and inevitably related to one another in the marine ecosystem as predator and prey; and because such marine resources are inevitably interconnected, significant conflicts would occur later in the 1970s and in the 1980s between the Magnuson Act and the MMPA and ESA. And yet, we find these interconnections were largely ignored during the legislative passage of these acts.

The Failure to Set Priorities Among Uses

A corollary of the single-purpose, sector-by-sector approach to ocean resources management is that no explicit priorities are established among various ocean uses. Each act, in effect, mandates the implementing agencies to fully promote and/or protect the particular use involved. Thus, the CZMA exhorts to "preserve, protect, develop, and where possible, to restore, and to enhance the resources of the Nation's coastal zone for this and succeeding generations." The Magnuson Act, in turn, mandates to "conserve and manage fishery resources found off the coasts of the United States . . . [and] to encourage the development of [underutilized] fisheries." The OCSLAA calls for "expedited exploration and development of the Outer Continental Shelf in order to achieve national economic and energy policy goals." In perhaps the strongest language, the MMPA finds that "population stocks should not be permitted to diminish beyond the point at which they cease to be a significant functioning element in the ecosystem of which they are a part"; therefore, they "should be protected and encouraged to develop to the greatest extent feasible commensurate with sound principles of resource management."

What this means, in effect, is that the decision-making arena associated with each of these laws is biased in favor of the particular use or resource at stake. Taking the OCSLAA as an example, while implementing agencies need to take into account the effects of oil development on other ocean users and on the marine environment, these reviews are inevitably tilted in favor of achieving the goal of expedited development of offshore oil resources.

What this means, too, is that there is no neutral public arena for resolving the conflicts that arise among various ocean users (such as oil developers, commercial fishermen, recreational boaters). A major result of the absence of a neutral public forum is that conflicts will most often end up being litigated; in some cases, being mediated, and in many cases, going unresolved.

New Management Concepts and Structures

In contrast to previous government mandates that stressed a service, and not a regulatory role (such as the service of collection and dissemination of fishery statistics), the legislation enacted in the 1970s all clearly called for management and regulatory actions. It established procedures to regulate activities in the oceans and coastal zone and to affect the behavior of other parties such as state governments, direct users, and public.

These laws also introduced innovative management concepts and structures. The Magnuson Act, for example, called for a new mode or philosophy of management, much broader than the biologically oriented theory of "maximum sustainable yield" on which most fishery management practices had previously been based. Implicit in the Magnuson Act was a more holistic view of management which considered the entire ecosystem, including the human system, involved in the exploitation of fishery resources. For example, the act calls for "managing stocks throughout their range," and for "multispecies management" (considering the ecological relations among species). Concern with the human systems involved in the exploitation of fish stocks and with equity questions pervades the act. The central concept of "optimum yield," for example, was meant to "provide the greatest benefit to the nation," and if allocations of fish needed to be made, such allocations had to be "fair and equitable to all [such] fishermen"; if limited entry had to be considered, the social, economic, and cultural implications had to be taken into account. Resources, moreover, were to be managed using the "best scientific data available," "in a cost-effective manner avoiding unnecessary duplication," and which allowed for a "multiplicity of options in regard to future use." In addition to pioneering a novel conceptual management framework, the Magnuson Act ushered in an innovative management structure in the form of the regional councils involving state, federal, public, and user-group involvement.

The MMPA and ESA both reflected a more holistic approach to management and a strong ecosystemic orientation. With these two acts, the federal government moved from a smattering of ad hoc legislation on a species-by-species basis to comprehensive policies that protected entire categories of species (marine mammals under the MMPA; threatened and endangered species under the ESA). In contrast to previous harvest-oriented management practices, which utilized the management standard of "maximum sustainable yield," these acts stressed multiple aesthetic, ethical, and ecological values associated with wildlife resources. In, particular, both acts emphasized not only the health and stability of wildlife species, but most important, the healthy condition of the entire ecosystems of which these species form a part (Lund 1980). The concept of "ecosystem" found in the MMPA and ESA, however, unlike that in the Magnuson Act, tended largely to ignore to role of humans in the ecosystem—a difference in approach that would ultimately lead to significant conflicts in implementation.

The concept of "consistency" in the CZMA, too, represented a significant innovation—for the first time, states could influence the conduct of federal activities inside and, under certain conditions, outside of state waters. Similarly, the stringent environmental review procedures incorporated in the OCSLAA represented a significant departure from previous management practices.

The Politics of Enactment

A number of the variables thought to be important in explaining the initiation of policies—perceived crisis or changing conditions; timely studies calling attention to a problem; the catalytic role of policy entrepreneurs—are all manifested in the enactment of the body of marine law in the 1970s. All of the acts highlighted—the CZMA, the MMPA, the Magnuson Act, the OCSLAA—were, in varying degrees, responses to perceived needs to protect and enhance resources that, for varying reasons, were in a critical state. The CZMA and the MMPA were reactions to threats to coastal resources and marine mammal species, respectively. The Magnuson Act, in turn, responded to a critical situation of threats to fishery stocks by foreign fishermen, while the OCSLAA sought to alleviate, in an environmentally sensitive manner, acute energy shortages.

By and large, policy studies played an important role in the initiation of all these acts, but most important, for the CZMA and the MMPA. In the case of the CZMA, the Stratton Commission had explicitly called for the creation of a coastal program, while much of the impetus for the MMPA came from a wide number of studies documenting the worldwide decimation of marine mammal species.

Policy entrepreneurs—both members of Congress and their staffs and interest-group advocates—played a key role in initiation. The presence of "slack resources" in the Congress (more staff, new subcommittees), and the fact that the "oceans" were a new, not previously "colonized" issue area, gave congress members and their staffs who were casting about for an issue a good opportunity for leadership attention.

An interesting difference between the enactment of this body of law and the experience in other policy areas is that in these cases the policy initiative came largely from Congress and external interest groups, not from the executive branch. The administrations involved—Nixon, Ford, and Carter—were, by and large, either generally indifferent or opposed to the new initiatives (Kitsos 1981). It was, in fact, the confluence of congressional interest, new opportunities for influencing Congress through the proliferation of subcommittees, and the mobilization of outside interests that was responsible for the ocean and coastal policy initiatives.

The 1970s witnessed a broadening of the interest-group participation within each marine sector, and, in this sense, a change in the traditional "iron triangles" that had previously characterized each issue area. In addition to more established conservation and wildlife groups, new groups (such as the Coastal States Organization, the National Federation of Fishermen, the Friends of the Earth, the Society for Animal Protective Legislation) mobilized to promote their favorite ocean and coastal issues. These new groups brought new concepts and new ways of doing things, often clashing in the process with the more established organization. The "iron triangles" prevailing within each issue area thus underwent considerable internal change.

However, within each issue area, a broadening of outlook was taking place. At the beginning of the decade, there was little interaction among different marine sectors, and in this sense, "iron triangles" within each sector remained relatively stable. Although marine mammal and fisheries issues are clearly intertwined and would become increasingly conflictive, there was little participation by fishing interests in debates over the MMPA. In a related example, although the energy industry would ultimately be considerably hampered by the provisions of both the MMPA and the ESA, energy groups were "asleep at the switch" and were little involved during the passage of the ESA (Manning 1990).

In effect, then, the single-purpose character of the body of law enacted in the 1970s was a clear result of the fact that single-issue groups, capitalizing on the interest of specific members of Congress and on the "multiple points of access" provided by the proliferation of congressional subcommittees, were able to prevail in isolated battles (Kitsos 1981). Each single interest was able to achieve its own law protecting or promoting its own preferred part of the marine system.

By the end of the 1970s, during the enactment of the OCSLAA, however, political interactions began to change and a considerable broadening or crossover among various ocean sectors became evident. Actors during this period began to broaden their scope and to more frequently interact (and attempt to influence) other marine sectors. The trend toward the more fluid, shifting, and less predictable situation of "issue networks" politics had begun.

FOUR

The 1970s to the 1990s

A Mixed Record of Success in Ocean Policy Implementation

This chapter first discusses the evolution of the political and policy context from the 1970s to the late 1990s, to put in perspective the progress of implementation of the major ocean and coastal programs enacted in the 1970s. It then examines, in closer detail, the evolution of policy in each of several sectors: coastal management, fisheries management, marine mammal protection, offshore oil and gas development, marine pollution control, and marine protected area management.

The Political and Policy Context of the Period

A flurry of activities on the part of federal agencies charged with implementing the new ocean and coastal acts ensued, as agencies worked to interpret the far-reaching statutes enacted in the 1970s, to put in place new decision processes, and to mobilize new resources. Many federal agencies were ill prepared to implement some of the innovative concepts contained in the new acts. Amidst these developments, the energy crisis of 1973–1974 torqued national ocean policy, for a time, toward the goal of energy independence, inducing, in the process, extensive conflicts between energy developers and the environmental interests that had been spawned earlier in the decade. Other conflicts took place as well, as various interest groups found themselves affected by the legislative actions made in other sectors. Litigation began to increase and escalated in the 1980s, particularly around offshore oil issues and fisheries/marine mammals conflicts.

In the 1970s, a clear move toward federal dominance in ocean policy matters had taken place, but in the 1980s, the pendulum swung more toward the states, in rhetoric, at least. The Reagan administration (1981–1989) brought with it a philosophy that the federal government was too big and that there was too much regulation of economic development activities, especially too much environmental regulation. The solution to this fundamental problem, in President Reagan's view, required a drastic shrinking of governmental programs, activities, and regulations. The administration called for a "New Federalism," which shifted government functions to state and local levels (often, however, without corresponding funding).

Many of the programs authorized in the 1970s were targeted for spending reduction, and some (such as coastal zone management and sea grants) were targeted for outright elimination. While these efforts were largely rebuffed through concerted action by Congress, the coastal states, and environmental organizations, the 1980s saw little growth in federal ocean programs, with the exception of a few legislative initiatives: the 1987 National Estuary Program (section 320 of the Clean Water Act),[1] the 1982 Coastal Barrier Resources Act (16 U.S.C. , 3501 *et seq.*),[2] the 1972 Ocean Dumping Act (Titles I and II of the Marine Protection, Research and Sanctuaries Act of 1972, as amended, 33 U.S.C., 1401 *et seq.*),[3] and the 1990 Oil Pollution Act (33 U.S.C., 2701 *et seq., inter alia*).[4] This period was typified by increasing conflicts, particularly with regard to offshore oil, with environmental interest groups ultimately successful in achieving a series of congressionally mandated moratoria on spending for offshore oil leasing activities.

Almost in reaction to the downturn in federal interest in the oceans, the late 1980s and the 1990s witnessed increased activism on the part of the coastal states and coastal communities, with, for example, a number of coastal states preparing integrated plans for the management of their offshore areas, with some of these (such as Oregon) going beyond the legal bounds (3 miles) of state jurisdiction (Bailey 1994). A great deal of experimentation on ocean, coastal, and estuarine management took place during this period, especially at the local level, with increased participation by a variety of stakeholders (private sector, environmental interests, state and local governments) in decisions about coastal and ocean resources. This period was also marked by the emergence of new management concepts (such as watershed management and ecosystem management) that emphasize the need to manage on the basis of natural systems rather than political jurisdictions (see, e.g., Imperial and Hennessey 1996).

The 1990s also witnessed profound changes in international relations as well as major changes in ocean and coastal management at the international level. The collapse of the Soviet Union, the growth in regional economic blocks (Europe, North America, East Asia), the globalization of economic activities, the growth in tensions between the

North (developed countries) and the South (developing countries) over issues related to environment and development, and the growing role of international trade agreements were all major factors that shaped foreign, and to a lesser extent domestic, policy in the 1990s. In the oceans arena, a major change to the Law of the Sea Convention, in the form of a new agreement revising the deep-seabed mining regime, made the LOS acceptable to major industrialized nations, such as the United States, which, in 1994, indicated its intention to seek Senate ratification of the treaty. The LOS treaty came into force in November 1994 following ratification by the sixtieth nation. In addition, the United Nations Conference on Environment and Development (the Earth Summit) in June 1992 brought major new paradigms to the fore—the concept of sustainable development, and major changes in ocean and coastal management thinking and practice through Chapter 17 (the oceans and coasts chapter) of Agenda 21 (the global blueprint on environment and development), and through the conventions on biological diversity and on climate change (Cicin-Sain and Knecht 1993).

Major new international accords were also made to halt the worldwide degradation of fisheries, such as the adoption of the 1994 agreement on straddling fisheries stocks and the FAO code of conduct for responsible fisheries. The United States participated in all of these international discussions, albeit sometimes reluctantly. While these advances in thinking at the international level had, in many cases, far-reaching impacts on the national behavior of other nations, by the end of the 1990s, they generally appeared to have little impact on U.S. domestic policy on oceans and coasts.

At the end of the 1990s, there is a growing consensus about the problems besetting single-sector management of the oceans, and the need to develop more integrated management approaches (Heinz Center 1998). Initiatives to create a national commission to provide a comprehensive examination of ocean policy and to create a national ocean council have become major factors in the discussion of future policy directions (Knecht, Cicin-Sain, and Foster 1998).

Politically, much of this period has been characterized by "divided" government, that is, one party holding the presidency while the other holds the congressional majority in one or both houses of Congress. In fact, as Table 4.1 suggests, in the period 1969 to 2000, the government was divided fourteen out of sixteen times. A situation of divided government, of course, makes the formulation and implementation of any policy more difficult, and requires that bipartisan support must be garnered in

TABLE 4.1 Incidence of "Divided" or "Unified" National Government, 1953–2001.

YEAR	PRESIDENT	CONGRESS	HOUSE (NUMBERS/PARTY)		SENATE (NUMBERS/PARTY)		GOVERNMENT
			MAJORITY	OPPOSITION	MAJORITY	OPPOSITION	
1953–1955	Eisenhower (R)	83rd	221 (R)	211 (D)	48 (R)	47 (D)	Unified
1955–1957	Eisenhower (R)	84th	232 (D)	203 (R)	48 (D)	47 (R)	Divided
1957–1959	Eisenhower (R)	85th	233 (D)	200 (R)	49 (D)	47 (R)	Divided
1959–1961	Eisenhower (R)	86th	283 (D)	153 (R)	64 (D)	34 (R)	Divided
1961–1963	Kennedy (D)	87th	263 (D)	174 (R)	65 (D)	35 (R)	Unified
1963–1965	Kennedy (D)/ Johnson (D)	88th	258 (D)	177 (R)	67 (D)	33 (R)	Unified
1965–1967	Johnson (D)	89th	295 (D)	140 (R)	68 (D)	32 (R)	Unified
1967–1969	Johnson (D)	90th	246 (D)	187 (R)	64 (D)	36 (R)	Unified
1969–1971	Nixon (R)	91st	245 (D)	189 (R)	57 (D)	43 (R)	Divided
1971–1973	Nixon (R)	92nd	254 (D)	180 (R)	54 (D)	44 (R)	Divided
1973–1975	Nixon (R)/ Ford (R)	93rd	239 (D)	192 (R)	56 (D)	42 (R)	Divided
1975–1977	Ford (R)	94th	291 (D)	144 (R)	60 (D)	37 (R)	Divided
1977–1979	Carter (D)	95th	292 (D)	143 (R)	61 (D)	38 (R)	Unified
1979–1981	Carter (D)	96th	276 (D)	157 (R)	58 (D)	41 (R)	Unified
1981–1983	Reagan (R)	97th	243 (D)	192 (R)	53 (R)	46 (D)	Divided
1983–1985	Reagan (R)	98th	269 (D)	165 (R)	54 (R)	46 (D)	Divided
1985–1987	Reagan (R)	99th	252 (D)	182 (R)	53 (R)	47 (D)	Divided
1987–1989	Reagan (R)	100th	259 (D)	176 (R)	55 (D)	45(R)	Divided
1989–1991	Bush (R)	101st	260 (D)	175 (R)	55 (D)	45 (R)	Divided
1991–1993	Bush (R)	102nd	267 (D)	167 (R)	56 (D)	44 (R)	Divided
1993–1995	Clinton (D)	103rd	261 (D)	173 (R)	56 (D)	42 (R)	Unified
1995–1997	Clinton (D)	104th	235 (R)	204 (D)	53 (R)	47 (D)	Divided
1997–1999	Clinton (D)	105th	228 (R)	206 (D)	55 (R)	45 (D)	Divided
1999–2001	Clinton (D)	106th	223 (R)	211 (D)	55 (R)	45 (D)	Divided

Source: Adapted from Ragsdale 1996.

order for policy proposals to be successfully enacted and implemented. Bitter partisanship between the president and the Congress was especially pronounced during the 104th Congress (1995–1997), which was marked by Newt Gingrich's "Contract with America" and the accession of Republicans to the majority in both houses of Congress. Such partisanship continued in the 105th Congress, culminating in the impeachment of President Clinton in the House in December 1998.[5]

During this period, the ocean policy landscape has changed from the domination of isolated "iron triangles," which predominated in the early 1970s, to the emergence of much more interaction among interests representing various ocean sectors—"issue networks" as Heclo (1978) calls them. This is especially true in the case of groups representing fishers and marine mammal protection who have been forced to deal with each other in resolving conflicts between marine mammals and fisheries. There have been efforts throughout this period to bring together environmental interests and offshore oil and gas interests, but relations among these groups remained largely wary and acrimonious at the end of the 1990s. Some programs, such as the National Estuary Program, have achieved some notable successes in bringing together a variety of stakeholders (representing all coastal sectors and government jurisdictions) in efforts to manage particular bodies of water; but all too often, the lack of implementation follow-through has blunted the effectiveness of these efforts. And although there has been growing admission among all ocean and coastal interests that more integrated management is needed, a powerful political coalition pushing for this perspective has yet to appear.

The 1973–1974 Energy Crisis and Its Impacts

Just as ocean programs enacted in the 1970s were coming on line, the energy crisis developed and legislative actions were taken to address it. The Arab oil embargoes of 1973–1974 took most Americans by surprise. Gasoline to power America's automobiles and oil to generate electricity have been viewed as essential to the modern American way of life and were very much taken for granted. Thus, the disrupted oil supplies focused public attention in a way that more diffuse environmental problems or the crisis of confidence in our educational system (during the *Sputnik* crisis) could not.

A number of the ocean and coastal laws enacted in the early 1970s, however, were not seen as helpful to the perceived energy problems. Laws to protect marine mammals and endangered species, to the extent that they blocked the development of energy resources, were seen as obstructionist, and efforts were made to soften or repeal them. Just as the ESA's protection of the snail darter interfered with construction of the Tellico Dam, the MMPA's protection of the bowhead whale was seen as slowing offshore oil exploration in the Arctic Ocean.[6] The CZMA, too, came under increasing attack as the federal consistency provisions were frequently called into play by coastal states seeking to ensure that coastal energy activities did not adversely affect their coastal zones. Several lawsuits were brought to determine how far coastal states were required to go in considering/accommodating the national interest in the siting of facilities of more than local significance in coastal areas.[7]

In 1976, Congress grafted the Coastal Energy Impact Program (CEIP) onto the CZMA in an effort to assist the states as they grappled simultaneously with completing their coastal zone management programs and were pressured to site additional energy-related facilities in their coastal zones (Eliopoulous 1982). But this program did little to ease the tension that was building between coastal management and coastal energy development. By 1978, Congress had completed work on a revised and environmentally sensitive approach to offshore leasing of the outer continental shelf for oil and gas development. The Outer Continental Shelf Lands Act Amendments (OCSLAA) of 1978 created a new decision-making process for offshore oil leasing and development in federal waters that ostensibly brought state and local governments into the process (P.L. No. 95-372, 1978). In fact, however, as implementation of the new legislation proceeded, it became clear that the secretary of interior was largely free to ignore state and local recommendations regarding the timing and size of OCS lease sales and specific exploration and development projects (Cicin-Sain and Knecht 1987).

During this period, the federal government encouraged coastal states and oil companies to consider the construction of offshore deepwater ports to service the supertankers carrying imported oil from the Middle East and elsewhere. Federal legislation was enacted in 1974 (P.L. No. 93-627, 1975) and one state, Louisiana, constructed an offshore terminal. The government also sought to foster innovative renewable energy projects such as those based on the concept of ocean thermal energy conversion (OTEC) (P.L. No. 96-283, 1980).

Although the Law of the Sea negotiations being conducted during this time attracted little public notice, this activity commanded the attention of ocean policy makers in the United States. The objectives of the United States in the Third United Nations Conference on Law of the Sea

(UNCLOS III) negotiations were varied: to stabilize international ocean law; to protect freedom of navigation to the maximum extent possible; to establish clear ownership of the resources of the 200-mile exclusive economic zones (EEZs); and to create a deep-seabed minerals regime that incorporated free-market principles insofar as possible and guaranteed licenses to qualified U.S. companies.

The oil embargoes that occurred during these protracted negotiations gave U.S. mining interests an opportunity to focus attention on a new problem. They argued that the nation, because of its lack of domestic sources, faced a potential shortage of strategic minerals such as cobalt, nickel, and manganese. According to this argument, a minerals cartel along the lines of OPEC could place the country in a vulnerable position. Although not widely supported, this "problem," when coupled with the growing impatience with UNCLOS III, was sufficiently compelling to push a domestic seabed mining bill through Congress and into law in 1980: the Deep Seabed Hard Mineral Resources Act (DSHMRA) (P.L. No. 96-283, 1980).

What impact, then, did the energy problem stream have on national ocean policy? One effect was to partially blunt the environmental character that had been impressed upon some aspects of ocean policy, most notably endangered species, marine mammals, and to a certain extent, coastal zone management. For example, an amendment to the Endangered Species Act (P.L. No. 93-205, 1973) created a council to consider exceptions to the requirements of the act under circumstances typically found in proposals for new energy facilities.[8] Also, as the Reagan administration gained power in the 1980s, it became clear that the balance between coastal resources protection and energy facilities siting would be set at a point favorable to development interests.

Onc might have expected that renewable ocean energy (for example ocean thermal energy conversion, or OTEC) would have received a sustained boost from the emergence of the energy problem, but this was not the case. Advocates of a strong federal role in the development of this technology ran up against forces in the incoming Reagan administration that believed commercializing OTEC technology was best left to the private sector. Although a small program in NOAA was created to implement the regulatory and licensing provisions of the domestic seabed mining legislation (DSHMRA) and also of the OTEC licensing act (Ocean Thermal Conversion Act [42 U.S.C., 9101 et seq.]), this activity was hardly of a scale that would signal a new turn in national ocean policy.

The Reagan Years and Governmental Retrenchment

The stance of the Reagan administration (1981–1989) toward ocean and coastal policy and programs was determined by the administration's basic antigovernment philosophy: The federal government was too big and there was too much regulation, especially environmental regulation, of economic development activities. Thus, the major problem perceived by the Reagan administration was the inhibiting effect the federal government had on the economy. The solution to this fundamental problem, according to the new administration, was a drastic shrinking of governmental programs, activities, and regulations.

The perspective of the Reagan administration was radically different from that of previous administrations. The ocean and coastal "problems" identified by the Stratton Commission, under President Johnson, required governmental action, and the Nixon, Ford, and Carter administrations sought legislation and established federal programs to address these problems and opportunities. The new administration, however, believed that free market forces would allocate the nation's resources (including ocean and coastal resources) more efficiently, and that industry should be freed from a vast array of governmental regulation in order to allow the market to operate. This basic reform would be accomplished by depriving federal civilian programs of funds through tax cuts and by rigorous budgetary control exercised by the Office of Management and Budget (OMB) to reduce spending, eliminate programs, and rescind federal regulations. Ocean and coastal programs were targeted for deep reductions by the OMB (Archer and Knecht 1987).

Programs such as coastal zone management (CZM) and Sea Grant were targeted for elimination; remaining programs were to be cut significantly. OMB was instructed to take charge of the regulatory review process and to reduce the burden of federal regulation on economic activity by modifying or rescinding existing regulations and prohibiting new ones (U.S. President 1982). Environmental regulations, in particular, were closely scrutinized. Meanwhile, the exploration and development of the oil and gas resources of the outer continental shelf were to be substantially expanded by offering vast new ocean areas for leasing. Environmental concerns associated with the development of these resources received less attention in the Reagan administration, and the efforts of coastal states to participate in the management of offshore energy and mineral resources and to share in the revenues flowing from their development were resisted. Funding for ocean science and research for nondefense purposes was also reduced.

These policies were initially quite successful in forcing a reduction in the funding of federal ocean and coastal programs. Certain programs such as CEIP and OTEC were defunded. NOAA's "ocean" activities were deemphasized, and the administration sought to "privatize" other NOAA activities, including national weather services, or to "contract out" agency functions. However, opposition from Congress, the coastal states, environmental organizations, and the public soon asserted itself, and proposals to eliminate particular programs such as CZM and Sea Grant were rejected.

As an additional curb on federal authority, the new administration adopted the philosophy of "New Federalism" in support of the interests of the states vis-à-vis the federal government (Domestic Policy Council 1986). In the case of ocean and coastal programs, however, the administration opposed the efforts of coastal states to regulate offshore energy and mineral development activities and defended federal prerogatives in these areas as vigorously as any previous administrations (Eichenberg and Archer 1987). In this respect, growing coastal state interest in managing offshore resources was viewed by the administration as part of the "problem" inhibiting economic development of these resources.

Internationally, the administration also opposed restrictions on ocean development activities. In the President Reagan's view, the oceans were a "frontier" and should be open for exploitation by pioneer entrepreneurs, much as the western lands of the United States had been developed. The administration rejected the impending Law of the Sea (LOS) treaty because it would have restricted access to and the right to develop seabed minerals by U.S. industry. In an attempt to protect U.S. ocean mining claims, the administration, asserting that seabed mining was a high-seas freedom, sought agreements with like-minded countries whereby such claims would be recognized and competing claims resolved under an interlocking system of national legislation (Deep Seabed Hard Mineral Resources Act 1980). Finally, to assert sovereign rights over the resources of the ocean adjacent to its shores, the United States proclaimed its 200-mile EEZ (U.S. Presidential Proclamation of 10 March 1983).

Problems in the ocean and coastal management system began to emerge during the early 1980s. One example was the dispute over the development of offshore energy resources between coastal states and environmental groups on the one side, and the oil industry and the Department of the Interior on the other. Other problems included conflicts between fishers and the oil industry over access to ocean areas and interference in each other's activities; between industry and environmentalists over drilling in or near sensitive natural habitats for seabirds and marine mammals; between state and federal agencies over state authority under the CZMA's federal consistency doctrine; and among federal agencies that had different and sometimes conflicting legal mandates respecting ocean resources (Eichenberg and Archer 1987). The dilemma posed by conflicting legal authorities was well illustrated when local governments were thwarted in their efforts to enforce air quality standards onshore under the Clean Air Act because they were unable to control air emissions from offshore oil and gas projects in federal waters, which are solely regulated by Interior under the OCSLAA.

These and other problems highlighted by the dispute over offshore energy development were strong indications that the system was unable to provide an integrated approach to ocean and coastal management or to resolve serious ocean and coastal use conflicts. Further, the system allowed only limited coastal state participation in decision making (*Secretary of the Interior v. California* 1980). Yet, there was no general agreement about how to address these problems, and the administration steadfastly opposed any initiatives to do so.

Growing Capacity and Action by the Coastal States

Important developments were also taking place during this period at the state level. Stimulated in part by their involvement in the development and operation of coastal zone management programs and, in part, by federally encouraged activities such as oil leasing in nearby federal waters, most coastal states developed considerably increased capacity in dealing with a variety of coastal and ocean issues (King 1986; Cicin-Sain 1990). Participation in the national CZM program required coastal states to acquire expertise on a range of topics, including coastal planning, wetlands protection, management of beaches and erosion, public access, coastal hazards, and special area management.

At about the same time that most coastal states were starting work on their CZM programs, the great acceleration in oil and gas leasing in federal waters began. Coastal states adjacent to these leases felt a strong need to comment on the proposed oil and gas developments, which they believed had the potential to affect their communities in important ways. Indeed, input from state governors had been formally built into the new decision-making process of the OCSLAA of 1978. Furthermore, the federal consistency process of the CZMA of 1972 also gave the coastal states with federally approved CZM programs some legal leverage over federal actions such as offshore leasing, if the states judged the federal action to be inconsistent with state coastal policies.

The participation of coastal states was also incorporated into the management of fisheries in federal waters under the terms of the Fishery Conservation and Management Act (FCMA) of 1976, as participants in the regional fishery management councils. This meant that state competence in marine fishery matters had to expand beyond the relatively narrow band of state waters to include the much broader EEZ. In addition, fishery management plans (FMPs) coming out of the new federal FCMA-mandated process could, under certain circumstances, come under the terms of the federal consistency provisions of the CZMA.

Exercise of state prerogatives under CZMA's federal consistency provisions and coastal state involvement in the EIS reviews brought the coastal states into a wide range of issues beyond oil and gas and fisheries, requiring an increase in expertise in areas such as the disposal of dredged materials, building of bridges to barrier islands, designation of floodplains and high-hazard zones, port and harbor development and the maintenance of navigational channels.

In 1980, the CZMA was amended to include a set of nine national goals for coastal zone management and to require states to deal more effectively with an enumerated set of issues that included increased specificity in protecting significant natural resources, reasonable coastal-dependent economic growth, improved protection of life and property in hazardous areas, and improved predictability in governmental decision making. The need to demonstrate increased attention to these issues also contributed to the growth of state competence (and action) relative to them (Coastal Ocean Policy Roundtable 1992). Realizing that strong state interests existed in the offshore areas beyond state waters, some coastal states, most notably Oregon, North Carolina, Hawaii, California, and Florida, began to expand the geographic scope of their planning and policy making to include adjacent federal waters (Hawaii Ocean and Marine Resources Council 1991). Oregon, for example, after an extensive study of the state's interest and activities in its territorial sea, created an ocean planning process that eventually led to the designation of an "ocean stewardship zone," a band of water varying in width from approximately 30 to 80 miles from the shore in which the state asserted strong interest especially related to fishing and environmental concerns. North Carolina, Hawaii, California, and

Florida all completed ocean plans for offshore areas adjacent to their presently recognized boundaries. These efforts demonstrate that the competence to prepare multisectoral (integrated) plans and policies in some states currently resides to a greater degree at the state level than within the federal government (Clark and Whitesell 1994).

Taken together, then, a number of forces and factors in the 1970s, 1980s, and 1990s conspired to significantly increase the capacity of the states to implement coastal and ocean policy. Some of this growth in competence came about in reaction to federally stimulated ocean activities that were seen as threats, such as oil and gas development, but much of it was brought about by the partnerships created by the legislation enacted in the early 1970s. Coastal states were given the lead role in the national CZM program, with federal agencies providing support and consistency; states were made virtual partners with the federal government in marine fisheries management; states were given important (though not determinative) roles in oil and gas development in the outer continental shelf; and states were permitted to begin planning and policy making in the adjacent federal waters under the CZM program if they so chose. As a result, by the mid-90s, states with important coastal and ocean interests were in a much better position to defend and protect those interests than had been the case earlier (Hershman 1996).

The 1983 Exclusive Economic Zones Proclamation and the 1988 Territorial Sea Proclamation

In contrast to the preceding Carter administration, the Reagan administration was skeptical of the Law of the Sea negotiations going on under U.N. auspices. Shortly after the administration took office in early 1981, it ordered a comprehensive review of the pending agreement. Although the United States had been a major player in shaping the negotiations up to that point, this review made it clear that the treaty as it then stood was unacceptable to the administration. Subsequently, the United States was one of only four nations[9] to vote against the adoption of the treaty in April 1982, but was nonetheless quick to take advantage of some of its provisions.

The 1983 Proclamation of the U.S. Exclusive Economic Zone (EEZ)

Arguing that the concept of national 200-mile zones in which coastal states had sovereign rights over resources and economic uses had become customary international law, the Reagan administration, in a

March 1983 proclamation, claimed a 200-mile EEZ for the United States. In its customary international law argument, the United States claimed that the concept of the 200-mile EEZ had been essentially agreed upon in the LOS negotiations by 1976, and that a number of other nations had unilaterally declared 200-mile zones of their own, thus indicating that the EEZ concept had become fully accepted by the world community even before the LOS treaty was finalized and approved.[10]

The 1988 Proclamation of a 12-Mile Territorial Sea In the past, the United States had always been a staunch supporter of the narrowest possible territorial sea—3 miles. In the earlier LOS negotiations, only when the broadening of the territorial sea to 12 miles was coupled with the guarantee of free transit through straits used for international navigation, would the United States agree to the change. Even so, six years elapsed after the approval of the LOS treaty before President Reagan, in 1988, claimed a 12-mile territorial sea for the United States. Press releases at the time made it clear that the action was taken primarily for national security purposes. The United States said that it wanted to be able to keep foreign submarines at least 12 miles from the entrances to sensitive naval complexes.

The 1983 proclamation creating a new 200-mile EEZ around the United States and its territories and the 1988 proclamation expanding the territorial sea of the United States from 3 to 12 miles significantly changed the relationship between the nation and the surrounding ocean (Cicin-Sain and Knecht 1985). The expansion of the territorial sea quadrupled the ocean area over which the United States claimed sovereignty, while the EEZ claim declared "sovereign rights" for the United States over all living and nonliving resources in the EEZ—an area about equal to the country's land area.

Although proclaiming a higher level of involvement by the United States with the surrounding ocean, these proclamations provide only the bare bones of a framework for governing these new areas. Neither proclamation spells out how resources in these areas and the ocean space itself are to be governed. Significant questions remain; for example: (1) Do governments (federal, state, and local) have duties and responsibilities toward the ocean areas newly under the jurisdiction of the United States above and beyond those spelled out in existing statutes (such as the Magnuson Act and the OCSLAA)? (2) What roles should the coastal states play in the governance of these areas? The expanded territorial sea areas, after all, are immediately adjacent to existing state waters, which are included within state CZM boundaries. (3) Will an overall plan be needed for the conservation and development of the nation's new EEZ area? Should a regional approach be considered for such planning? What role should the coastal states play in such planning and management?

Coastal Management: Successes and Shortcomings

Initiating the Program

As discussed in Chapter 3, the assignment of responsibility for implementation of the CZMA was not clear until the very last stages of the legislative debate. Because most observers had been expecting coastal management to go to the Interior Department, a significant degree of advanced planning was based on that assumption and an administrative unit was ready to take on the job. NOAA, on the other hand, had the benefit only of a relatively short implementation plan prepared several months earlier at the request of NOAA administrator Robert M. White. The plan[11] was influenced to a significant degree by limited coastal zone management activities already started by several coastal states (Rhode Island, Washington, Oregon, California, Minnesota, Michigan, and Wisconsin).

The early months of effort of NOAA's newly created CZM task force were devoted to assembling a staff with the necessary expertise, preparing draft rules and regulations for the financial grants portion of the program, working with the thirty-five eligible coastal states and territories to begin to interact with the federal government on CZM matters, and preparing the request for a supplemental appropriation to fund the new program, especially the initial round of planning grants to the states.

Because the Nixon administration still hoped for comprehensive national land use legislation that would encompass the new CZM program, it did not permit an appropriations request to go to Congress in fiscal years 1972 or 1973, frustrating efforts to get the state grants phase of the program underway promptly. It was only when CZM advocates Senators Warren Magnuson and Ernest Hollings made it clear that the Nixon administration's emergency energy legislation (the 1973 Arab oil embargo had just started) would be not acted upon until funding was requested to begin the CZM program, that funding was finally requested and appropriated, initially at a level of $12 million.

The authorizing legislation—the CZMA—was short and cast in general terms and, regarding the standards that state CZM programs must meet for federal approval and funding, entirely process oriented. Substantive standards were not included. Furthermore, the overall purpose of the state coastal zone management programs that the federal CZM program was supposed to facilitate and encourage was somewhat ambiguous at best and contradictory at worst. For example, Section 303 (1) (of the act) states: "to preserve, protect, develop, and where possible, to restore or enhance, the resources of the Nation's coastal zone for this and succeeding generations."

The legislative history of the CZMA was not very helpful when it came to writing rules and regulations to interpret the statute. One of the main arguments during debate of the act had been the degree to which the resulting state CZM programs were to be protectionist in their goals (a position strongly supported by groups such as the Natural Resources Defense Council) as opposed to actively promoting development that was perceived to be in the national interest, most notably, deciding when and under what conditions energy facilities could be sited on the coast, meeting needs that are more than local in nature. NOAA took the position that the legislation called neither for strict protection in every instance (though protection of the unique resources of the coastal zone was certainly called for), nor for the siting of proposed energy facilities, but that the legislation was principally aimed at rational and informed management of the coastal zone.

Two years after the first funding became available in 1974, the first state CZM program was approved—the state of Washington program in 1976 (see Table 4.2). Washington was one of the states with state CZM legislation (the 1971 Shoreline Management Act) that pre-dated the federal program. The following year saw federal approval of the programs of two additional states, California and Oregon.

Tangling with Oil and Gas Interests

Shortly after the approval of the California program, the American Petroleum Institute and several other oil interests sued the director of the federal CZM program, arguing that the California CZM program, with its constraints and restrictions on offshore oil and gas activities, did not meet the national interest standards contained in the CZMA. Ultimately, the Ninth Circuit Court of Appeals ruled in favor of the NOAA decision to approve the program, which, in effect, validated the approach taken by NOAA in attempting to set an appropriate balance between resource protection and energy resource development (*API v. Knecht*, note 7).

TABLE 4.2 State Coastal Zone Management Programs: Year of Federal Approval.

STATE	YEAR OF APPROVAL	STATE	YEAR OF APPROVAL
Washington	1976	Louisiana	1980
Oregon	1977	Mississippi	1980
California	1978	Connecticut	1980
Massachusetts	1978	Pennsylvania	1980
Wisconsin	1978	New Jersey (remaining segments)	1980
Rhode Island	1978		
Michigan	1978	Northern Marianas	1980
North Carolina	1978	American Samoa	1980
Puerto Rico	1978	Florida	1981
Hawaii	1978	New Hampshire (ocean and harbor segment)	1982
Maine	1978		
Virginia	1986	New York	1982
Maryland	1978	New Hampshire (remaining segments)	1988
New Jersey (bay and ocean shore segment)	1978	Texas	1997
Ohio	1997	Georgia	1998
Virgin Islands	1979	Minnesota	1998 (pending)
Alaska	1979	Indiana	1998 (pending)
Guam	1979		
Delaware	1979		
Alabama	1979	*Nonparticipating*	
South Carolina	1979	Illinois	

Source: Data from NOAA, Office of Ocean and Coastal Resource Management, 1998.

Coastal energy issues played important roles in the early implementation of the CZM program in several other ways as well. In 1976, the CZMA was amended (see Table 4.3 for a listing of important substantive amendments to the CZMA) to add a new coastal energy impact program (CEIP) to the overall CZM effort. Depending on the level of oil and gas activity in adjacent federal waters, coastal states were eligible under this new program for a series of grants and loans to assist them in dealing with the impacts of energy activity that affected their coastal zones.

Use of the federal consistency provisions of the CZMA by several coastal states in an attempt to stop oil and gas development in certain sensitive locations off their shores also became very controversial in the late 1970s and early 1980s. Ultimately, a case involving this conflict

TABLE 4.3 Major Amendments to the Coastal Zone Management Act.

Coastal Energy Impact Program Amendments of 1976[a]	• Redefines "coastal zone" under the CZMA of 1972, and defines, under such act, "outer continental shelf energy activity," "energy facilities," and "coastal energy activity." • Makes changes in the Management Program Development Grants, under the CZMA, by adding the following requirements for the coastal zone management program that a state is to develop and maintain under the act: (1) The program is to include a general plan for the protection of, and access to, public beaches and other coastal areas of environmental, recreational and historical, aesthetic, ecological, and cultural value; and (2) the state coastal zone management program is to include a process for the planning for energy facilities likely to be located in the coastal zone and for the planning for, and management of, the anticipated impacts from any energy facility. • Increases the maximum federal share of the costs of the development phase of a coastal zone management program to 80 percent from 66⅔ percent and extends, by one year, the time during which a coastal state may receive such grants for development of a program before it must have an approved program in order to continue to receive grants under the act. • Increases the maximum federal share for administrative costs of the ongoing state program operation to 80 percent from 66⅔ percent. • Requires a coastal state to establish an effective coordination and consultative mechanism between a designated state coastal zone agency and local governments within such state. Requires states to consider any applicable interstate energy plans or programs in the planning for and siting of energy facilities in the coastal zone of such states.

continued

TABLE 4.3 (*continued*)

Coastal Energy Impact Program Amendments of 1976[a] (*continued*)	• Requires each federal lease to be submitted to each state with an approved coastal zone management program for a determination by that state as to whether or not the lease is consistent with such state's program. • Directs the secretary of commerce to administer and coordinate a coastal energy impact program. • Requires pursuant to this program the provision of financial assistance to meet the needs of coastal states and local governments. Provides for formula grants to coastal states. Specifies the purposes for which such grants may be used. Requires the secretary to make such grants if the secretary finds that the coastal zone of such state is being, or is likely to be, significantly affected by the siting, construction, expansion, or operation of new or expended energy facilities. • Requires the making of loans to assist any state or local unit of government to provide new or improved public services required by coastal energy activity. • Establishes the Coastal Energy Impact Fund for the purpose of making payments under the Coastal Energy Impact Program.
Coastal Zone Management Improvement Act of 1980[b]	• Amends the CZMA of 1972 to declare that it is the national policy to provide for: (1) the protection of natural resources within the coastal zone; (2) the management of coastal development; (3) priority consideration to coastal dependent uses and orderly processes for siting major facilities related to national defense, energy, fisheries development, recreation, ports and transportation; (4) public access to the coasts for recreation purposes; (5) the coordination and simplification of procedures in order to ensure expedited governmental decision making for the management of coastal resources; (6) continued consultation and coordination with affected federal agencies; (7) the giving of timely and effective

continued

TABLE 4.3 *(continued)*

Coastal Zone Management Improvement Act of 1980[b] ***(continued)***	opportunities for public participation in coastal-management decision making; (8) assistance in the redevelopment of deteriorating urban water-fronts and ports, and sensitive preservation and restoration of historic, cultural, and aesthetic coastal features; and (9) assistance to support comprehensive planning, conservation, and management for living marine resources. • Encourages the coastal states to amend their coastal management programs to provide for: (1) the inventory and designation of areas that contain one or more resources of national significance; and (2) specific and enforceable standards to protect such resources. Specifies that if the secretary determines that a coastal state has failed to make satisfactory progress in such activities by September 30, 1984, the secretary shall not make any grants to such state. • Allows the secretary to make grants to any eligible coastal state to assist that state in meeting one or more of the following objectives: (1) the preservation or restoration of specific areas of the state; (2) the redevelopment of deteriorating and underutilized urban waterfronts and ports that are designated in the state's management program areas of particular concern; and (3) the provision of access to public beaches and other public coastal areas and to coastal waters. • Directs the secretary to withdraw approval of the management program of any coastal state, and to withdraw any financial assistance available to that state if the secretary determines that the coastal state is failing to adhere to, and is not justified in deviating from: (1) the management program approved by the secretary; or (2) the terms of any grant or cooperative agreement funded under this act, and refuses to remedy the deviation.

continued

TABLE 4.3 *(continued)*

Coastal Zone Management Reauthorization Act of 1986[c]	• Amends the CZMA to phase-down the federal share of the development, implementation, and operation of coastal zone management programs form 80 percent to 50 percent over a three-year period. Directs any state receiving federal funds to promptly notify the secretary of commerce of any proposed program changes. Requires the secretary to notify the state of its approval or disapproval of such changes within thirty days of notice from the state of such proposed changes. • Establishes the National Estuarine Reserve Research System (the System). Authorizes the secretary to designate an estuarine area as a national estuarine research reserve upon certain findings. Requires the secretary to develop guidelines for research within the System. Requires the secretary to promote and coordinate the utilization of such reserves. Authorizes the secretary to make grants to coastal states for coordinating the utilization of such reserves. Authorizes the secretary to make grants to coastal states for acquisition or operation of such reserves or to a coastal state or other entity to support research and monitoring within a reserve. Sets per-reserve grant limits. Directs the secretary to periodically evaluate the operation and management of each reserve and its research. Authorizes the secretary to suspend a reserve from eligibility for financial assistance or withdraw its designation as a national reserve upon certain findings. Outlines information to be included in annual reports concerning such reserves. • Repeals federal provisions that establish (1) research and technical assistance grants for coastal zone management, and (2) the Coastal Zone Management Advisory Committee and certain other positions. • Adjusts the authorization of appropriations for administrative grants to states for coastal resources management programs and for the establishment of national estuarine sanctuaries.

continued

TABLE 4.3 *(continued)*

1990 Amendments[d]	• Requires states with approved coastal zone management programs to submit coastal nonpoint pollution control programs to the secretary of commerce and the administrator of the EPA for approval. Sets forth requirements of such programs and review and approval procedures. Withholds specified coastal management and water pollution control assistance from states failing to submit approved programs. Requires the secretary and the administrator to provide technical assistance to coastal states and local governments for implementing such programs.
	• Directs the secretary to review the inland coastal boundary of each state program and evaluate whether such boundary extends inland to the extent necessary to control land and water uses having a significant impact on state coastal waters. Requires the secretary to recommend modifications to such boundaries.
	• Requires that each federal agency activity within or outside the coastal zone that affects any land or water use or natural resource of the coastal zone to be carried out in a manner that is consistent to the maximum extent practicable with the enforceable policies of approved state management programs.
	• Authorizes appropriations of: (1) sums not exceeding $750,000 for each fiscal year between 10/1/90 and 9/30/93; (2) sums not exceeding $42,000,000 for the FY ending 9/10/91; $48,890,000 for the FY ending 9/30/92; $58,870,000 for the FY ending 9/30 93; $67,930,000 for the FY ending 9/30/94; $90,090,000 for the FY ending 9/30/95, as may be necessary for grants under section 306, 306A, and 309; (3) sums not exceeding $6,000,000 for the FY ending 9/30/91; $6,270,000 for the FY ending 9/30/92; $6,552,000 for the FY ending 9/30/93; $6,847,000 for the FY ending 9/30/94; $7,155,000 for the FY ending 9/30/95, as may be necessary

continued

TABLE 4.3 (*continued*)

1990 Amendments[d] (***continued***)	for grants under section 315; (4) such sums, not to exceed $10,000,000 for each of the fiscal years occurring during the period beginning 10/1/90 and ending 9/30/95 as may be necessary for activities under section 310 and for the administrative expenses incident to the administration of this title; except that expenditures for such administrative expenses shall not exceed $5,000,000 in any such fiscal year.

[a]http://thomas.loc.gov. Bill Summary and Status for the 94th Congress. Public Law 94-370.
[b]http://thomas.loc.gov. Bill Summary and Status for the 96th Congress. Public Law 96-464.
[c]http://thomas.loc.gov. Bill Summary and Status for the 99th Congress. Public Law 99-272.
[d]http://thomas.loc.gov. Bill Summary and Status for the 101st Congress. Public Law 101-508.
Source: Prepared with assistance from Rosemarie Hinkel.

(*Secretary of Interior v. California*) reached the Supreme Court, and in 1984 the Court ruled, in a divided opinion, that the federal consistency provisions do not apply to federal activities outside the state's coastal zone, that is, to oil and gas leasing in federal waters, which was widely interpreted as a serious blow to the federal consistency process and to the CZM program as a whole (Fitzgerald 1985; Eichenberg and Archer 1987).

Meanwhile, additional states were completing CZM programs and receiving federal approval and implementation grant funding. By 1980, sixteen states were in this category with an additional thirteen states still developing their programs (Mitchell 1982, p. 269). The year 1980 also saw amendments to the act that, in addition to reauthorizing it for an additional five years, clarified its purposes by adding nine national goals and added language encouraging coastal states to strengthen selected aspects of their CZM programs. These changes were, in part, in response to a report by a 1980 General Accounting Office report that had been critical of some facets of the CZM program (U.S. GAO 1980).

The period from 1981 to 1987 was a difficult one for the CZM effort and a number of other environmental programs at the national level (Hildreth and Johnson 1985). The Reagan administration sought to reduce the size of the federal government and to reduce federal regulations and their impact on the nation's economic activities. New program administrators were appointed with these goals in mind, and the CZM program was no exception (Archer and Knecht 1987). Strong efforts in

Congress and in the coastal states, however, thwarted these attempts and the program survived, but much of the innovation during this period took place within the coastal states and their communities rather than at the federal level.

Expanding the Focus

After this relatively long "low" period, the CZM program got a needed boost by a remarkable set of amendments that advocates were able to get through the Congress in 1990 (Archer 1991). These changes and additions reinvigorated the program at a time when it was very badly needed. In addition to authorizing a new type of grant for program improvements (section 309) and restoring program development grants (for a limited period) (section 305) to accommodate the few states not in the program, the amendments strengthened the federal consistency provisions by clarifying that they indeed apply to federal activities occurring outside the coastal zone, such as oil and gas leasing, overturning the 1984 Supreme Court decision in *Secretary of Interior v. Watt.* Also, states were mandated to add new non-point-source pollution management programs to their CZM programs within the next three to four years—called the 6217 provision after the section of the amended act (The Coastal Zone Act Reauthorization Amendments of 1990).[12] All told, the 1990 amendments revitalized the program and put it on a much more secure footing for the decade of the 1990s.

Of the five states without approved CZM programs in 1990 (Texas, Ohio, Indiana, Illinois, and Minnesota), four of them took advantage of the renewed program development grants and resumed their CZM program efforts. Of these, by the end of 1997, the programs of both Texas and Ohio had been approved and funded with the Minnesota and Indiana programs nearly completed. As of early 1999, only Illinois (with its shoreline of 59 miles) remains outside the national CZM program. The level of federal funding of the CZM program by fiscal 1998 had risen to about $50 million a year, exclusive of funding for the National Estuarine Research Reserve (CSO 1998, p. 1). Of course, having virtually all of the U.S. shoreline under federal approved CZM programs does not necessarily mean that the goals, on the ground, of the program are being achieved.

Some Assessments of the CZM Program

Evaluation of the impacts of the CZM program is difficult at best, and consequently, relatively little information of this kind exists. Most assessments to date have involved the use of "process" indicators (number of new state laws and regulations, strength of enforcement, better mapping, etc.), and not "on the ground" indicators of outputs and program performance. A study was prepared in 1996 on the "perceptions" of the performance of state CZM programs among coastal interest groups, academics, and state CZM managers. The results suggest that state CZM programs are perceived as doing a reasonably good job of resource protection and public access, a moderately good job of dealing with natural hazards in the coastal zone, and only a fair job of managing coastal development (Knecht, Cicin-Sain, and Fisk 1996, 1997).

The most comprehensive evaluation of the CZM program has recently been completed by a team of external consultants and results are now available (Hershman et al. 1999). This study assessed the effectiveness of four aspects of the program: protecting estuaries and coastal wetlands; protecting beaches, dunes, bluffs, and rocky shores; seaport development; and the redevelopment of urban ports. As expected, the study revealed a severe shortage of on-the-ground outcome data. For example, for only two of twenty-nine states and territories were sufficient on-the-ground outcome data available to conclusively judge the effectiveness of programs to protect wetlands and estuaries. In ten other states, sufficient outcome data were available to make probable determinations of effectiveness. For these twelve states, 56 percent rated their wetlands protection program high, 39 percent rated moderate, and only 6 percent rated low in effectiveness. In the case of beaches, dunes, bluffs, and rocky shores, virtually no data on on-the-ground outcomes were available; hence, the assessment had to rely on process indicators and, especially, case examples. In the seaport aspect of the study, the analysis revealed that only about half of the states had dealt with port issues in their programs, and only a fraction of these in a comprehensive way. Twelve port-active states were identified and studied in greater detail. The urban waterfront revitalization part of the study revealed that about 300 such projects had been undertaken in the coastal states and territories studied, many of which benefited from their state CZM programs in various ways. In addition to amassing some very valuable information on the tools and techniques currently being used by the states and territories in their CZM programs, the study clearly demonstrated the urgent need for more systematic collection of on-the-ground outcome data to allow more accurate assessment of state CZM program performance over time.

The CZM program has been very effective in diffusing coastal management practices to practically all of the U.S. coastal zone in the coastal states and territories. More than 97 percent of the U.S. shoreline is now under the CZM program; all but one of the thirty-five eligible states and territories are participating (see Figure 4.1). No state or territory has dropped out of the program in its twenty-seven-year history.

The CZM program stimulated coastal management action in the states and territories, where, in most cases, none had existed before. Established focal points that permit an integrated view of the coastal zone and its management needs now are present in all of the participating states and territories.

The flexibility built into the act to allow each state to define its own structure and process for implementing its CZM program has meant that, in effect, there are thirty-five different variations of coastal management in the United States, albeit all meeting certain procedural standards established and enforced by the national OCRM office. This represents a good example of creative federalism at work—which has allowed for a great deal of innovation to take place and for "lesson learning" to be shared among the states. The coastal states have developed a strong network of relationships through the Coastal States Organization, and lessons learned in one state may be easily diffused to other states.

Implementation of the act's federal consistency provision, although sometimes controversial and conflictive, has been particularly effective in ensuring that federal actions that affect state coastal zones are consistent with state coastal policies. The federal consistency provision has worked as one of the few mechanisms available for achieving intergovernmental coordination and policy harmonization among federal, state, and local governments on coastal issues.

Although because of the absence of baseline data on the status of coastal zones at the time of inception of coastal programs, it has been difficult to demonstrate empirically the accomplishments of the program, it is clear that the CZM program has, in thousands of instances around the U.S. shoreline, prevented inappropriate coastal development, fostered public access to the coast, served to protect fragile coastal resources such as wetlands, and protected the public from coastal hazards. As an experienced practitioner in the field, Peter Douglas (executive director of the California Coastal Commission) has noted, "the greatest accomplishment of the CZM program is what has not been built in the coastal zone . . ." (quoted in Lester 1996b).

The CZM program also faces a number of continuing problems and challenges. All too often, little scientific knowledge (from both the natural and the social sciences) has been built into the CZM programs, and this is one area in which the program needs to improve. Our survey of

FIGURE 4.1.
Map of States Participating in the Coastal Zone Management Program.
Source: NOAA Office of Ocean and Coastal Resource Management 1998.

coastal managers, interest groups and academics (Knecht, Cicin-Sain, and Fisk 1996, 1997) pointed "greater use of science" as a major area in need of improvement in CZM programs. As noted earlier, assessment of the effectiveness and impact of the CZM program continues to be difficult due to a lack of systematic collection of data on on-the-ground outcomes. Enhancing the use of scientific information in state CZM programs specifically aimed at establishing appropriate monitoring and evaluation systems should work to address this shortcoming.

State CZM programs are sometimes of low visibility, especially in states where no specialized coastal permit legislation and/or no state coastal commissions or councils exist (i.e., "networked" programs [Born and Miller 1988]). In many states (e.g., Massachusetts), the established CZM program has been successful in working with other types of coastal management programs established later (such as the National Estuary Program); in others, the program has not adapted accordingly and works in isolation of more recently established programs in the coastal zone.

Although generally good relations exist between coastal states programs and NOAA's Office of Ocean and Coastal Resource Management, there continues to be a lack of a formal workable mechanism to coordinate coastal and ocean activities among federal agencies and offices at the federal level, which obviously affects state coastal managers. Although the OCRM office has often worked to play a catalytic role in bringing federal ocean and coastal agencies together to achieve more coherent national policy, its somewhat low level within the bureaucratic hierarchy (four levels down the Department of Commerce ladder of authority) makes it difficult for it to play a central leadership role in federal interagency proceedings.

With a few exceptions (such as Oregon, California, Hawaii, Maine, Mississippi, Florida, and North Carolina), state CZM programs continue to be focused primarily on traditional shoreland uses in the coastal zone. Extending the programs to encompass ocean uses in the territorial sea (and in the EEZ to the extent that state interests are involved) and bringing other ocean sectors such as fisheries under the ambit of CZM have generally been slow.

Fisheries Management: Fisheries Depletion Overshadows Positive Institutional Changes

As discussed in Chapter 3, the Fishery Conservation and Management Act (FCMA) of 1976 ushered in a new era of fisheries management. Novel features of the act were (1) the creation of a new Fishery Conservation zone (to 200 miles offshore); (2) a new management system was put into place in the form of eight new regional fishery management councils charged with preparing and implementing fishery management plans for fisheries in their region; (3) new national standards for fisheries management were promulgated; and (4) foreign fishing was eliminated unless permitted through the conclusion of Governing Fishery Agreements between the United States and foreign governments. Specific measures for the development of U.S. fisheries were not included in the 1976 act, nor were measures aimed at addressing the problems of fisheries spanning multi-state jurisdictions.

Initial Implementation Challenges

Following passage of the FCMA, the first few years were spent on implementing its very complex process of fishery management plan development involving the regional councils and their committees, and subsequent review by the federal level (National Marine Fisheries Service and the secretary of commerce). This proved invariably to be a very protracted and often frustrating process. Implementation of the FCMA involved the mobilization of new actors and roles at the regional level, the working out of new standards and rules for managing fisheries, and the redefinition of relationships between national and regional fishery authorities. Figure 4.2 provides a summary of the FMP development

DEPARTMENT OF COMMERCE
Office of the Secretary

National Oceanic and Atmospheric Administration (NOAA)
Office of the Administrator
Office of Policy and Planning

National Marine Fisheries Service
Office of the Assistant Administrator for Fisheries
Office of Policy and Planning
Office of Resource Conservation and Management Plan Review Division

Fishery Policy Group (informal group)

Office of NOAA General Counsel

National Level

Regional Level

Southwest Regional Office (Los Angeles)

Southwest Fisheries Research Center (La Jolla)

PACIFIC FISHERY MANAGEMENT COUNCIL
- Council Members (based in Portland)
- Council Staff (based in Portland)
- Scientific and Statistical Committee (based in Portland)
- Anchovy Advisory Panel (worked primarily in Los Angeles)
- Anchovy Plan Development Team (worked primarily in La Jolla)

STATE OF CALIFORNIA
- California Department of Fish and Game (based in Sacramento)
- California Fish and Game Commission (based in Sacramento)

OTHER ADMINISTRATIVE AGENCIES
- Department of State
- Department of the Interior
- U.S. Fish and Wildlife Service
- U.S. Coast Guard
- Environmental Protection Agency

FIGURE 4.2.
Decision-Making Process for the Development of Fishery Management Plans under the Magnuson Act
Source: Adapted from Cicin-Sain and Orbach 1986, p. 351.

process—in the case of the anchovy fishery off California, a relatively simple example since only one state was involved—showing the many national and regional participants.

Implementation of the FCMA put a heavy burden on the main federal agency charged with implementation—the National Marine Fisheries Service (NMFS), which was ill equipped for the task. Prior to 1976, NMFS had basically been a service and research organization in charge of collecting and analyzing fishery statistics and managing services to fishermen (such as awarding loans for vessel construction). NMFS had no real authority over the management of fisheries. The agency was populated largely with fishery biologists, with most of its personnel resources (66 percent) found in regional research centers around the country (Chandler 1988; see Figure 4.3). These research scientists typically conducted fishery assessment studies and, much like academic scientists, were interested in and motivated by the advancement of scientific knowledge about fisheries, peer review of their work, and publications.

Much of the challenge of implementing the FCMA thus lay in how to move the agency away from its traditional research and service mission to a new management and regulatory role. Much of the time would be spent on reorganizing the agency—several times—in an attempt to achieve this shift. Implementation of the major concept of optimum yield required perspectives other than fishery biology; hence, social scientists such as economists and anthropologists came to be new participants in the fisheries management process, in both the scientific and the statistical committees set up by each council, and in the NMFS itself.

Another major implementation challenge lay in the changing relationship between NMFS and its major client group—commercial fishers. This relationship, originally very amicable, underwent considerable change after 1976. At the time the act passed, U.S. fishers were elated at their triumph over foreign fishers. Elation, however, was soon followed by dismay as they began to realize the magnitude of the domestic

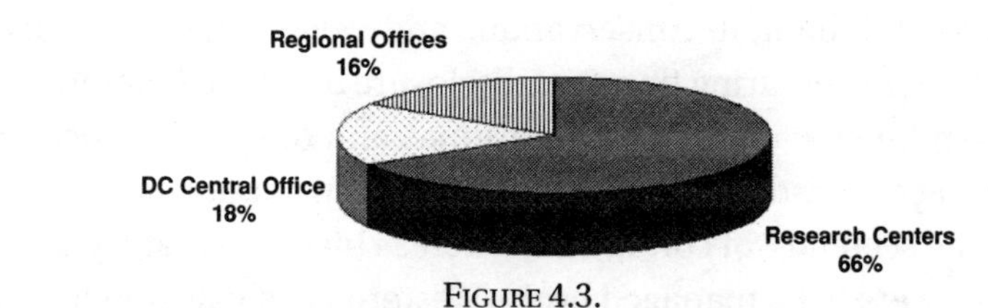

FIGURE 4.3.
Regional Personnel Breakout, National Marine Fisheries Service.
Research Centers vs. Regional Offices Total = 2,310 Full-Time Equivalents
Source: Adapted from Chandler 1988, p. 23.

regulatory regime that had been set up; as immortalized by a fisherman who spoke at one of the first FCMA hearings held in Astoria, Oregon, in 1977; "First we had the Russians, now we have the bureaucrats." Acrimonious relationships followed in the initial period of implementation as commercial fishers strove to control the new decision-making roles established by the FCMA: membership on the regional councils, the scientific and statistical committees, and the industry advisory panels set up for each fishery management plan. Eventually, by the mid-1980s, the commercial fishers would come to dominate most of the councils, and be relatively supportive of the domestic fishery management regime (see Table 4.4).[13]

Other factors that affected the implementation of the act were frequent reorganizations and changes in direction—from 1976 to 1999 the NMFS has had eight different directors, each, of course, wishing to put his own imprint on the organization. The agency was also given new responsibilities related to other laws, such as the MMPA, and ESA, which often turned out to conflict with the agency's fishery management responsibilities. As noted earlier, too, during the Reagan administration, the program was proposed for elimination several times and was only maintained through congressional action.

Amendments to the Fishery Conservation and Management Act

Since it was enacted in 1976, the FCMA has been amended more than a dozen times, with major substantive amendments occurring in 1980, 1986, 1990, and 1996 (see Table 4.5). In particular, the 1980 amendments (which changed the act's name to the Magnuson Fishery Conservation and Management Act) reduced allocations of foreign harvest, raised foreign fishery fees, and provided 100 percent observer coverage of foreign fishing operations in an effort to encourage the expansion of the domestic fish harvesting and processing industries.

The 1986 amendments to the act established sovereign rights over fish in the EEZ (following the 1983 EEZ proclamation), established new requirements for habitat conservation, and revised the FMP decision process (e.g., by requiring financial disclosure by council members). The 1990 amendments, in turn, brought tunas and other migratory species under the act's jurisdiction—thenceforth Atlantic tunas were to be managed by the secretary of commerce, whereas highly migratory species in the Pacific were to be managed by the Western Pacific Regional Fisheries Management Council. In addition, the 1990 amendments required tuna products to be accurately labeled as to the effect of tuna harvesting on

TABLE 4.4 Composition of the Regional Fishery Management Councils, 1985–1987

COUNCIL	COMMERCIAL		RECREATIONAL		ENVIRONMENTALIST		CONSUMER		ACADEMIC/ SCIENTIST/ CONSULTANTS		OTHER		TOTAL COUNCIL APPOINTMENTS
	1985	1987	1985	1987	1985	1987	1985	1987	1985	1987	1985	1987	
New England	8	7	2	3	—	—	—	—	1	1	—	—	11
Mid-Atlantic	5	3	5	5	—	—	—	—	—	1	2	3	12
South Atlantic	1	1	5	4	—	—	—	—	—	1	1	1	8
Caribbean	—	1	2	1	—	—	—	—	1	—	1	1	4
Gulf of Mexico	5	5	4	4	2	2	—	—	—	—	—	—	11
Pacific	4	4	1	3	1	1	—	—	1	—	1	—	8
North Pacific	6	5	—	—	—	—	—	—	—	—	1	2	7
Western Pacific	3	3	3	3	3	—	—	—	1	1	1	1	8
TOTAL	32	29	22	23	3	3	—	—	4	3	8	7	69

Source: Chandler 1988.

Note: State and federal officials hold 41 seats on the eight regional fishery management councils, bringing the total number of council members to 100. Totals differ because vacancies existed on the following councils: South Atlantic, Caribbean, and Western Pacific.

TABLE 4.5 Major Amendments to the Fishery Conservation and Management Act.

1978 Amendments[a]	• Amends the FCMA to include within the definition of fishing under the act, fish processing at sea. • Directs the secretary of the treasury, in cooperation with the secretaries of commerce and state, to report to Congress and to the president on all allocations of U.S. fish to foreign nations and all trade barriers imposed by such nations. • Revises certain fishing permit requirements for specified foreign fish-processing ships at sea in certain U.S. fisheries. • Disallows foreign processing of U.S.-harvested fish if U.S. fish processors have adequate capacity to process such fish. • Prohibits transfer of U.S.-harvested fish to any foreign ship, within the U.S. fishery conservation zone, not having the fishing permit required by such act.
Packwood-Magnuson Amendment of 1979[b]	• Authorizes appropriations for fiscal years 1980–1982. • Defines "certification" as a determination made by the secretary of commerce that foreign vessels are conducting fishing operations or engaging in trade or taking that diminishes the effectiveness of the International Convention for the Regulation of Whaling for the purpose of reducing a nation's allocation of the total allowable level of foreign fishing within the U.S. fishery conservation zone. Requires the secretary of state to reduce the allocation of a nation so certified by not less that 50 percent.
American Fisheries Promotion Act of 1980[c]	• Amends the FCMA to set forth a formula for the reduction of the total allowable level of foreign fishing, if any, with respect to any U.S. fishery for each harvesting season after the 1980 harvesting season.

continued

TABLE 4.5 (*continued*)

American Fisheries Promotion Act of 1980[c] (*continued*)	• Directs the secretary of state, upon a determination by the secretary of commerce that a portion of the optimum yield for a harvesting season will not be harvested by U.S. vessels, to allocate such portion for use by foreign fishing vessels during the harvesting season or the immediately succeeding season. • Directs the secretary of commerce to establish a program requiring that a U.S. observer be stationed aboard each foreign fishing vessel while that vessel is engaged in fishing in the fishery conservation zone, effective October 1, 1981. • Requires the secretary of commerce, in consultation with the secretary of state, to establish a schedule of fees to be paid by the owner or operator of any foreign fishing vessel for which a permit is issued pursuant to this act. • Declares that the FCMA shall be cited as the Magnuson Fishery Conservation and Management Act (MFCMA).
1982 Amendments[d]	• Amends the MFCMA to permit a foreign ship to process fish within the internal waters of a state only if (1) such ship is under the flag of a country with which the United States has a fishing treaty, and (2) the ship's owner or operator has applied for and received the governor's permission to engage in such activities.
Act to Improve Fishery Conservation and Management of 1983[e]	• Amends the MFCMA to require, as a condition of each Governing International Fishery Agreement, that the foreign nation (1) pay for required U.S. observer costs, including related monitoring and data expenses; and (2) agree to a binding commitment to comply with allocation provisions. • Allows foreign vessels to engage in recreational fishing within the fishery conservation zone and state waters, subject to applicable federal and state requirements.

continued

TABLE 4.5 *(continued)*

Act to Improve Fishery Conservation and Management of 1983[e] (*continued*)	• Eliminates the imprisonment penalty for illegal fishing in the fishery conservation zone. Authorizes "fair market value" payments in lieu of forfeiture of fish. • Requires the secretary of commerce to appoint to Regional Fishery Management Councils only persons with fishery resource backgrounds. • Authorizes appropriations through fiscal year 1985 for carrying out the act.
1984 Amendments[f]	• Amends the MFCMA to authorize the secretary of state to allocate to foreign fishing vessels the yield of a fishing season that exceeds the amount that will be harvested by U.S. vessels. Authorizes the secretary of state to determine the allocation among foreign nations of the total allowable level of foreign fishing. Requires such an allocation to be based on, among other factors: (1) whether and to what extent such nation imposes import barriers or otherwise restricts the market access of U.S. fish and fishery products, particularly fish and fishery products for which the foreign nation has requested an allocation; and (2) whether and to what extent such nation is cooperating with the United States in both the advancement of existing and new opportunities for fishery exports from the United States and the advancement of fisheries trade through the purchase of fish and fishery products from U.S. fishermen.
1986 Amendments[g]	• Amends federal law concerning fishery conservation to define "exclusive economic zone" for purposes of exercising sovereign rights to fishery resources in such zone. States that the United States shall exercise sole fishery management authority (except with regard to highly migratory fishes) within the EEZ. Provides that the United States shall exercise exclusive fishery management authority over all anadromous fishes

continued

TABLE 4.5 (*continued*)

1986 Amendments[g] **(*continued*)**	throughout their migratory range beyond the EEZ (except when such fishes are in a foreign nation's territorial sea or EEZ) and all continental shelf fishery resources beyond the EEZ. • Provides that fishing permits issued to foreign vessels to permit fishing within the U.S. fishery zone shall be valid only for one year, with renewal required thereafter. • Requires members of each regional fishery management council to be knowledgeable and experienced with regard to the conservation and management of the fishery resources of the geographic area concerned. Prohibits the governor of a state from submitting names of individuals to the secretary of commerce for appointment to such councils unless the governor has first consulted with commercial and recreational fishing representatives of the state regarding such individuals. Prohibits an individual's appointment to a council position until such individual complies with certain financial disclosure requirements. Provides a three-year term for each voting member appointed. • Authorizes each council to comment on a proposed state or federal agency action that may affect the habitat of a fishery resource under its jurisdiction. • Requires voting members of a council to disclose any financial interest held in any fish-harvesting, processing, or marketing activity over which such council has jurisdiction. • Requires the secretary to ensure that those persons dependent upon the fisheries within the jurisdiction of the Regional Fishery Management Councils for their livelihood are fairly represented as voting members of the council.

continued

TABLE 4.5 (*continued*)

Marine Mammal Protection Act Amendments of 1988[h]	Affect MFCMA in that they: • Direct the secretary of commerce to compile and publish lists of fisheries in three categories based on frequency of incidental taking of marine mammals by vessels in those fisheries. • Direct the secretary of commerce to grant an exemption for a vessel engaged in a fishery identified under either of the two categories of frequent or incidental taking of marine mammals, upon receipt of a completed registration form. • Establish a 240-day grace period after the enactment of this act after which owners of vessels in fisheries where there is either frequent or occasional taking of marine mammals must (1) have registered with the secretary to obtain an exemption for each vessel; (2) ensure that the decal or other physical evidence of exemption is displayed on or in the possession of the master of each vessel; and (3) comply with specified reporting requirements. • Direct the secretary to review information on the incidental taking of marine mammals and evaluate the effects of such taking on the affected population stocks. • Direct the secretary, upon finding that such taking is having an immediate and significant adverse impact on a marine mammal population stock, or such taking results in the annual killing of higher-than-specified numbers of Steller sea lions and North Pacific fur seals, to consult with appropriate Regional Fishery Management Councils and state fishery managers and prescribe emergency regulations to prevent any further taking. • Direct the secretary, upon finding that such taking is not having an immediate and significant adverse impact but will likely have such an

continued

TABLE 4.5 (*continued*)

Marine Mammal Protection Act Amendments of 1988[h] (*continued*)	impact over a period of time longer than one year, to request the appropriate Regional Fishery Management Council or state to initiate, recommend, or take such action within its authority as it considers necessary to mitigate the impact, including adjustments to requirements on fishing times or areas, or imposition of restriction on the use of vessels or gear.
Fishery Conservation Amendments of 1990[i]	• Amend the MFCMA to replace provisions excluding highly migratory fish from the U.S. assertion of sovereign rights and exclusive fishery management authority with provisions requiring the United States to cooperate, directly or through international organizations, with nations involved in fisheries for highly migratory species to ensure conservation and promote optimum utilization of such species throughout their range, both within and beyond the EEZ. Bring tunas and other highly migratory species under the MFCMA. Atlantic tunas to be managed by secretary of commerce; highly migratory species in the Pacific to be managed by the Western Pacific Regional Fishery Management Council. • Set the total allowable level of fishing by a foreign fishery subject to U.S. exclusive fishery management authority at the portion of the optimum yield of such fishery that will not be harvested by vessels of the United States. • Directs the secretary of state to (1) evaluate the effectiveness of each existing international fishery agreement that pertains to highly migratory species; (2) initiate negotiations to obtain access for U.S. vessels fishing for tuna species within the EEZs of other nations; (3) report to the Congress on the evaluation and the negotiations; and (4) negotiate for international fishery agreements on highly migratory species as necessary to correct inadequacies identified by the evaluation.

continued

TABLE 4.5 (*continued*)

Fishery Conservation Amendments of 1990[i] (*continued*)	• Directs the secretary of commerce, in consultation with the secretary of state and the appropriate Regional Fishery Management Council, to establish reasonable foreign fishing permit fees that apply nondiscriminatorily to each nation. • Directs the secretary of commerce, through the secretary of state and the secretary of the department in which the Coast Guard is operating, to seek international agreements to implement specified findings, policies, and provisions of this act, in particular an international ban on large-scale drift-net fishing. • Dolphin Protection Consumer Information Act declares that it is a violation of the Federal Trade Commission Act for any producer, importer, exporter, distributor, or seller of any tuna product that is exported from or offered for sale in the United States to include on the label the term "dolphin safe" or any other term or symbol that falsely claims or suggests that the tuna was harvested using a method of fishing that is not harmful to dolphins if the product contains tuna harvested (1) on the high seas by drift-net fishing, or (2) in the eastern tropical Pacific Ocean by purse-seine nets that do not meet dolphin-safe requirements of this act.
Sustainable Fisheries Act of 1996[j]	• Key provisions include: preventing overfishing and ending overfishing of currently depressed stocks; rebuilding depleted stocks; redefining "optimum yield" (it must be based on MSY and only revised downward, not upward, "by any relevant social, economic, or ecological factor"); reducing bycatch and minimizing the mortality of unavoidable bycatch; designating and conserving essential fish habitat; reforming the approval process for fishery management plans and regulations; and reducing conflict of interest on regional councils.[k]

continued

TABLE 4.5 *continued*)

Sustainable Fisheries Act of 1996[j] **(*continued*)**	• Modifies foreign fishing prohibitions and requirements. Provides for an international agreement on bycatch reduction standards and measures. • Requires national fishery conservation and management standards to (1) provide for the sustained participation of fishing communities and minimize adverse economic impacts on those communities; (2) minimize bycatch and its mortality; and (3) promote the safety of human life at sea. • Modifies Fishery Management Council requirements regarding composition, operations, jurisdiction, disclosure of financial interest, and other matters. • Modifies fishery management plan required and discretionary contents. • Prohibits, until October 1, 2000, individual fishing quota programs unless approved before January 4, 1995. • Directs the secretary to establish an advisory panel, conduct surveys and workshops, and take other actions regarding pelagic longline fishing vessels that participate in fisheries for Atlantic highly migratory species. Authorizes the secretary to implement a related comprehensive management system. • Authorizes a fishing capacity reduction program if certain requirements are met. Makes participation voluntary. Provides for funding, including authorizing an industry fee system if approved by an industry referendum. Mandates establishment of a task force to study and report on the federal role in subsidizing the expansion and contraction of fishing capacity and otherwise influencing aggregate capital investments in fisheries.

continued

TABLE 4.5 (*continued*)

Sustainable Fisheries Act of 1996[j] **(*continued*)**	• Appropriations for the purposes of carrying out the provisions of the act are not to exceed the following sums: (1) $147,000,000 for FY 1996; (2) $151,000,000 for FY 1997; (3) $155,000,000 for FY 1998; and (4) $159,000,000 for FY 1999.

[a]http://thomas.loc.gov. Bill Summary and Status for the 95th Congress. Public Law 95-354.

[b]http://thomas.loc.gov. Bill Summary and Status for the 96th Congress. Public Law 96-61.

[c]http://thomas.loc.gov. Bill Summary and Status for the 96th Congress. Public Law 96-561.

[d]http://thomas.loc.gov. Bill Summary and Status for the 97th Congress. Public Law 97-191.

[e]http://thomas.loc.gov. Bill Summary and Status for the 97th Congress. Public Law 97-453.

[f]http://thomas.loc.gov. Bill Summary and Status for the 98th Congress. Public Law 98-623.

[g]http://thomas.loc.gov. Bill Summary and Status for the 99th Congress. Public Law 99-659.

[h]http://thomas.loc.gov. Bill Summary and Status for the 100th Congress. Public Law 100-711.

[i]http://thomas.loc.gov. Bill Summary and Status for the 101st Congress. Public Law 101-627.

[j]http://thomas.loc.gov. Bill Summary and Status for the 104th Congress. Public Law 104-297.

[k]For an overview of the SFA, see the June 1997 issue of *SFA Update,* Office of Sustainable Fisheries, NMFS, NOAA.

Source: Prepared with assistance from Rosemarie Hinkel.

dolphins and made other changes to the FMP process in the direction of promoting greater accountability. The 1996 amendments to the act, the passage of the Sustainable Fisheries Act, represented a major change in the language.

Achievements under the Magnuson Act: Successes, Problems, Continuing Issues

By 1995, the new process of fisheries management had produced a good number of tangible outputs—by this time, the councils had implemented thirty-four FMPs for various fish and shellfish resources, and eleven additional plans were in various stages of development (Buck 1995). Some of the plans addressed individual or closely related species (e.g., FMP for red drum in the South Atlantic Council), while others, such as the FMP for the Gulf of Alaska, addressed larger species assemblages inhabiting similar habitats.

What have been the effects and impacts of the new system of fishery management brought about by the Magnuson Act? To answer this question, it is first useful to describe the extent to which, and when, evaluations of the success of the fisheries management regime have taken place. As is often the case with national legislation, there have been no overall assessments of what results were being achieved, as no specific evaluation procedures were built into the legislation. Thus, it is often the perceptions of how well things are working that are more important than actual impacts since information on actual impacts is notably lacking (Mazmanian and Sabatier 1981). In the 1980s, most stakeholders appeared relatively content with the progress being made under the Magnuson Act, once the initial issues surrounding putting the new system in place had been settled. As already mentioned, after initially opposing implementation of the domestic management regime, commercial fishers accommodated to the process, and in fact, came to dominate the membership of most of the councils. Starting in the mid-1980s, however, a number of reports—from inside NMFS as well as from other parts of the government and from the scientific community—began to warn about fisheries depletion.

A NMFS study comparing the condition of fishery stocks in 1975 with that in 1985 was the first to show fisheries decline. While underscoring the difficulties of evaluating progress achieved under the Magnuson Act (due to the lack of the information and methods needed to estimate time-bound targets for restoring depleted resources and the difficulty of coping with the normal fluctuations that many fisheries undergo even under ideal circumstances), Finch (1985) showed that eight species groups were unchanged, eleven were in an improved condition, and six were in a deteriorated condition. In the late 1980s and early 1990s, a number of other evaluations of the fishery management regime created under the Magnuson Act were conducted. Some were the result of congressional initiative, initiative by the NOAA administrator, initiative by the OMB, the inspector general, the GAO, the regional fisheries councils, conservation groups such as the Center for Marine Conservation, and professional groups and academics (see, for example, U.S. GAO 1983; NOAA 1986; Sutinen and Hanson 1986; Branson, Larson, and Miller 1986; Wise 1991; Gordon 1986).

The various studies of the MFCMA prepared in the late 1980s and early 1990s cited several major benefits, foremost of which is the regional approach to fisheries management and its strong emphasis on shared management—including state, federal, and public participation. Second, the act rationalized the system of fishery management by establishing national standards and criteria for evaluating management approaches and developing detailed procedures for planning. Third, it brought overfishing by foreigners under control and initiated the trend toward full domestic utilization. Fourth, it provided greater economic and recreational opportunities for domestic fishers as a result of the decrease in foreign fishing. Fifth, the act highlighted the importance of research and scientific information in the fishery decision-making process and the need to make decisions on the basis of the best scientific information available.

Among these benefits or successes, perhaps the clearest one is the significant decline of foreign fishing and the attendant rise in domestic fishing, readily evident in Figure 4.4. As noted by Buck (1995), foreign fisheries harvest from the U.S. EEZ declined from about 3.8 billion pounds in 1977, to zero in 1992. Concomitantly, domestic offshore catch increased from about 1.56 billion pounds in 1977 to more than 6.32 billion pounds in 1993.

Successes notwithstanding, various studies of the FCMA conducted in the late 1980s and early 1990s also identified several major problems. The first and foremost problem is that overfishing persists in certain fisheries. Similarly, although optimum yield (OY) is theoretically a good concept, its implementation has been used in a number of cases to justify overallocation. As William F. Gordon, past director of the National Marine Fisheries Service, put it, "In practice, OY has increasingly served

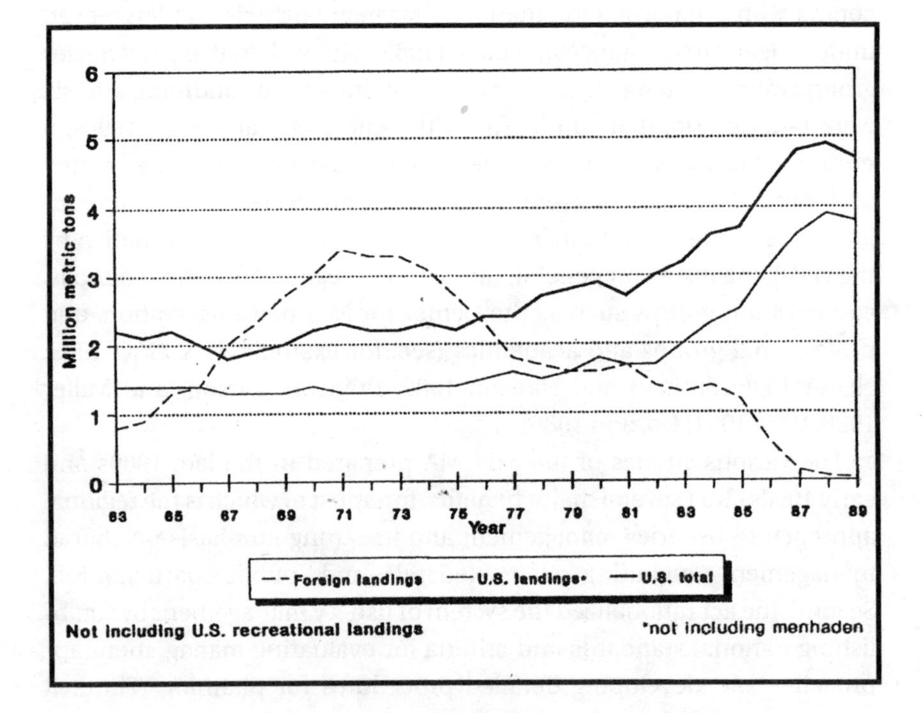

FIGURE 4.4.
U.S. and Foreign Landings from the U.S. 200-Mile Zone, 1963–1989.
Source: Wise 1991, p. 3. Copyright © Center for Marine Conservation. Reprinted with permission.

as an arena for user competition for short-term advantage, which has too often demonstrated the reverse—that what is good for the fisherman at the moment is often bad for the fish" (Gordon 1986, p. 10).

Second, the Magnuson system has been criticized for being too cumbersome, convoluted, and expensive. Some argue that the responsibility and authority of the federal government and the councils is ambiguous, and that, therefore, accountability is not clear. Others point to the great expense involved in running this system. One analysis, for example, estimated that the total cost to taxpayers of running the system was $223 million in 1985, with a net cost of $180 million when the $40–$45 million generated by the system in foreign fishing fees was subtracted. This net cost translated into about $0.05 per pound of fish produced from federal waters in 1985 (Gutting 1986).

Third, fishery information has often been inadequate because of the lack of long-term research commitments; the inadequate correlation between federal research agendas and council management information needs; and the insufficient attention to habitat and ecosystem relationships. Fourth, domestic users do not contribute specifically to the cost of services provided them. Fifth, domestic processing capacity has not increased sufficiently. The total U.S. fishing industry and the economy have not realized the full value of fishery resources. Sixth, recreational fishing, which takes place mostly in state waters, is managed with uneven degrees of thoroughness and, in some cases, not at all. Seventh, the Magnuson Act has not addressed underlying problems in fisheries management related to the absence of well-defined property rights over fishery resources, even though some experimentation has occurred with limited entry and with individual transferable quotas. Eighth, in an effort to keep as many fishers fishing as possible, the Magnuson Act has continued the practice of inefficiency in fisheries management (i.e., the imposition of detailed and cumbersome regulations), which thwarts the development of new technology and discourages innovation. Thus, a skeptic might conclude that the bottom line might be "*Plus ça change, plus c'est la même chose*"—a convoluted management system is producing outputs similar to those that prevailed before the 1970s.

Among the problems cited above, the issue of overfishing and fisheries depletion has been confirmed and further highlighted in other studies. The 1991 NMFS report *Our Living Oceans* and a follow-up report in 1992 acknowledged that overfishing was a national problem and showed that sixty-seven fish species were overutilized. Fisheries depletion was particularly pronounced in New England, where there has been an 80 percent decline in stocks of haddock, cod, and flounder since the 1960s (U.S. DOC, NMFS, 1991 1992) (see Figure 4.5).

Environmental Groups Enter Fisheries Management

Evidence of fisheries depletion drew the attention of major environmental groups that had, up until this point, been largely absent from fisheries discussions. In the 1980s, major environmental interests, such as the Natural Resources Defense Council and the American Oceans Campaign, had largely been preoccupied with offshore oil policy, in a full-out attempt to block new leasing and development in frontier areas offshore California, Oregon, Washington, North Carolina, and Florida. The environmental groups and their allies (such as the coastal states) were able to stop offshore oil development in these areas through the congressional imposition of yearly moratoria on spending for development of these areas by the

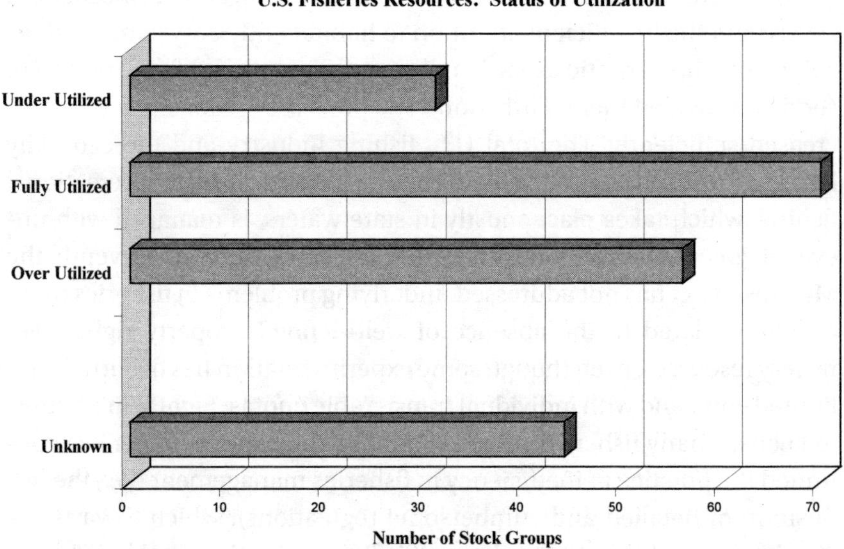

FIGURE 4.5.
Numbers of Stock Groups Classified by Their Status of Utilization for Stocks Under the Purview of National Marine Fisheries Service.

Source: Adapted from U.S. DOC, NMFS 1995b, p. 10.

Notes: A fishery resource is defined as *fully utilized* when the amount of fishing effort used is about equal to the amount needed to achieve long-term potential yield (LTPY or the maximum long-term average yield that can be achieved through conscientious stewardship) and the resource is near its LTPY stock level. The resource is considered *overutilized* when more fishing effort is employed than is necessary to achieve LTPY. A fishery resource is classified as *under utilized* when more fishing effort is required to achieve LTPY.

Department of Interior. By the early 1990s, fresh from winning the offshore oil battles, the environmental interests sought a "next issue" to tackle, and fisheries became their first priority. Conservation groups such as the Center for Marine Conservation had already began to operate in the fisheries arena and had been a sponsor of one of the major reports showing problems in the fishery management regime (Wise 1991).

The environmental interests organized a number of workshops and meetings in the period 1990–1992 to define and assess fisheries management issues. At these meetings, the concept of privatization of fisheries (through limited entry, individual transferable quotas, etc.) was at first embraced as a potential panacea for fisheries depletion, and later largely abandoned, as the full implications of attendant socioeconomic impacts were revealed (Center for Marine Conservation and World Wildlife Federation [CMC and WWF] 1994; Anderson 1995).

A major analysis published in 1997 further demonstrated the plight of America's fisheries, providing a detailed account of the status of fish stocks in each of eight marine regions around America's coasts (NRDC 1997). In the late 1990s, a number of other participants, such as the Pew and Packard Foundations, joined the debate on fisheries management, strongly promoting the cause of conservation. Pew, for example, established Seaweb, a public information effort designed to raise the awareness of decision makers, influential media people, and the public to the alarming decline of U.S. fishery resources.

The 1996 Sustainable Fisheries Act

Anticipating a "window of opportunity" in the upcoming reauthorization of the Magnuson Act in 1993, several major environmental and recreational interest groups—the Center for Marine Conservation, the Audubon Living Oceans Program, Greenpeace, the National Coalition for Marine Conservation, and the World Wildlife Fund—came together to create a coalition "to put the 'C' for 'conservation' back into the Magnuson Act" (Iudicello, Burns, and Oliver 1996, p. 341). By 1995, the coalition—called the Fish Conservation Network—had grown to include more than a hundred conservation, environmental, recreational, and commercial fishing groups (Iudicello, Burns, and Oliver 1996, p. 342). From 1993 to 1996 these groups engaged in negotiations designed to produce an acceptable draft of a revised Magnuson Act for consideration by the Congress.

As noted by Suzanne Iudicello of the Center for Marine Conservation and her colleagues, organizers of the negotiations, debate during the

1996 amendment process focused on issues of stewardship over public resources, waste, conflicts of interest, resistance by industry to regulation, the cost to the taxpayers of free access to public resources, subsidies for their development and extraction, and finally, the cost of bailout and restoration after the system fails. "For the first time, these themes were raised in a context of public interest and public resources, rather than constituent casework" (Iudicello, Burns, and Oliver 1996, p. 340).

In 1996, the Magnuson Act was amended and renamed the Sustainable Fisheries Act (SFA). It passed the House by a resounding bipartisan vote of 388 to 37. The 1996 amendments to the Magnuson Act represented a major change: Key provisions of the act included preventing overfishing and ending overfishing of currently depressed stocks; rebuilding depleted stocks; reducing bycatch and minimizing the mortality of unavoidable bycatch; designating and conserving essential fish habitat; reforming the approval process for fishery management plans (FMPs) and regulations; reducing conflicts of interest on regional councils; and establishing user fees (U.S. DOC, NMFS 1997, p. 1). As noted by Joel Gay in *Pacific Fishing*, "[F]or two decades, the focus had been the high-speed development of the Exclusive Economic Zone. The new bill puts expansion into neutral, throws capitalization into reverse, and orders the protection of fish, fisheries, and habitat full speed ahead" (1996, p. 26).

The Sustainable Fisheries Act made a number of changes and additions to the national standards put forth by the Magnuson Act. A major change was in the definition of optimum yield. As noted by Jim Gilmore of the American Factory Trawler Association (reported in Gay 1996), "'[T]he only thing worth a damn in the whole bill,' is a one-word change in the definition section. In the past, 'optimum yield' was defined as the 'maximum sustainable yield' from the fishery 'as modified by any relevant social, economic, or ecological factor.' The new bill defines optimum yield as the maximum sustainable yield 'as reduced by' those same factors" (p. 26).

Another significant change enacted by the SFA required NMFS to conduct several activities to describe, identify, conserve, and enhance essential fish habitat (EFH). Essential fish habitat was defined as "those waters and substrate necessary to fish for spawning, breeding, feeding, or growth to maturity." Under the EFH amendments, fisheries management plans were required to (1) describe and identify EFH and adverse impacts to EFH for the fishery, (2) minimize the adverse effects of fishing on EFH to the extent practicable, and (3) identify other actions to encourage the conservation and enhancement of EFH (U.S. DOC, NMFS 1997, p. 3).

The SFA also established three new national standards regarding fishing communities, by-catch, and safety at sea. National Standard 8 mandated that the value of fishery resources to fishing communities must be taken into account by conservation and management measures. Such consideration is required in order to sustain the participation of fishing communities in the fishery and to reduce adverse economic impacts on those communities. National Standard 9 called for the minimization of bycatch and bycatch mortality "to the extent practicable." National Standard 10 directed conservation and management measures to encourage the safety of human life at sea (U.S. DOC, NMFS 1997, p. 3).

It remains to be seen whether the SFA will be successful in solving the problems it was created to address. While one of the principal changes in the act is the redefinition of overfishing and the commitment to rebuild stocks, National Standard 8 requires management and conservation measures to minimize adverse economic impacts on fishing communities. Is the emphasis on conservation and restoration of stocks compatible with consideration of fishing communities' needs? Or must one take precedence over the other? If a stock is in grave danger of being depleted, fishing effort must be either greatly reduced, which is detrimental to dependent fishing communities, or allowed to continue, which is detrimental to the fish stocks. The identification of essential fish habitat for every fishery in the United States is an enormous task requiring many resources and personnel. The act places many new responsibilities, which are not always particularly well defined, upon the regional fishery management councils, which already have many duties to conduct with a limited number of resources (Heinz Center 1998, p. 2). Finally, one of the primary characteristics of fishery management is the immense amount of uncertainty surrounding the dynamics of many fisheries. While the SFA undeniably was enacted with good intentions, its operational success is not guaranteed (see further discussion in Chapter 5).

Marine Mammal Protection: Has This Single-Purpose Law Been the Most Successful?

As discussed in Chapter 3, the MMPA is essentially a prohibitive and narrowly focused policy that established a detailed regulatory system for the protection of marine mammals, generally giving these animals—through the moratorium on their take—priority over other resources and uses of the marine environment. In the case of some marine mammal species that are depleted (such as the California sea otter), the protective mantle of the Endangered Species Act and its process for creating recovery plans for threatened and endangered species has added an additional layer of protection.

Conflicts with Other Resources and Uses

Implementation of the MMPA has been fraught with conflicts with many other marine resources and activities. Some of the most serious have involved interactions with commercial and recreational fisheries, but significant conflicts with other uses and activities have occurred as well. Among the fishery–marine mammal conflicts, a major controversy that had animated the initial passage of the act in 1972 was the negative impact of tuna-fishing operations on dolphins. As dolphins tend to swim above schools of yellowfin tuna, the fishing method of "setting on dolphins" had been used by the tuna industry since the 1960s; the tuna purse-seine nets encircled both the tunas and the dolphins, causing extensive dolphin mortality in the course of fishing operations. Seals and sea lions, too, have come into conflict with salmon and steelhead fisheries, particularly on the Pacific Coast. A well-publicized case involved a sea lion nicknamed "Herschel," which had taken up residence along the fish ladders built into the Ballard Locks in Lake Washington in Washington State, eating the salmon and steelhead trout swimming up the ladders to spawn (U.S. DOC 1994a).

Protection of sea otters in California has come into bitter conflict with shellfish fisheries. This is an especially interesting story that has involved extensive debate over the meaning of optimum sustainable population (OSP) and the time frame involved in such calculations. Prior to the colonization of California, sea otter herds extended from Alaska to the tip of the Baja California peninsula. In the 1800s, much of the sea otter population was taken by fur traders from England, Russia, and other countries for sale and exchange to the Chinese court. By the early 1900s, the population was thought to be extinct, but in 1911, a small band of sea otters was found off Big Sur, California (just south of Monterey), and became protected under California law. As the sea otter population recovered, it began to extend southward, threatening the profitable shellfish fisheries, primarily the one for abalones, which had developed in the central California coast during the twentieth century. Lacking a protective layer of blubber, sea otters have a voracious appetite and typically eat the equivalent of up to one-quarter of their body weight per day, their favorite food being the economically profitable abalone. Bitter disputes ensued over the adverse economic impacts of the expanding range of the sea otter, which was resulting in the closing down of the shellfish fisheries.

The sea otter protection groups argued that the OSP goal should be interpreted as restoring the otters to their original precolonization range, while the fishers contended that the goal of OSP must be tempered to take into account the fact that profitable fisheries had grown up in the same area since the state had been populated (Cicin-Sain 1982b).

Recreational activities have also come into conflict with marine mammals. Manatees in Florida, for example, have often been harmed or killed by recreational boats; it is estimated that of the 1,000–2,000 manatees left in Florida's coastal regions, 150–200 die each year, largely as a result of being struck by motor boats and their propellers (Kubasek et al. 1995; U.S. DOC, NMFS 1994a). Marine mammals such as whales and dolphins have at times suffered from the great interest they evoke from the whale-watching industry and other recreational modes of viewing and enjoying marine mammals. For example, commercial cruises to observe and "feed wild dolphins" operating in Texas, South Carolina, and Florida were found to harm dolphins by making them reliant on food offered by humans and decreasing their ability to survive in the wild (U.S. DOC, NMFS 1994d, 1995c, 1998).

Conflicts between offshore oil development and marine mammal protection have also ensued. Concerns with the effects of noise emanating from offshore oil platforms in the Arctic on the bowhead whale have led to restrictions on the industry's operations (Bolze 1990). The special sensitivity of sea otters to oil spills (lacking a layer of fat, sea otters must constantly groom themselves to keep bouyant—in the event of an oil spill, they would ingest oil from their coats and die from the toxic effects), led to placing several offshore oil tracts in central California off-limits to development (Cicin-Sain 1982b). In the Gulf of Maine, seals and sea lions have come into conflict with salmon and lobster aquaculture operations, causing economic loss for the industry. Another type of conflict has arisen over concern with the interaction between marine scientific experiments involving low-frequency sounds and their possible negative effects on marine mammals. In this case, the Department of Defense provided funding in 1993 to the Scripps Institution of Oceanography for a thirty-month large-scale study to determine whether travel times of low-frequency sounds across ocean basins can be used to detect changes in temperature associated with global climate change (MMC 1996). The experiment involved the establishment of low-frequency transmitters to be installed and operated periodically in deep-ocean areas off Hawaii and California. Environmental groups and others rallied against this experiment, claiming that such acoustic signals would harm marine mammals, especially whale populations. Marine mammals have also been adversely affected by commercial shipping operations; in particular, there are concerns about the harm that the depleted northern right whale is suffering from strikes from ships and other ship-associated hazards in Atlantic waters.

Amendments to the MMPA

The MMPA has been amended eight times (see Table 4.6), with especially important (and complex) amendments occurring in 1988 and 1994. Two major themes characterize the MMPA amendments: establishing the details of a regime managing marine mammal–fisheries interactions, and extending MMPA-like protection to marine mammals found in the waters of other nations.

As discussed in Chapter 3, the original 1972 act allowed some exceptions and exemptions to the moratorium: takings for scientific research and public display, takings by native Alaskans for subsistence or creation of traditional handicrafts, and, most significant, taking of marine mammals in the course of commercial fishing operations. As reported in Alker (1996), "Although the 'immediate goal' was to approach zero mortality, Congress believed that 'present technology was not adequate to the task' and some flexibility would have to be accepted" (p. 536). Initially, the MMPA gave commercial fisheries a two-year exemption from the moratorium; subsequently in 1974, the NMFS gave the tuna industry a general permit allowing the U.S. purse-seine tuna fleet to take an unlimited number of dolphins until December 31, 1995. This permit was challenged by litigation (*Committee for Humane Legislation, Inc. v. Richardson*) and NMFS was required to set population-specific quotas (after determining the OSP of each species and the impact of the regulations on the species' ability to reach its OSP) (Young, Irvin, and McLean 1997, p. 62). Starting in 1976, NMFS established incidental take quotas for dolphins, implemented several changes in tuna purse-seine fishing operations (such as the "backdown procedure" that allows the dolphins to swim out of the purse seine, and the use of the "Medina panel" [strips of fine mesh installed in the purse seine that prevent dolphins from entangling their snouts and flippers in the net]) (Young, Irvin, and McLean 1997). Incidental take authorizations through formal general permit procedures were utilized in the case of salmon–Dall's porpoise interactions (Young and Iudicello 1997).

TABLE 4.6 Major Amendments to the 1972 Marine Mammal Protection Act.

1976 Amendments[a]	• Amends the MMPA to conform the ocean areas defined in the act to the fishery conservation zone.
1978 Amendments[b]	• Authorizes appropriations to carry out the act during fiscal years 1979, 1980, and 1981.
1981 Amendments[c]	• Exempts purse-seine tuna from the zero-mortality goal.
	• Allows incidental take of small numbers of marine mammals by U.S. nontuna commercial fishermen, provided that the population is not depleted and the taking has a negligible impact on the stock. Such take is to be systematically monitored. Also establishes similar conditions for taking of small numbers of nondepleted species by U.S. citizens engaged in specified activities (other than commercial fishing) within specific geographic areas.
	• Establishes conditions under which management authority for certain species of marine mammals may be transferred to a state.
	• Authorizes appropriations for the Departments of Commerce and Interior and the Marine Mammal Commission for fiscal years 1982, 1983, and 1984.
1984 Amendments[d]	• Requires secretary of commerce to obtain documentation from foreign nations wishing to import yellowfin tuna into the United States that such foreign tuna-fishing operations are conducted (1) under a regulatory program for governing the incidental take of marine mammals that is comparable to the U.S. program, and (2) with an average rate of incidental taking that is also comparable to the U.S. program.
	• Requires secretary of commerce to carry out a program of research on marine mammal populations to monitor the effect yellowfin tuna fishing is having on them.

continued

TABLE 4.6 (*continued*)

1986 Amendments[d] **(*continued*)**	• Authorizes appropriations for fiscal years 1985 through 1988 to the Departments of Commerce and Interior for activities under Title I of the act and to the Marine Mammal Commission for activities under Title II of the act.
1986 Amendments[e]	• Allows U.S. citizens to hunt for specified species of depleted fish if the administrator of NOAA determines that such hunting will not have an unmitigable adverse impact.
1988 Amendments[f]	• Amends the MMPA to provide for an interim exemption period for commercial fishing operations (other than commercial yellowfin tuna fishing) from specified provisions of the act governing the incidental taking of marine mammals. Sets forth requirements that supersede such provisions during the interim period (from the enactment of this act until October 1, 1993). Makes such exemptions available only to owners of vessels that (1) are vessels of the United States, and (2) have valid fishing permits issued by the secretary of commerce under the MFCMA. • Declares the immediate goal that the incidental kill or serious injury of marine mammals permitted in the course of commercial fishing operations be reduced to insignificant levels approaching a zero mortality and serious injury rate. • Directs the secretary of commerce to compile and publish lists of fisheries in three categories based on frequency of incidental taking of marine mammals by vessels in those fisheries. • Directs the secretary of commerce to place observers on from 20 to 35 percent of exempted vessels for each fishery identified as having frequent taking of marine mammals, in order to obtain statistically reliable information on species and numbers of marine mammals incidentally taken in the fishery. Directs the secretary

continued

TABLE 4.6 (*continued*)

1988 Amendments[f] (***continued***)	to implement an alternative observation program if fewer than 20 percent of the fishing operations will be monitored. Set forth guidelines for determining the distribution of observers among fisheries and vessels. Sets forth priorities for allocating observers among fisheries when the required level of coverage cannot be met.
	• Directs the secretary of commerce to consult with the secretary of the interior before taking actions or making determinations for the interim exemption period that affect or relate to species or population stocks of marine mammals for which the secretary of the interior is responsible under the act.
	• Regulates provisions relating to the taking of porpoise in a tuna fishery. Prohibits the secretary of commerce from finding that the regulatory program of a foreign nation is comparable[g] to the U.S. program unless it has met specific standards relating to such taking, for purposes of importation embargo provisions. Requires any intermediary nation from which yellowfin tuna or tuna or tuna products will be exported to the United States to certify and provide reasonable proof that it has acted to prohibit their importation from any nation from which direct export to the U.S. is banned under such embargo, within 60 days after the effective date of such a ban.
	• Directs the secretary, through the secretary of state, to initiate (1) negotiations with foreign nations for treaties to protect marine mammals, and (2) discussions with foreign nations whose vessels harvest yellowfin tuna with purse seines in the eastern tropical Pacific Ocean to conclude international arrangements to conserve marine mammals taken incidentally in the course of harvesting.
	• Directs the secretary of commerce to convene annual meetings with representatives of

continued

TABLE 4.6 (*continued*)

1988 Amendments[f] **(*continued*)**	conservation and environmental organizations, the commercial tuna fishing industry, and other interested persons to discuss results of efforts to reduce incidental mortality and serious injury of marine mammals and to develop plans for such efforts during the next year. • Extends through fiscal year 1993 the authorization of appropriations to the Department of Commerce, the Department of Interior, and the Marine Mammal Commission to carry out specified duties under the act. • Secretary of commerce required to review and transmit recommendations to Congress before January 1992. • Conditions and procedures established for the secretaries of commerce and interior to review the status of populations to determine if they should be listed as depleted. • Preparation of conservation plans for any species listed as depleted required. Plans required to be modeled after recovery plans developed pursuant to the Endangered Species Act. • Conditions under which permits may be issued to take marine mammals for the protection and welfare of the animals, including importation, public display, scientific research, and enhancing the survival or recovery of a species listed. • Reward system under which the secretary of the treasury can pay up to $2,500 to individuals providing information leading to convictions for violations of the act established. • Other nations' programs of dolphin protection required to be comparable to that of the United States. • The rate of incidental take of dolphins cannot be more than 2 times that of the U.S. tuna fleet and 1¼ after 1990 and thereafter.

continued

TABLE 4.6 *(continued)*

1988 Amendments[f] *(continued)*	• "Intermediary" nations that export yellowfin tuna to the United States must provide proof that these products did not originate from a country without a porpoise protection program. • Provided for 100 percent observer coverage on U.S. tuna vessels in the eastern tropical Pacific.
1990 Amendments[h]	• Amends the MMPA regarding fish or products containing fish harvested by nations whose fishing vessels engage in high-seas drift-net fishing, shall require that the government of the exporting nation provide documentary evidence that the fish or fish product was not harvested with a large-scale drift net in the South Pacific Ocean after July 1, 1991, or in any other waters of the high seas after July 1, 1992. Also, in the case of tuna or a product containing tuna harvested by a nation whose fishing vessels engage in high-seas drift-net fishing, requires that the government of the exporting nation provide documentary evidence that the tuna or tuna product was not harvested with a large-scale drift net anywhere on the high seas after July 1, 1991.
1992 Amendments[i]	• Amends the MMPA to direct the secretary of commerce, in consultation with the secretary of the interior, the Marine Mammal Commission, and individuals with knowledge and experience in marine science, marine mammal sciencc, marine mammal veterinary and husbandry practices, and marine conservation, including stranding network participants, to establish the Marine Mammal Health and Stranding Response Program.
1994 Amendments[j]	• Amends MMPA with respect to the moratorium and exceptions on the taking and importing of marine mammals to allow the issuance of permits for (1) scientific research, public display,

continued

TABLE 4.6 *(continued)*

1994 Amendments[j] ***(continued)***	photography for educational or commercial purposes, or enhancing the survival of a species or stock; or (2) importing polar-bear parts taken in sport hunts in Canada. • Allows the issuance of permits for non-commercial-fishing operations within specific geographic regions for the incidental, but not intentional, taking by harassment of small numbers of marine mammals of a species or stock; and (2) will not have an unmitigable adverse impact on the availability of such species or stock. • Authorizes a take reduction plan, if necessary, to reduce incidental taking during commercial fishing operations. • Describes regulations to accompany the general authorization for the incidental commercial taking of marine mammals. Establishes the immediate goal that the incidental kill or serious injury of marine mammals in the course of commercial fishing operations be reduced to insignificant level approaching a zero mortality and serious injury rate within seven years. • Directs the secretary to establish a program to monitor incidental mortality and serious injury takes during the course of commercial fishing operations that include observers on certain vessels. • Requires the establishment of take reduction plans to reduce incidental mortality or serious injury to insignificant levels approaching zero within five years. Authorizes the establishment of take reduction teams to facilitate the development of such plans. • Authorizes the secretary of commerce to enter into cooperative agreements with Alaska Native organizations to conserve marine mammals and provide co-management of subsistence use by Alaska Natives.

continued

TABLE 4.6 (*continued*)

1994 Amendments[j] (*continued*)	• Allows a state to apply to the secretary of commerce for the intentional taking of individually identifiable pinnipeds that are having a significant negative impact on the recovery of certain salmonid fishery stocks. Authorizes the secretary to establish a Pinniped-Fishery Interaction Task Force to advise on responding to such application.
	• Authorizes appropriations: (1) Department of Commerce: $12,138,000 for FY 1994; $12,623,000 for FY 1995; $13,128,000 for FY 1996; $13,653,000 for FY 1997; $14,2000,000 for FY 1998; and $14,768,000 for FY 1999. (2) Department of the Interior: $8,000,000 for FY 1994; $8,600,000 for FY 1995; $9,000,000 for FY 1996; $9,400,000 for FY 1997; $9,900,000 for FY 1998; and $10,296,000 for FY 1999. (3) Marine Mammal Commission: $1,500,000 for FY 1994; $1,600,000 for FY 1995; $1,650,000 for FY 1996; $1,650,000 for FY 1997; $1,700,000 for FY 1998; and $1,750,000 for FY 1999.

[a] http://thomas.loc.gov. Bill Summary and Status for the 94th Congress. Public Law 94-265.

[b] http://thomas.loc.gov. Bill Summary and Status for the 95th Congress. Public Law 95-316.

[c] http://thomas.loc.gov. Bill Summary and Status for the 97th Congress. Public Law 97-58.

[d] http://thomas.loc.gov. Bill Summary and Status for the 98th Congress. Public Law 98-364.

[e] http://thomas.loc.gov. Bill Summary and Status for the 99th Congress. Public Law 99-659.

[f] http://thomas.loc.gov. Bill Summary and Status for the 100th Congress. Public Law 100-711.

[g] For details on comparability standards, see Young, N.M., Irvin, W.R., and McLean, M.L. "The Flipper Phenomenon: Perspectives on the Panama Declaration and the "Dolphin Safe" Label." *Ocean and Coastal Law Journal*, vol. 3:57, 1997 pp. 68–69.

[h] http://thomas.loc.gov. Bill Summary and Status for the 101th Congress. Public Law 101-627. H.R. 2061.

[i] http://thomas.loc.gov. Bill Summary and Status for the 102nd Congress. H.R. 3486, S. 1898.

[j] http://thomas.loc.gov. Bill Summary and Status for the 103rd Congress. Public Law 103-328, S. 1636.

Source: Prepared with assistance from Rosemarie Hinkel.

Major amendments to the MMPA in 1988 and 1994 represented efforts to develop a detailed regime for systematically addressing fishery–marine mammal interactions, not only in the two cases noted above but in all other cases of significant interaction. The 1988 amendments provided for an information-gathering program and an Interim Exemption

Program for Commercial Fisheries, a five-year interim regime for governing the interaction between marine mammals and fisheries. The exemption was designed to allow NMFS time to increase its data gathering, observations, and research into marine mammal–fishing interactions (with the active participation of the fishing industry through the maintenance of fish logs and other measures), while allowing commercial fishing to continue (Young and Iudicello 1997). The act required placing observers on fishing vessels involved in Category I fisheries[14] to monitor the interactions. The 1988 amendments also directed the MMC to develop recommended guidelines to the secretary of commerce, who would in turn provide a suggested regime to Congress to replace the exemption provision after its scheduled expiration in October 1993. The MMC issued its report in 1990, providing guidelines to NMFS to govern the incidental take of marine mammals after 1993 (Young and Iudicello 1997). NMFS, after a series of consultations with affected groups, delivered a description of its proposed regime to Congress in November 1992.

In anticipation of the 1993 authorization "window," a coalition of environmental groups, animal welfare groups, commercial fishing associations, and Native Alaskans began meeting in 1993, assisted by a professional mediator, the Keystone Center (Young and Iudicello 1997, p. 176). The major issues during these negotiations were, from the conservation community side, that the MMPA would continue to impose a moratorium on taking marine mammals, with the goal of reducing incidental lethal take of marine mammals in commercial fishing operations to insignificant levels approaching zero mortality and serious injury rate; and on the part of the commercial fishing groups, to avoid a burdensome management regime that would apply across the board to all fisheries regardless of level of interaction between marine mammals and fisheries (Young and Iudicello 1997).

The 1994 amendments to the act established a new system for governing the incidental take of marine mammals in commercial fisheries. The 1994 amendments added three major sections to the MMPA: (1) requirements on stock assessments, status determinations, and calculations of the stock's potential biological removal level (PBR); (2) requirements for fishermen; and (3) establishment of a process for addressing interactions between pinipeds and fishery resources (Young and Iudicello 1997, p. 189). The most significant changes to the act, according to Young and Iudicello, were the new Sections 117 and 118, which incorporated the new regime to govern incidental takes of marine mammals during commercial fishing operations. Take reduction plans were to be developed for each "strategic stock," defined as one for which the "level of direct-caused mortality exceeds the PBR," which is declining and likely to be listed as a threatened species under the ESA within the foreseeable future, or which is already listed as threatened or endangered under the ESA or designated as depleted under the MMPA (p. 205). Take reduction teams for (1) Gulf of Maine harbor porpoises, (2) Pacific offshore cetaceans, (3) Atlantic offshore cetaceans, (4) Atlantic large baleen whales, and (5) Mid-Atlantic coastal gillnets have been convened to develop plans.

The Tuna-Dolphin Issue: Reducing Mortality

In concurrent developments, starting in the mid-1980s, efforts to reduce dolphin mortality were expanded from the U.S. tuna fleet to also encompass foreign fishing activities in the eastern tropical Pacific (MMC 1996, p. 92). The U.S. tuna fleet had traditionally dominated purse-seine tuna fishing in this area, but starting in the 1980s, the U.S. tuna fleet began moving its operations to the western Pacific (Pacific Islands region) for several reasons, among them the MMPA restrictions and the 1983–1984 El Niño event. During this period, it is reported that a number of U.S. vessels "reflagged"—switched countries and sailed under foreign flags (Brower 1989, reported in Kubasek et al. 1995). During the same period, the tuna fleets from other nations operating in the eastern tropical Pacific (such as Mexico and Venezuela) doubled or tripled in size (Kubasek et al. 1995). As the eastern tropical Pacific tuna fishery shifted to foreign control, so did the problem of incidental take of dolphins; the number of dolphins killed by foreign fleets grew greatly—from the estimated 22,980 in 1984 to 39,642 in 1985, and 112,482 in 1986 (MMC 1996).[15]

Despite the decline of the its tuna fleet in the eastern tropical Pacific, the United States remained an important market for tuna caught in that area. Therefore, it began to use the MMPA to affect the behavior of other nations. In the 1984 amendments to the act, foreign fleets and intermediary countries (countries in which tuna is transhipped or processed) that wanted to sell tuna to the United States had to show that the producing country had environmental programs for protection of dolphins similar to those of the United States. If not, a ban was placed on the import of their tuna products. This requirement was applied more rigorously in the 1988 amendments to the act, which established that the kill rate (dolphin mortality per set) by foreign fleets could be no more than twice (and subsequently no more than one and a quarter) the kill rate of the U.S. fleet (Kubasek et al. 1995).

In the meantime, faced with domestic consumer boycotts as well as with the possible advantage of jumping on a "green labeling" bandwagon, the major tuna canners—Starkist, Van Camp Seafood, and Bumblebee—announced in 1990 that they would adopt "dolphin safe" purchasing practices and begin labeling their tuna as dolphin safe (Young, Irvin, and McLean 1997). As dolphin mortality continued at a fairly high level even after the MMPA amendments, Congress enacted the Dolphin Protection Consumer Information Act of 1990, which called for labeling tuna products as "dolphin safe" (not responsible for the killing of dolphins) (Young, Irvin, and McLean 1997).

Concurrently, in response to continued deaths of dolphins by foreign fleets, U.S. courts ordered a number of embargoes against various harvesting nations that did not have comparable dolphin conservation programs (see e.g., a 1990 lawsuit filed by the Earth Island Institute against the Department of Commerce, reported in Young, Irvin, and McLean 1997). Mexico's tuna products were subsequently embargoed by the United States. In response, Mexico filed a challenge in 1991 with a dispute resolution panel of the General Agreement on Tariffs and Trade (GATT) alleging that the tuna ban imposed on its fleet under the MMPA represented a violation by the United States of the GATT. The panel found in favor of Mexico, ruling that the embargo clause violated the GATT because it applied to animals and resources outside the jurisdiction of the United States. (Kubasek et al. 1995). In the end, the United States decided to attempt to negotiate a new multilateral treaty with Mexico, and consequently Mexico did not have the full GATT council take up the panel decision.

The panel's decision in favor of Mexico called into question the MMPA's embargo procedures and highlighted the need to work in a multilateral fashion. The subsequent passage of the International Dolphin Conservation Act in 1992 amended the MMPA with the creation of Title III, which called for the establishment of a global moratorium prohibiting tuna harvesting using the purse-seine method, and authorized the secretary of state to enter into international agreements that establish this moratorium. Any nation acting in accordance with this act would have the embargo provisions imposed upon them by the MMPA lifted.

By 1995, however, all Latin American nations fishing for yellowfin tuna were embargoed under the MMPA (Young, Irvin, and McLean 1997), and tuna markets had shifted considerably, with much of the tuna catch from the eastern tropical Pacific going to non-U.S. markets (e.g., Europe, Latin America, and, especially, Mexico). Because of this market shift, denial of access to the U.S. tuna market no longer provided nations fishing in the ETP with a powerful incentive for dolphin protection. These factors led the United States to heighten efforts at multilateral negotiations with the other nations operating in the eastern tropical Pacific. These resulted in the Panama Declaration in 1995, endorsed by some conservation groups (such as the Center for Marine Conservation) as a "historic international agreement to protect dolphins and biodiversity in the ETP" (Young, Irvin, and McLean 1997, p. 98). The Panama Declaration is expected to lead to an international agreement that provides protection for dolphin populations, further decreases dolphin mortality, gives protection to the ETP ocean ecosystem, and manages and conserves the tuna fishery more effectively. On May 21, 1998, the United States and seven Latin American nations (Colombia, Costa Rica, Ecuador, Mexico, Nicaragua, Panama, and Venezuela) signed the "Agreement on the International Dolphin Conservation Program." The agreement allows the United States to lift the embargo on tuna imports from participating nations, and will enter into force upon ratification by four nations (U.S. DOS, 1998).

Key Ingredients to Successful Implementation

As discussed in Chapter 3, the major driving forces that had led to the enactment of the MMPA were dolphin mortality from tuna-fishing operations, the decline of whales, and the Canadian harp seal and Pribiloff seal hunts. When we examine the status of these marine mammal populations twenty-seven years after enactment of the MMPA, we find a good measure of success. With regard to dolphins, Table 4.7 shows a decline in mortality for U.S. vessels from 368,600 dolphin deaths in 1972 to 0 deaths in 1995, and for non-U.S. vessels, a decline in mortality from 55,078 in 1972 to 3,274. However, MMC staff note that as a result of high levels of mortality in past years, at least two of the stocks are now severely depleted and are likely to remain depleted for many years to come despite those much lower mortality rates (Twiss 1998). With regard to northern fur seals, the commercial harvest has been eliminated and subsistence harvest continues at a modest level, as shown in Table 4.8. With regard to whales, the effects of the MMPA moratorium on halting the decline of whale populations are not as easily discerned because of the global range of many whale species and the international regime for whale conservation being overseen by the International Whaling Commission (IWC). While many species of great whales remain severely depleted due to earlier overharvesting, at least one of them, the North

TABLE 4.7 Estimated Incidental Kill of Dolphins in the Tuna Purse-Seine Fishery in the Eastern Tropical Pacific Ocean, 1972–1995[1].

YEAR	U.S. VESSELS	NON-U.S. VESSELS
1972	368,600	55,078
1973	206,697	58,276
1974	147,437	27,245
1975	166,645	27,812
1976	108,740	19,482
1977	25,452	25,901
1978	19,366	11,147
1979	17,938	3,488
1980	15,305	16,665
1981	18,780	17,199
1982	23,267	5,837
1983	8,513	4,980
1984	17,732	22,980
1985	19,205	39,642
1986	20,692	112,482
1987	13,992	85,185
1988	19,712	61,881
1989	12,643	84,403
1990	5,083	47,448
1991	1,002	26,290
1992	439	15,111
1993	115	3,601
1994	106	4,095
1995	0	3,274[2]

Source: Adapted from Marine Mammal Commission 1996, p.100.

[1]These estimates based on kill per set and fishing effort data are provided by the National Marine Fisheries Service and the Inter-American Tropical Tuna Commission. They include some, but not all, seriously injured animals released alive.

[2]Preliminary estimate.

Pacific (California) gray whale has recovered to the point that it was removed from the endangered species list by the United States in 1994 (*Marine Conservation News* 1995). A moratorium on the commercial taking of most whale species, put in place by the IWC in 1984, currently remains in effect. Certain whaling nations, most notably Japan and

TABLE 4.8 Subsistence Harvest Levels for Northern Fur Seals in the Pribilof Islands, 1985–1995[1].

YEAR	ST. PAUL	ST. GEORGE	TOTAL
1985	3,384	329	3,713
1986	1,200	124	1,324
1987	1,710	92	1,802
1988	1,145	113	1,258
1989	1,340	181	1,521
1990	1,077	164	1,241
1991	1,645	281	1,926
1992	1,482	194	1,676
1993	1,518	319	1,837
1994	1,616	161	1,777
1995	1,265	260	1,525

Source: Adapted from Marine Mammal Commission 1996, p. 40.

[1]Data provided by the National Marine Fisheries Service, Alaska Region.

Norway, however, continue to take limited numbers of minke whales from Arctic and Antarctic populations that do not appear to be endangered (*New York Times* 1999).

Other marine mammal populations, such as the California sea otter, have also shown good signs of recovery as a result of protective measures established by the MMPA and ESA. As noted in Table 4.9, for example, California sea otter populations grew about 70 percent from 1982 to 1995 (although post-1995 surveys have shown population declines [Twiss 1998]). Other successes are noted by William Aron (1988): "The California sea lion, harbor seal, and northern elephant seals all appear to be rapidly increasing and in many areas are close to, if not greater than, their abundance prior to exploitation . . . The California sea lion population was as low as a few thousand and is now about 175,000. Harbor seals number 400,000 and northern elephant seals now number 100,000 from a low of a few hundred" (p. 103).

Notwithstanding such successes, a number of marine mammal species found in U.S. waters are still faced with serious threats. This includes marine mammal species and populations listed as endangered or threatened under the ESA: West Indian manatee, southern sea otter, Hawaiian monk seal, Guadalupe fur seal, Steller sea lion, and a number

TABLE 4.9 California Sea Otter Population Counts by the Fish and Wildlife Service and the California Department of Fish and Game, 1982–1995.

YEAR	SEASON	TOTAL
1982	Spring	1,346
	Fall	1,338
1983	Spring	1,251
	Fall	1,226
1984	Spring	1,304
	Fall	—
1985	Spring	1,360
	Fall	1,221
1986	Spring	1,570
	Fall	1,201
1987	Spring	1,650
	Fall	1,367
1988	Spring	1,724
	Fall	—
1989	Spring	1,864
	Fall	1,599
1990	Spring	1,680
	Fall	1,636
1991	Spring	1,941
	Fall	1,661
1992	Spring	2,101
	Fall	1,715
1993	Spring	2,239
	Fall	1,805
1994	Spring	2,359
	Fall	1,845
1995	Spring	2,377
	Fall	2,190

Source: Adapted from Marine Mammal Commission 1996, p. 50.

of whale populations (northern right whale, bowhead whale, humpback whale, blue whale, finback whale, sei whale, and sperm whale) (adapted from MMC 1996, p. 4). Other species not listed but that nonetheless received special attention in 1995, according to the Marine Mammal Commission, included harbor seals in Alaska, Pacific walruses, gray whales, harbor porpoises in the Gulf of Maine, beluga whales, and polar bears. At the same time that some species of marine mammals remain at risk, other marine mammal populations, such as seals and sea lions have grown to very large populations. According to the *National Fisherman* (March 1995), for example, California sea lions are increasing in population around 10 percent each year, and are "chewing the steelhead [salmon] population down to the nub of extinction" (p. 9).

Part of the reason for the successes achieved through the MMPA can be attributed to the very committed interest groups that have doggedly followed the implementation of the act and have, on many occasions, intervened at key points in the implementation process: American Cetacean Society, American Humane Association, Animal Protection Institute, Defenders of Wildlife, Friends of the Sea Otter, Greenpeace, Humane Society of the United States, Center for Marine Conservation, International Wildlife Coalition, National Audubon Society, Northwind Undersea Institute, Oceanic Society, Sierra Club, Society for Animal Protective Legislation, Whale Center, World Wildlife Fund, among others (Young and Iudicello 1997). Another reason for the act's successes is the role that has been played by the Marine Mammal Commission (MMC) and its scientific advisors. Although the MMC does not have regulatory authority over marine mammals, it has acted as an official monitor or watchdog over marine mammal policy, conducted studies and assessments, and has rendered its opinion on matters affecting public policy on marine mammals. Federal agencies are required to respond to the commission recommendations within 120 days (Hoffman 1989). It also has, as mandated in section 204 of the act, duly reported to Congress once a year on the progress in the act's implementation—something that doesn't happen as rigorously or predictably in most other areas of U.S. ocean policy. The independent and scientifically credible oversight role played by the MMC is one that other areas of U.S. ocean policy may want to emulate (Chapter 7 returns to this point in a discussion of the desirability of creating a national ocean council).

Notwithstanding the successes, problems remain in this area of U.S. ocean policy. As we discuss further in Chapter 5, we fear that the 1994 amendments to the act, which created a highly complex process for studying and eliminating the most detrimental interactions with fisheries, are difficult and costly to implement. Others decry the inadequate performance of NMFS as a marine mammal protector, saddled as the agency is with its fishery management responsibilities under the

Magnuson Act (Kubasek et al. 1995). Yet others note that while the act has been successful in its own sector, it has created serious problems for other ocean sectors; for example, in causing the continued rise in certain marine mammal populations that are no longer at risk, such as seals and sea lions on the Pacific Coast. Whether it is appropriate to use the MMPA as an instrument of foreign policy is also a question that we raise again at the end of Chapter 5.

All in all, however, the MMPA has probably been the most successful of all the programs in achieving the goals of the original legislation. This is a bit ironic insofar that it is, by far, the most single-purpose and single-minded of all U.S. ocean laws—involving a cessation on the take of marine mammals notwithstanding impacts on other sectors, albeit with some exceptions. The success of the MMPA, at the expense of other interests, illustrates that success in one part of the U.S. ocean governance system, while fundamental to that sector or resource, can well lead to problems in other sectors and can produce an overall U.S. ocean policy that is less than the sum of its parts.

Offshore Oil and Gas Policy: Continued Policy Stalemate in Most of the Country

Implementation of the Outer Continental Shelf Lands Act Amendments has always been very controversial, although not equally controversial everywhere. In the western Gulf of Mexico, especially in Louisiana and Texas, offshore oil development has generally been a welcomed activity. In the early years of the program, OCS activity was largely confined to offshore Louisiana, which had already established a considerable infrastructure for offshore oil development, having already landed 35 million barrels of oil by the time of the first federal lease sale in 1954 (Kitsos 1994b). The coastal economy of Louisiana, and of Texas to some extent, thus evolved with the offshore oil and gas industry. As Kitsos notes, "... there was a symbiotic compatibility among the key actors involved in the OCS program: the Department of the Interior; the oil and gas production industry; the myriad offshore service and supply companies; and the State and its coastal communities. The industry was critical to the economic development of the area, and, quite naturally, was integrated into the State's political system" (p. 38). In contrast, as the federal government began to attempt to develop offshore resources in frontier areas—California, Oregon, Washington, Alaska, the west coast of Florida, the Mid-

Atlantic, and New England—it soon ran into significant opposition by the coastal states and environmental groups. Concerns centered around the impacts of offshore oil and gas development on other uses of the marine environment, such as commercial and sports fishing, tourism and recreation, vessel traffic, military operations, subsistence hunting and fishing activities, and protection of marine mammals (Cicin-Sain and Tiddens 1989). In addition to competition with other users, opposition has also focused on the potential adverse environmental impacts of offshore oil development, especially the effects of seismic surveys, drilling, and the discharge of effluent waters and cuttings on the marine environment. In Alaska, major concerns have centered around the safety of the oil development, and especially the fear of oil spills, the effects on marine mammals, and the impacts on fisheries of interest to Native Alaskans.

As Kitsos (1994b) notes, underlying many of these concerns has been the fear that a way of life would be lost—"unlike in Louisiana, many public officials and citizens in frontier coastal communities perceived that they did not need, indeed they did not want, a new and potentially frightening industry that would bring in people and pollution and, when the oil was gone, leave little of value in its place" (p. 39).

The extent of the offshore oil controversy is related to the value of the offshore resources and to the fact that there is a disjuncture in the distribution of the costs and benefits arising from the exploitation. Most of the benefits accrue to the national level, consumers, and the federal treasury, while coastal states and communities adjoining the development bear most of the potential risks and costs of the development. The OCS program contributes about 18 percent of U.S. domestic production of oil and 27 percent of U.S. production of natural gas (as noted in Figures 4.6 and 4.7, most of these resources come from the Gulf of Mexico). The program has generated very sizeable revenues to the federal government over time—$120 billion since the inception of the program in 1953 (Quarterman 1998). During the years of highest OCS production (1979–1984), the OCS represented the largest source of federal revenues after individual and corporate income taxes (Kitsos 1994b). In contrast to the situation prevailing in mineral exploitation on land (where states receive 50 percent of the revenues generated from mining), because of the federal ownership of the OCS under the OCSLAA, the bulk of the revenues generated has gone to the federal government, with only a small portion being returned to the states.

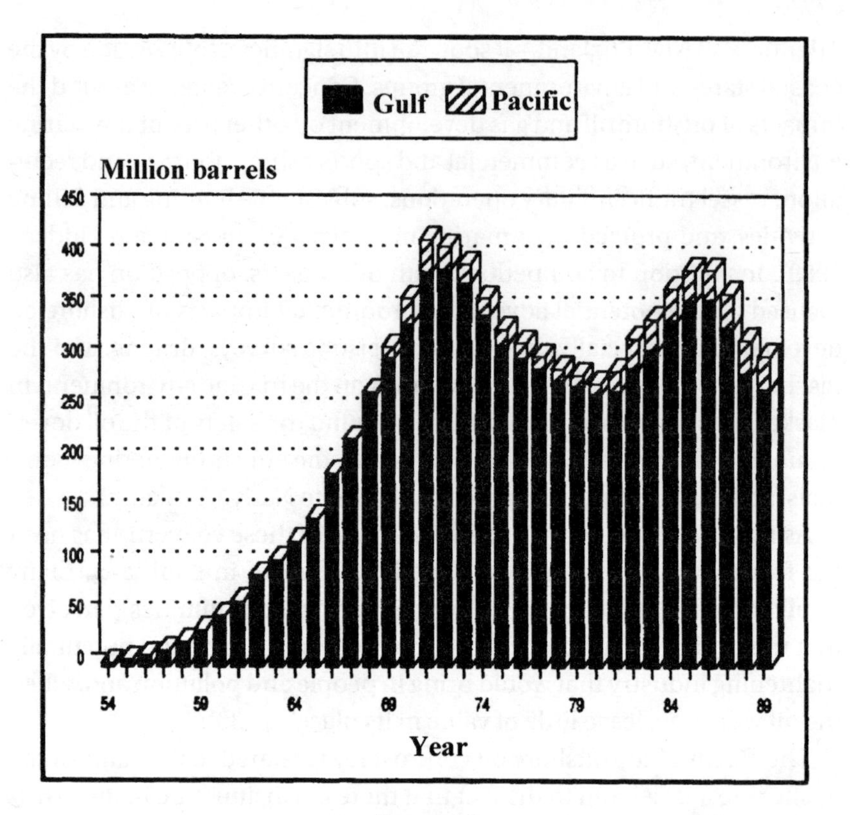

FIGURE 4.6.
National OCS Oil and Condensate Production, 1954–1989 (showing relative amounts produced in the Gulf of Mexico and Pacific OCS regions).

Source: U.S. Department of Interior, Minerals Management Service.

A Troubled History from the Outset of Implementation

As noted by Lester (1996a), the "balance" between development and environmental safeguards and between the interests of federal and state governments that all parties to the negotiations of the 1978 amendments thought had been reached was shattered soon after enactment. After a brief period of attempted balance under President Carter and Secretary of Interior Cecil Andrus, responsibility for the program was passed to President Reagan and to his secretary of interior, James Watt. Within two weeks after taking over the Department of Interior, Watt reversed Andrus's decision not to develop sensitive areas off northern California. In 1982, Watt adopted an "areawide" approach to leasing the OCS. This was a marked departure from the "tract selection" program that had been in operation up to that point and had made only a limited number of offshore lands available for lease. Secretary Watt announced that practically the entire OCS (1 billion acres) would be made available for leasing.

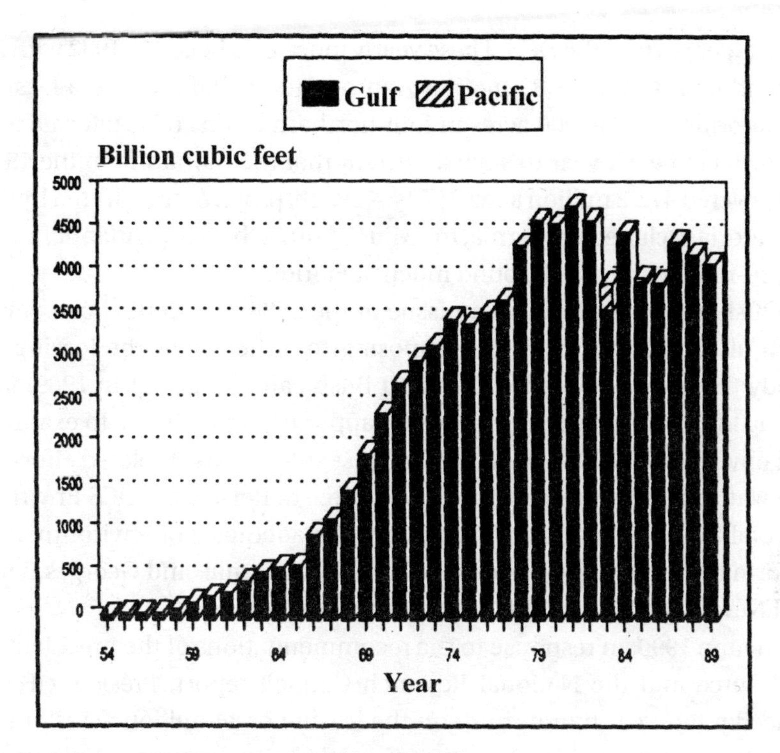

FIGURE 4.7.
National OCS Gas Production, 1954–1989 (showing relative amounts produced in the Gulf of Mexico and Pacific OCS regions).

Source: U.S. Department of the Interior, Minerals Management Service.

In the first eighteen months of the areawide program, over 265 million acres were offered for lease (GAO 1985). This took place while the Reagan administration's budget-cutting initiatives also proposed the elimination of federal funding for the CZMA and the Sea Grant programs, and of the Coastal Energy Impact Program. The coastal states and the environmental groups flocked to Washington to protest the inconsistency of these policies—that is, opening up of the OCS to development while eliminating many of the programs designed to protect the OCS from the effects of development (Armitage 1984; Kitsos 1994b).

Coastal states, environmental groups, and the Congress reacted swiftly to Secretary Watts's leasing plan. Realizing, after the fact, that the OCSLAA, as passed, had given the secretary of interior, perhaps inadvertently, vast discretionary power with regard to the leasing program, James Watts's opponents mobilized to add language to the Interior Department appropriation bill to prohibit the secretary from spending any money on leasing activities in specified parts of the U.S. outer continental shelf. This approach proved highly effective in preventing leasing from larger and larger parts of the OCS. These yearly moratoria began in 1982 with an amendment to the FY 82 Interior Appropriations bill that put a leasing moratorium on 736,000 acres in four northern basins off California and "snowballed each year to such an extent that the moratoria in the 1992 bill covered 472.2 million acres" (Kitsos 1994b, p. 36). Areas covered by the moratoria included California, the Mid-Atlantic, the North Atlantic, and a 20–30-mile buffer zone around much of Florida.

Offshore leasing became an issue in the 1988 presidential campaign with all candidates basically proposing to either defer the leasing or study the issue. Soon after President Bush came to power in 1989, fulfilling his own campaign promise, he appointed a task force to examine the environmental issues present in these three areas. A related effort by the National Research Council's Committee to Review the OCS Environmental Studies Program and to assess the adequacy of environmental information for OCS decisions in Florida, California, and Georges Bank (off New England) was also conducted.

In June 1990, in response to the recommendations of the presidential task force and the National Research Council report, President Bush placed a ten-year moratorium on the leasing of 84 million acres of the OCS (U.S. DOI MMS 1993), areas off northern and southern California, southwest Florida, Georges Bank, Washington, and Oregon. Observers speculated that the presidential action may well have been tied to the political fallout from the Exxon Valdez oil spill in Alaska, which had occurred only six weeks after the presidential task force had been appointed (Kitsos 1994b). Although not an OCS accident, the oil spill focused renewed public attention on the potential environmental impacts of offshore oil development and oil transportation. The moratoria were further expanded, this time by the Congress, to include a 50-mile buffer stretching from Rhode Island south to Maryland, and by the Outer Banks Protection Act, which prohibited new leasing off North Carolina's coast and delayed the approval of a drilling permit for a proposed Mobil Oil development until environmental and socioeconomic information had been reviewed by a panel of scientists (Kitsos 1994b).

As discussed in Chapter 5, in June 1998, President Clinton further extended the existing moratorium through June 30, 2012, and designated parts of the eastern Gulf of Mexico and all national marine sanctuaries off-limits to offshore oil development. The result is that, as shown in the map in Figure 4.8, new offshore oil activity is prohibited in most of the U.S. outer continental shelf, with the exception of areas in the Gulf of Mexico, Alaska, and on existing leases in south central California.

FIGURE 4.8.
Status of the Federal OCS Leasing Program.

Source: Adapted from U.S. Department of Interior, Minerals Management Service 1998 (agency handout).

This period also saw battle over the interpretation of the federal consistency provisions of the Coastal Zone Management Act and their application to federal lease sales. NOAA had always taken the position that federal lease sales, even though they occurred beyond a state's coastal zone, could directly "affect" a state's coastal zone and the resources therein. The Minerals Management Service of the Interior Department had always taken exception to NOAA's interpretation but was unable to secure a change in NOAA's regulations. Ultimately, the applicability of the federal consistency doctrine to leasing was decided in connection with a lawsuit over several tracts off California that were proposed to be leased as a part of lease sale 53. In *Secretary of Interior v. California* (464 U.S. 312 [1984]), the Supreme Court held that OCS lease sales were exempt from consistency review under section 307(c)(1) of the CZMA (Van de Kamp and Saurenman 1990). Following a concerted effort by environmental groups and coastal states, the 1990 amendments to the CZMA overturned the Supreme Court decision and made it clear that oil and gas lease sales were subject to the requirements of 307(c)(1) (Archer 1991).

Thomas R. Kitsos, a chief staffer for many years with the House Merchant Marine and Fisheries Committee, and in the late 1990s with the Department of Interior, attributed environmental groups' and coastal states' hostility to the offshore oil program to a number of "attitudinal mistakes" and "policy errors" by the implementing agency, the Minerals Management Service. The first attitudinal mistake was "the failure to acknowledge the power of state governments in our federal system." (Kitsos 1994b, p. 37) "For inexplicable reasons, the agency appears to ignore history and [the] inherent jurisdictional conflict between the federal government and coastal states, and, more importantly, eschews efforts to moderate this history by any meaningful attempts to establish more of a partnership with the coastal states (p. 37)." The second attitudinal mistake, in Kitsos's view, was the failure or refusal to recognize the variability among the coastal states, and the fact that what might "sell well" in the Gulf of Mexico might not be at all acceptable in California, Washington, or Oregon (p. 38). The third attitudinal mistake, Kitsos noted, was the "failure to recognize the growing and substantial power of the coastal states–environmentalist coalition in opposition to offshore oil development, and the pursuit of an adversarial rather than a cooperative relationship with the coalition (p. 39)." These attitudinal errors led the agency, in Kitsos's view, to three major "policy errors": the failure to support consistent offshore and coastal policies (i.e., pushing for rapid offshore oil development while at the same time cutting back on coastal protection programs); the failure to support, in the 1980s, congressionally endorsed, state-supported, and House-passed OCS revenue-sharing legislation; and the failure of not taking the 1990 presidential policy on the OCS to its logical conclusion—that is, long-lasting reform of the program (Kitsos 1994b).

The OCSLAA has been amended twice. The OCSLA amendments of 1985 resolved a dispute between the federal government and the states over how revenues from production in the 3- to 6-mile zone should be distributed. Since development in federal waters in this area may drain state oil reservoirs, this issue had been litigated and revenues derived from exploitation of this zone had been placed in escrow. The amendments mandated that 27 percent of all revenues from production within 3 miles seaward of the federal-state boundary be given to the states, and set up a schedule for distribution of the revenues placed in the escrow account. In 1994, the act was amended again, this time not in relation to the offshore oil and gas resources, but rather in reference to sand, gravel, and shell resources. These amendments authorized the secretary of interior to negotiate agreements (rather than conduct a competitive lease sale) for the use of these resources in projects undertaken by federal, state, or local governments for such activities as shore protection and beach or coastal wetlands restoration.

Agency Adaptations

In response to the political turmoil and to the criticisms made by the National Research Council, the Minerals Management Service made significant changes in its modus operandi, starting in the 1990s, with the goal of achieving orderly development of offshore resources through consensus, not direct confrontation. As noted by Minerals Management Service Director Cynthia Quarterman (1998), the agency has endeavored to foster a cooperative relationship with affected states and localities through efforts such as the following:

- Resolution of litigation concerning leases in Alaska and in the southeast part of the Gulf of Mexico that led to their extinguishment.
- The establishment, in central California, of a tri-county forum involving local governments and other stakeholders in a joint assessment effort—the California Offshore Oil and Gas Energy Resource Study, aimed at forecasting the issues and potential impacts associated with additional offshore oil development in the region.
- The establishment of an Alaska Regional Stakeholders Task Force to address issues of concern, particularly issues related to the subsistence fishing and hunting of Native Alaskans and issues related to commercial fishers.
- Providing greater attention to socioeconomic issues in the agency's Environmental Studies Program.
- *De facto* acceptance of the congressional moratoria by not including any of the areas covered in the congressional moratoria in the department's five-year plan (U.S. DOI, MMS 1995).
- Efforts to negotiate with the governor of Florida over the proposed Chevron USA development of natural gas in the Destin Dome Block 56 Unit off the Florida Panhandle. These efforts, however, have met with little success because of the perceptions of Floridians that their west coast tourism industry could be adversely affected.

Some Successes, but Largely a Policy Stalemate

The OCS program has clearly been successful in the Gulf of Mexico and has provided significant benefits nationally, in the form of energy resources, jobs (it is estimated that approximately 85,000 people are employed directly in the offshore oil and gas industry, with an equal number employed in supporting jobs [YOTO 1998, D-3]), and significant revenues to the federal government. Many new environmental safeguards incorporated into the 1978 OCSLAA have been put into place, apparently with good success. The YOTO report on offshore oil and gas development (YOTO 1998, D-8) notes that there has not been a spill larger than 1,000 barrels from an OCS platform or rig since 1980—that is, OCS operators have produced some 5.5 billion barrels of oil of which only 0.001 percent has been spilled.

Development in frontier areas has been blocked by an effective coalition of environmental groups and coastal states through the adoption of moratoria on the appropriations for the program. The conflicts and animosity generated through these battles have been very costly, not only in terms of the time and resources spent in dealing with these issues and the negative tenor they set for the resolution of related ocean policy issues, but in foregone development opportunities as well. It is estimated that the OCS contains about 19 percent of the nation's proven gas reserves, 15 percent of the proven oil reserves, and more than 50 percent of the nation's remaining undiscovered oil and gas reserves (YOTO 1998, D-7). The potential development of these resources in a paced and orderly manner is very much hampered by the current OCS policy stalemate.

Significant questions must still be addressed if this stalemate is to be resolved: What measures may be taken to restart the program, to rebuild trust between the states and the federal government? How can the states be brought into the OCS decision-making process in an equitable way? What kind of revenue-sharing system should be put in place and how can this be accomplished? Through what kinds of approaches can paced development of offshore resources be accomplished? Through what kinds of programs and approaches can adverse effects on coastal communities be mitigated? These questions are discussed in Chapter 5.

Controlling Water Pollution: Unresolved Issues

Water pollution became one of the leading environmental issues in the late 1960s and early 1970s in the United States. Events such as the infamous fire on the Cuyahoga River in Ohio during June 1969 graphically demonstrated the seriously degraded nature of certain national waterways. Two years later, a report entitled *Water Wasteland* was released by a task force initiated by Ralph Nader. The report emphasized the poor condition of many U.S. waters and was subsequently confirmed by official government sources. Also in 1971, the President's Council on Environmental Quality reinforced many of the findings of *Water Wasteland* in its *Second Annual Report.* As public attention was repeatedly drawn to contaminated shellfish beds and fish kills, it became increasingly obvious that the issue of water pollution needed to be addressed (Adler et al. 1993, pp. 5–6).

The Federal Water Pollution Control Act of 1972, also known as the Clean Water Act (CWA) was enacted to "restore and maintain the chemical, physical and biological integrity of the Nation's waters" (Clean Water Act §§ 101 [a]). The CWA established three national goals as a means of achieving this objective. The first goal called for the termination of pollutant discharge into navigable waters. The original date for attainment of the "no discharge" goal was 1985. The second goal called for an intermediate goal of water quality that provided for the health of fish, shellfish, and wildlife as well as for recreation in and on the water. The original date for attainment of the "fishable and swimmable waters" goal was July 1, 1983. The third goal called for the prohibition of toxic pollutant discharges in toxic amounts (CWA §§ 101 [a] [1]-[3]).

Successes in Managing Point Sources of Pollution

Under the 1972 Clean Water Act, the EPA began to take steps toward curbing point sources of pollution by establishing the National Pollutant Discharge Elimination System (NPDES). Point-source pollution, as its name indicates, comes from an identifiable point, such as a pipe releasing a factory's waste chemicals into a nearby stream. The NPDES mandates that every discharge of pollution into U.S waters from a point source requires a permit. In general, technology-based standards are used in setting permit discharge limits. However, for water bodies in which technology-based limits (limits based on the use of the best available technology) fail to maintain water quality, the permit discharge limits are based instead on water quality standards. Such water quality standards are aimed at preserving specified uses of the particular water body and are set by the states with the assistance of the EPA (YOTO 1998, E-36).

In addition to creating the NPDES, the CWA required states to generate water quality standards for both in-state and interstate waters and to identify all substandard waters. States were required to calculate the amount of pollution reduction required in order to bring substandard waters into compliance with the new regulations, and to include these requirements in permits. Should individual states fail to perform the new duties imposed by the act, the EPA was to assume responsibility for carrying out any requirements neglected by the states (Adler et al. 1993, pp. 8-9).

Since its enactment in 1972, the CWA has met with a certain degree of success in reducing pollution to the nation's waterways. In 1972, it was estimated that 30 to 40 percent of evaluated waters were "fishable and swimmable." According to the 1998 Clean Water Action Plan, state monitoring data indicate that 60 to 70 percent of assessed waters now meet state water quality objectives. In 1973, 85 million people were served by adequate sewage treatment facilities. By 1998, that number had increased to 173 million. Furthermore, national standards for industrial discharges have led to an estimated annual reduction in pollutant discharges of 108 million pounds of conventional pollutants and 24 million pounds of toxic pollutants (Clean Water Action Plan 1998).

Continuing Thorny Problems in the Management of Non-Point-Source Pollution

While considerable progress has been made in reducing point-source pollution to the nation's waterways, U.S. waters continue to be plagued by non-point-source pollution. Pollution that does not originate from a readily identifiable source poses a serious obstacle to regulators.

Non-point-source pollution occurs in the form of runoff from both agricultural and urban areas and from atmospheric deposition. Agricultural runoff contributes chemicals from pesticides and herbicides and excess nutrients from fertilizers and animal wastes to waterways and coastal waters. Chemicals can contaminate shellfish, and nutrients can lead to excessive algal growth that reduces dissolved oxygen levels and places aquatic life at risk. Overabundant nutrients also have the potential to trigger toxic algal blooms. Agricultural runoff also leads to increased sedimentation of waterways and is responsible for 88 percent of the total suspended solids entering marine waters. Increased sediment loads reduce light penetration, much to the detriment of many aquatic plants (YOTO 1998, E-8).

Urban areas are also major sources of non-point-source pollution. As much as 363 million gallons of oil from land, municipal, and industrial wastes annually contribute to the degradation of estuaries. Street runoff from a city of 5 million people can contain as much oil as a large tanker spill. Urban areas also contribute more than 2 million pounds of cadmium, copper, and zinc to U.S. waters annually (YOTO 1998, E-9).

While the role that non-point-source pollution plays in the degradation of the nation's waterways has gained increasing attention during the last decade, Congress acknowledged the severity of non-point-source pollution as early as 1972. Both EPA and the states have been required to develop thorough plans aimed at the regulation of water pollution from both point and nonpoint sources since the 1972 passage of the act. Sections 102(a), 201(c), 208, 303, and 305(b) of the CWA as passed in 1972 are all concerned with the evaluation, control, and reduction of non-point-source pollution. Unfortunately, there has been inadequate implementation of many of these regulations (Adler et al. 1993, pp. 9, 183).

In 1987, Congress took new action to control non-point-source pollution by enacting Section 319 of the Clean Water Act, aimed at making states identify waters that are suffering from non-point-source pollution and undertake comprehensive programs to decrease and eliminate such pollution. In effect, Section 319 brought together the various provisions related to control of non-point-source pollution that were previously scattered throughout the act. Section 319(a)(1)(C) added to the existing provisions the mandate that non-point-source pollution be reduced "to the maximum extent practicable." While certain watersheds throughout the country have witnessed a marked improvement in water quality since the inception of Section 319, there has not been significant progress in the reduction of non-point-source pollution nationally (Adler et al. 1993, pp. 185–186).

Congress moved to address non-point-source pollution again in 1990 during the reauthorization of the Coastal Zone Management Act. In the 1990 Coastal Zone Act Reauthorization Amendments (CZARA), the states' Coastal Zone and Clean Water Act Section 319 programs were combined to form a coastal nonpoint pollution control program. Jointly administered by EPA and NOAA, the CZARA approach focuses on the reduction of non-point-source pollution by employing land-use measures implemented through the CZMA and CWA (Archer 1991, p. 218). Specifically, Section 6217 of the CZARA requires that coastal states with federally approved coastal zone management programs create and institute specific non-point-source management measures. Should initial control measures fail to meet water-quality objectives within a designated time, states are to develop and implement stricter measures. Once EPA and NOAA have approved Section 6217 programs, they are combined with CWA Section 319 management programs and are thus eligible to receive Section 319 funds (U.S. EPA 1995).

Public awareness of the non-point-source pollution was considerably heightened starting in 1996 as a response to the flare-up of the toxic microbe *Pfiesteria piscicida*, which bloomed in several Mid-Atlantic estuaries and their tributaries, resulting in fish kills and disease, and also, according to some scientists, posing some threats to human health (Barker 1997). The *Pfiesteria* blooms have mobilized public debate on different ways of controlling non-point-source pollution, particularly that emanating from chicken-producing operations found especially in Mid-Atlantic states such as Delaware and Maryland. Maryland is moving to become the first state in the nation to force poultry companies to be responsible for the waste generated by their operations (Goodman 1999).

Despite the progress made by various programs in reducing pollution to the nation's waterways, 40 percent of the waterways still fail to meet the "fishable and swimmable waters" goal of the CWA (Clean Water Action Plan 1998). In light of this fact, President Clinton and Vice President Gore announced the Clean Water Action Plan in February 1998. This plan emphasizes collaborative strategies focusing on watersheds and their dependent communities. In order to implement the new initiative, the president proposed a budget of $568 million in new resources for Fiscal Year 1999 and a total increase of $2.3 billion over five years. The Clean Water Action Plan contained more than a hundred major action items in the categories of new watershed focus, protecting public health, controlling polluted runoff, providing incentives for private land stewardship, restoring and protecting wetlands, protecting coastal waters, expanding citizens' right to know, and enhancing federal stewardship (Clean Water Action Plan 1998).

Issues in the Reauthorization of the Clean Water Act

When the CWA came up for reauthorization in 1993, three broad issues dominated the debate: pollution prevention, protection and restoration of aquatic ecosystems, and successful implementation of the act. Each of these issues contained a number of important subissues. Pollution prevention encompassed reducing and eliminating toxic pollutants, preventing non-point-source pollution, the permitting of irrigation return flows and feedlots, diminishing the flow of urban stormwater runoff, and preventing combined sewer overflows. Significant topics falling under the category of protecting and restoring aquatic ecosystems included restoring urban watersheds, protecting the nation's wetlands, strengthening the National Estuary Program, identifying and safely disposing of contaminated sediments, preventing clean waters from becoming degraded, and protecting biological integrity. Issues related to the implementation of the CWA included enlarging communities' right to know, creating standards for beach closures and advisories, enforcing the Clean Water Act, and adequately funding clean water programs (Clean Water Network 1993). When reauthorization of the CWA failed repeatedly in the Congresses since 1993, these issues went unresolved. A number of them have since been incorporated in the Clean Water Action Plan of 1998.

The National Estuary Program: Good Concept, Slow Implementation

While the significance of the nation's waterways in general has always been recognized, an awareness of the particular importance of estuaries has developed largely within the last decade, although two important studies were undertaken in 1970: the National Estuary Study and the National Estuarine Pollution Study. The importance of estuaries and the devastating effects that both point- and non-point-source pollution was having on them gained attention during the 1980s. As a result, Congress developed the National Estuary Program (NEP) under Section 320 of the 1987 amendments to the Clean Water Act in order to address the difficult problems that led to the degradation of the estuarine systems throughout the nation (Clean Water Network 1993, p. 26).

Under the NEP, the EPA, working with state governors, designates certain estuaries of "national significance" for participation in the program. Once an estuary is designated for the program, the EPA convenes a management conference with participants representing relevant federal commissions and agencies, the governor and pertinent state commissions and agencies, private companies, organizations, and citizens. The conference addresses the uses that affect the integrity (chemical, physical, and biological) of the estuary. The objective of the conference is to develop a Comprehensive Conservation and Management Plan (CCMP) that will reestablish and maintain both the water quality and resources of the estuary (Briefing Papers on the CWA Reauthorization 1993, p. 26). By 1999, twenty-eight estuary programs were working to safeguard the health of some of the nation's most important coastal waters (see Figure 4.9).

The NEP seems to present a comprehensive and effective method for managing the nation's estuaries; however, while there is no lack of innovative ideas and concepts behind the program, the concepts have not always been translated into success. Several problems have caused the NEP to suffer from slow implementation. First, although the NEP requires the development of a Comprehensive Conservation and

FIGURE 4.9.
Map of National Estuary Program Sites.
Source adapted from U.S. EPA National Estuary Program. 1999 (at http://www.epa.gov).

Management Plan, once an estuary is designated for the program, there is no specific structure or process prescribed to implement the plan. Second, the EPA has not coordinated the relevant environmental agencies consistently for each estuary. The EPA's general level of commitment to each NEP has fluctuated as well. Third, the process of CCMP development has sometimes lacked sufficient citizen involvement. Finally, many states have not been able to implement their CCMPs due to lack of funds resulting from state budget shortages and inadequate federal support (Clean Water Network 1993, pp. 26–27).

The NEP also seems to suffer from the more general difficulties associated with implementing "ecosystem management." The concept of ecosystem management is guided by several principles: providing an increased role for science in management; using natural system boundaries and not political ones for management; developing increased cooperation and coordination among agencies; and fostering larger roles for public participation and education. Unfortunately, implementing ecosystem management has proven very complex (Korfmacher 1998, p. 207; Cicin-Sain, Knecht and Bradly, in preparation). In a detailed study of the implementation of the CCMP in the Albemarle-Pamlico Estuarine Study (part of the NEP in North Carolina), Korfmacher (1998) identified seven "paradoxes of ecosystem management." The Albemarle-Pamlico Estuarine Study was found to suffer from these paradoxes, and it is likely that they apply to a greater or lesser extent to the NEP as a whole.

1. *Geographical scale.* While a large scale helps to ensure the consideration of all impacts on an estuary, it also detracts from effective public participation and problem solving on a local level.

2. *Scope.* While a broader scope interests and involves more stakeholders, it also increases the difficulty of reaching any type of agreement on the issues at hand.

3. *Consensus.* In order to ensure continued implementation of management measures, consensus for the measures within and outside the program is necessary. However, arriving at consensus on any issue is extremely time consuming and at odds with the need to achieve initial success rapidly in order to maintain support for the program. (We would add, too, that the process of achieving consensus can result in pressure to arrive at a "lowest common denominator" solution, which is amenable to all parties but which may not be sufficiently tough to make a real difference in the health of the estuary [Cicin-Sain, Knecht, and Bradly, in preparation]).

4. *Obtaining useful science.* According to interviews conducted by Korfmacher, scientists maintained that an inverse relationship exists between scientists' status or reputation and their willingness to focus their efforts on specific, applied questions.

5. *Timescale.* While ecosystem management necessarily functions over a very long period of time, slow progress can cause a program to lose momentum resulting in a concomitant loss of public interest and participation.

6. *Independent institutions.* While credibility of a program attempting to critique existing problems and approaches is enhanced if the program is independent of extant management agencies, creating a new and separate organization makes it difficult to sustain involvement and implementation commitments from existing agencies.

7. *Political support.* While strong political support enhances any program's chances for success, it also makes the program vulnerable to shifts in political leadership. (pp. 205–208).

Progress has undoubtedly been made in controlling certain types of water pollution in the United States, but much still remains to be done. The goals of the CWA of 1972 have yet to be accomplished. Non-point-source pollution continues to plague national waters. The NEP faces difficulties in implementation, and the effectiveness of the 1998 Clean Water Action Plan remains to be seen.

Marine Protected Areas: Putting the Program on a Stronger Footing

As reviewed in Chapter 3, 1972 saw the creation of two significant elements of what was to become a major part of the U.S. program in marine protected areas—the enactment of Title III of the Marine Protection, Research, and Sanctuaries Act authorizing the designation of national marine sanctuaries, and the enactment of Section 315 of the Coastal Zone Management Act authorizing the establishment of estuarine sanctuaries in connection with the CZM program. These elements of U.S. national ocean policy have had a long gestation period. Twenty-six years after enactment, some observers feel that the U.S. marine and estuarine protection program is still a very modest one given the magnitude and nature of ocean and coastal resources under U.S. control (CMC 1998; Klee 1999).

The slow start of this program can be attributed to at least three factors: (1) Neither the administration nor the Congress were interested in providing funding during the initial years of the program; (2) the goals of these new initiatives were somewhat uncertain; and (3) offshore oil and gas interests voiced early strong opposition to the program. The Marine Sanctuaries Program and the estuarine research reserves (as they are now called) have been administered by the same office in NOAA since their initiation, but are discussed separately because of the very significant differences between their goals and objectives and the manner in which they have been implemented.

The National Marine Sanctuaries Program

The basic goal of the National Marine Sanctuaries Program is to protect ocean areas because of the unique or significant resources that they contain (Foster and Archer 1988). Resources can be judged significant for their "conservation, recreation, ecological, historic, research, educational or aesthetic qualities" (MPRSA § 1431[a][2]). Soon after enactment, it became clear that the new program was subject to widely different interpretations. Some people saw it as principally a protection or preservation program, while others regarded it as analogous to the multiple-use management system operated by the nation's national forest service (Tarnas 1988). This dichotomy, especially since it involved members of Congress close to the enactment process, was to cause the NOAA staff charged with its implementation to move cautiously in its initial work. Also contributing to the slow start was the fact that the funding to begin implementation of the program was neither requested by the administration nor appropriated by the Congress during the years immediately after passage of the legislation. The priority that NOAA gave to securing appropriations to begin implementation of the Coastal Zone Management Act, judged to be more urgent at the time, could also have been a factor.

Almost from the beginning, some groups suggested that the marine sanctuary designation could be used to slow or prevent certain kinds of ocean development, most notably offshore oil and gas (Hoagland and Eichenberg 1988). Naturally, this made oil companies very wary of the program and its goals; thus, the offshore oil industry also took an early interest in the implementation of the Marine Sanctuaries Program.

In response to the Arab oil embargo of 1973–1974, the Nixon administration announced a greatly expanded effort to lease areas of the U.S. outer continental shelf for oil and exploration and development. As oil companies and the federal government geared up for this expanded effort, some members of Congress made it clear that an effort to use the Marine Sanctuaries Program to frustrate the expanding offshore oil and gas program could result in the elimination of the Marine Sanctuaries Program itself.

Immediately after the passage of the CZM and MPRS acts in 1972, a task force was created in NOAA to oversee the initial implementation of the programs authorized in these two pieces of legislation, specifically, the CZM program, the Marine Sanctuaries Program, and the Estuarine Research Reserve Program. Although most of the initial effort was devoted to the CZM program, some time was also spent on initiating the two sanctuary programs. There was keen interest in getting the Marine Sanctuaries Program operational as soon as possible but in a way that would not endanger its continued existence. Early attention was focused on the possibility of designating the underwater resting place of the Civil

TABLE 4.10 Sanctuaries in the U.S. Marine Sanctuaries Program.

YEAR DESIGNATED	SANCTUARY	SIZE OF PROTECTED AREA (IN SQUARE MILES)	TYPE OF HABITAT	KEY SPECIES
1975	*Monitor* (North Carolina)	1	pelagic, open ocean, artificial reef (protection of Civil War sunken vessel)	amberjack. black sea bass, red barbier, scad, corals, sea anemones, dolphin, sand tiger shark, sea urchins
1975	Key Largo* (Florida)	100	coral reefs, patch and bank reefs, mangrove-fringed shorelines and islands, sand flats, seagrass meadows	coral reef and associated reef species
1980	Channel Islands (California)	1,658	kelp forests, rocky shores, sandy beaches, seagrass, meadows, pelagic, open ocean, deep rocky reefs	California sea lion, elephant and harbor seals, blue & gray whales, dolphin, blue shark, brown pelican, western gull
1981	Gulf of the Farallones (California)	1,255	coastal beaches, rocky shores, mud and tidal flats, salt marsh, esteros, deep benthos, continental shelf and slope	dungeness crab, gray whale, stellar sea lion, common murre, ashy storm-petrel
1981	Gray's Reef (Georgia)	23	calcareous sandstone, sand bottom communities, tropical temperate reef	northern right whale, loggerhead sea turtle, barrel sponge, angelfish, ivory bush coral, grouper, black sea bass
1981	Looe Key* (Florida)	5.32	coral reefs, patch and bank reefs, mangrove-fringed shorelines and islands, sand flats, seagrass meadows	coral reef and associated reef species

1986	Fagatele Bay (American Samoa)	0.25	tropical coral reef	crown-of-thorns starfish, blacktip reef shark, surgeon fish, hawksbill turtle, parrotfish, giant clam
1989	Cordell Bank (California)	526	rocky subtidal, pelagic, open ocean, soft sediment continental shelf and slope, seamount	krill, Pacific salmon, blue whale, humpback whale, Dall's porpoise, shearwater, albatross, rockfish
1990	Florida Keys (Florida)	3,674	coral reefs, patch and bank reefs, mangrove-fringed shorelines and islands, sand flats, seagrass meadows	brain and star coral, sea fan, loggerhead sponge, turtle grass, angelfish, spiny lobster, stone crab, grouper, tarpon
1992	Flower Garden Banks (Texas/Louisiana)	56	coral reefs, artificial reef, algal-sponge communities, brine seep, pelagic, open ocean	brain and star coral, manta ray, loggerhead turtle, hammerhead shark
1992	Monterey Bay (California)	5,328	pelagic, open ocean, sandy beaches, rocky shores, kelp forests, wetlands, submarine canyon	sea otter, gray whale, market squid, brown pelican, rockfish, giant kelp
1992	Hawaiian Islands (Hawaii)	1,300	humpback whale breeding, calving and nursing grounds, coral reefs, sandy beaches	humpback whale, pilot whale, hawaiian monk seal, spinner dolphin, green sea turtle, trigger fish, limu, cauliflower coral
1992	Stellwagen Bank (Massachusetts)	842	sand and gravel bank, muddy basins, boulder fields, rocky ledges	northern right whale, humpback whale, storm petrel, white-sided dolphin, bluefin tuna, sea scallop, northern lobster
1994	Olympic Coast (Washington)	3,310	pelagic, open ocean, sandy and rocky shores, kelp forests, seastacks and islands	tufted puffin, bald eagle, northern sea otter, gray whale, humpback whale, Pacific salmon, dolphin

*Looe Key and Key Largo Marine Sanctuaries were incorporated into the Florida Keys National Marine Sanctuary.

Source: Adapted from NOAA NOS. 1997. National Marine Sanctuaries Accomplishments Report.

War vessel, the ironclad *Monitor,* as the first marine sanctuary. In January 1975, twenty-seven months after passage of the legislation, the site was formally designated as the first national marine sanctuary in a ceremony at the Commerce Department. A second marine sanctuary, in Key Largo, Florida, about 100 times larger than the first but also relatively noncontroversial, was designated later in 1975. This sanctuary was put in place to protect the portion of a coral reef located in federal waters but adjacent to a state marine park, already established to protect the portion of the reef under state jurisdiction. (See Table 4.10 for a list of marine sanctuaries and their dates of designation.)

In May 1977, as a part of President Carter's "message on the environment," he cited the Marine Sanctuaries Program as one of the priorities of his administration. In response to his direction that the secretary of commerce gather information on areas of the outer continental shelf that were imminently threatened by development, NOAA received 169 site nominations for its list of recommended areas (LRA) for consideration as national marine sanctuaries. Of these, about half were added to the LRA and seven sites were selected as "active candidates" for designation. However, the process ultimately became mired in controversy because some of the sites added to the LRA in this process were very large (e.g., one covered most of the Beaufort Sea off Alaska and another Georges Bank off New England), and many of them seemed aimed principally at preventing oil and gas development. Also, timely public participation was not built into the process nor was there a mechanism for adequate public notification. Consequently, in 1982, NOAA proposed a new site-selection process that clarified the program's goals and operational policies, its site-identification criteria, the nomination and designation process, and the elements and nature of site-specific management plans (Tarnas 1988). After work by a consultant and eight regional resource evaluation teams, NOAA selected twenty-nine sites for inclusion in its new site evaluation list (SEL) in 1983 (Tarnas 1988). Amendments to the marine sanctuaries authorizing legislation in 1984 put into law the new NOAA procedures and increased the oversight role of the Congress.

Meanwhile, additional marine sanctuaries were being designated in the Channel Islands and Farallones off California; in Looe Key, Florida; and in Gray's Reef off Georgia, bringing the total to six by the end of 1981. Significant controversy surrounded the designation of the two large sanctuaries off California and, especially, their prohibition of new oil and gas activity. However, after a year-long study, the incoming Reagan administration allowed the prohibitions to stand. Notwithstanding this decision, the Reagan administration was generally hostile to the Marine Sanctuaries Program; only one very small sanctuary was designated between 1982 and 1988, the Fagatele Bay sanctuary in American Samoa.

During this period, advocates of the sanctuaries program, especially in the coastal states and environmental interest groups, grew impatient with the pace of the program and began to explore the route of congressional designation for marine sanctuaries. Proponents found sympathetic ears in both houses of Congress, which ultimately led to congressional designation of three new marine sanctuaries—the Florida Keys National Marine Sanctuary in 1990, the Monterey Bay National Marine Sanctuary in 1992, and the Hawaiian Islands Humpback Whale Sanctuary in 1992. While congressional designation of a marine sanctuary was a relatively quick process, it led to difficulties that could have been avoided in a more systematic approach to sanctuary designation. For example, the process devised in NOAA's regulations permits public and interest group participation in considering alternative boundaries and management plans as a part of the deliberations on the draft environmental impact statement prior to the formal designation of the sanctuary. Congressional designation prior to such discussion and debate could prematurely foreclose the exploration of other desirable management or boundary options. All three congressional designations were followed by relatively long periods of work by NOAA and state and local interests to fashion appropriate boundaries and management plans. For example, work on the Hawaiian sanctuary was not completed until six years after the congressional designation.

Recognizing growing external dissatisfaction, NOAA appointed a team to review the Marine Sanctuaries Program in November 1990. In its February 1991 report, Frank Potter, chair of the Marine Sanctuaries Review Team, presented recommendations that underscored the fundamental importance of the program and called for strengthened programmatic leadership, an elevation of the program's organizational location within NOAA, increased visibility, and a tenfold increase in annual funding (to $30 million) (Potter 1991). Some of these recommendations ultimately, found favor with the incoming Clinton administration, which assumed office in January 1993. The report also was favorably received in Congress and helped lend support to a threefold increase in appropriations for the Marine Sanctuaries Program (to $12 million), which began in 1994 and continued into 1999. Another substantial boost in appropriations is possible if Congress endorses and funds President Clinton's Lands Legacy Initiative announced in early 1999.

Assessment of the Program

Through 1998, fourteen marine sanctuaries had been designated, with one more (Thunder Bay, Michigan) undergoing review (see Table 4.10 and Figure 4.10).

Although the increases in budgets for the program that have occurred since 1990 are encouraging, the program remains underfunded, understaffed, and sometimes politically invisible in national ocean policy bureaucratic battles. As discussed in Klee (1999), the Center for Marine Conservation has noted some interesting comparisons with land-based protected areas regarding the funding and staffing of the program. In 1993 the National Park Service received $1.4 billion to support its mission, while the marine sanctuaries program received only $7.1 million. Yet, one sanctuary alone, the Monterey Bay National Marine Sanctuary, is twice the size of Yellowstone National Park (Sobel and Merow 1993 quoted in Klee 1999). With regard to staffing comparisons between land and ocean parks, Yellowstone National Park in 1993 had a budget of $18,247,000 and employed a staff of 508 employees, whereas Monterey Bay National Marine Sanctuaries had a budget of $548,000 and only 2 employees (Klee 1999). In addition, from a bureaucratic point of view,

FIGURE 4.10.
Map of U.S. Marine Sanctuary Sites.

Source: Adapted from data from NOAA, National Marine Sanctuary Program, 1998.

the National Marine Sanctuaries Program office is small, relatively unknown, and low in the bureaucratic hierarchy, in comparison to the large and well-known bureaucracies involved in protecting land areas, such as the National Park Service, the U.S. Forest Service, or the U.S. Fish and Wildlife National Wildlife Refuge System (Klee 1999).

The successes of the program, however, should be noted. The number of marine sanctuaries has doubled in the last decade, and the funding of the program has tripled since 1990. The program has successfully developed management regimes, including detailed zoning plans, for managing vast and complex ocean areas off the Florida Keys and Monterey Bay. The targeted resources within the present system of sanctuaries are clearly now receiving a higher level of protection than prior to sanctuary discussion. Through the program's many public education efforts, the awareness of the public of the importance of ocean resources, both in the sanctuaries and elsewhere, has no doubt been considerably heightened.

Continuing problems for the program include the following: The three recent congressional designations suggest continuing unrest with the administrative pace of the program. Some confusion seems to continue as to whether the program should be seen as a multiple-purpose ocean management program or a program aimed at protecting or preserving some unique aspect of particular ocean areas. Most important, in our view, no overall philosophy or vision appears to exist on the ultimate size of the program, its rate of growth, and its role in protecting the marine biodiversity resources of the U.S. EEZ. The latter problem is touched upon again in Chapter 5.

The National Estuarine Research Reserve Program

The Coastal Zone Management Act of 1972 authorized the National Estuary Research Reserve Program as a mechanism for developing a national system of estuarine laboratories. The program sought to both identify a national system of estuaries representative of those found along the U.S. coastline and strengthen coastal zone management through a program of applied research in estuarine reserves. As mentioned earlier, the implementation of this program got off to a slow start due to the lack of funding and personnel devoted to it in the early years, given the perceived pressure to get the larger and more visible CZM program underway as quickly as possible. Even so, the necessary procedures and regulations were prepared on a timely basis, and the first estuarine sanctuary—the South Slough Estuarine Sanctuary in southern Oregon—was formally designated in 1974.

TABLE 4.11 Estuaries in the National Estuarine Research Reserve System (NERRS).

YEAR DESIGNATED	ESTUARY	APPROXIMATE SIZE OF RESERVE AREA (IN ACRES)	TYPE OF HABITAT	KEY SPECIES
1974	South Slough (Oregon)	4,500	upland forests, freshwater wetlands and ponds, salt marshes, tidal flats, eelgrass meadows, open water	Port Orford cedar, bald eagle, great blue heron, elk, eelgrass, Dungeness crab, ghost shrimp
1976	Sapelo Island (Georgia)	6,100	salt marshes, maritime forests, dunes and beaches, freshwater ponds, sloughs	cordgrass, live oak, osprey, great blue heron, wild turkey, bald eagle, loggerhead turtle
1978	Rookery Bay (Florida)	9,400	mangrove forests, barrier beaches, coastal dry-zone scrub, pine flatwoods, seagrass beds, tropical hardwood hammocks, open shallow waters	red, white, and black mangroves, bottle-nose dolphin, manatee, white ibis
1979	Apalachicola Bay (Florida)	24,600	forested floodplains, open water, oyster bars, salt marshes, barrier islands, freshwater marshes	tupelo gum, bald cypress, American oyster, white shrimp, osprey, black bear, Atlantic sturgeon
1980	Elkhorn Slough (California)	1,400	oak woodlands, salt marshes, grasslands, mudflats, freshwater ponds, Monterey pine grove, coastal scrub	shorebirds, great blue heron, great egret, smooth hound and leopard sharks, birds of prey
1980	Old Woman Creek (Ohio)	600	upland forests, old-field succession, streams, freshwater marshes, swamp forests, barrier beach/Lake Erie	American water lotus, bald eagle, eastern fox snake, migratory waterfowl

1980	Narragansett Bay (Rhode Island)	4,950	open water, eelgrass meadow, tidal flats, salt marshes, freshwater wetlands and ponds, upland fields, upland forests	bluefish, striped bass, winter flounder, osprey, great blue heron, whitetail deer
1980	Padilla Bay (Washington)	110,700	open water, seagrass meadows, tidal flats and sloughs, salt marshes, upland forests, upland meadows	seagrass, Dungeness crab, salmon, black brant, bald eagle, peregrine falcon
1981	Jobos Bay (Puerto Rico)	2,800	fringing and basin mangrove forests, seagrass beds, mangrove channels, salt flats, lagoons, coral reefs, subtropical dry forests	West Indian manatee, hawksbill sea-turtle, yellow-shouldered blackbird, brown pelican
1982	Hudson River (New York) (4 components)	4,800	open water, subtidal meadows, tidal flats, tidal marshes, tidal freshwater wetlands, mixed forest	grass shrimp, herring, blue crab, snapping turtle, muskrat, osprey, bald eagle
1982	Tijuana River (California)	2,500	dunes and beaches, mudflats, salt marshes, riparian, coastal sage, uplands	light-footed clapper rail, California least tern, least Bell's vireo, salt marsh bird's beak, cordgrass
	(North Carolina) (4 components)			
1982 1991	Currituck Banks, Rachel Carson, Zeke's Island Masonboro Island	10,000	salt/brackish/freshwater marshes, sandy beaches, shrub thickets, maritime forests, mud/sand flats, subtidal vegetation, oyster bars	cordgrass, hard clam, blue crab, flounder, loggerhead sea turtle
1984	Wells (Maine)	1,600	fields and forests, tidal rivers, swamp forest, salt marshes, dune forest, beach	whitetail deer, spartina grasses, slender blue flag, snowy egret, soft-shell clam, winter flounder, piping plover

continued

TABLE 4.11 Estuaries in the National Estuarine Research Reserve System (NERRS) (*continued*)

YEAR DESIGNATED	ESTUARY	APPROXIMATE SIZE OF RESERVE AREA (IN ACRES)	TYPE OF HABITAT	KEY SPECIES
	Chesapeake Bay (Maryland) (3 components)			
1985	Monie Bay	4,800	forested uplands, tidal freshwater marshes, tidal brackish water marshes, open waters, coastal grasslands	peregrine falcon, bald eagle, sora rail, Baltic rush, anglepod, wild rice
1990	Jug Bay, Otter Point Creek			
1986	Weeks Bay (Alabama)	3,000	tidal flats, freshwater marshes, salt marshes, forested swamps, upland forests	black needlerush, brown pelican, great blue heron, shrimp, sea trout, blue crab, redfish, American alligator, red bellied turtle
1988	Waquoit Bay (Massachusetts)	2,250	salt ponds, salt marshes, pine and oak forest, barrier beaches/dunes, open water	piping plover, least tern, sandplain gerardia, alewife, winter flounder, blue crab, osprey
1989	Great Bay (New Hampshire)	5,280	upland fields and forests, salt marshes, mudflats, rocky intertidal, eelgrass beds, channel bottom and subtidal	bald eagle, rainbow smelt, American oyster, horseshoe crab, large salt marsh aster, osprey
1991	Chesapeake Bay (Virginia) (4 components)	4,400	tidal waters, marshes and flats, submerged aquatic vegetation beds, swamps	migratory neotropical birds, sea grasses, all stages of life of commercially important fishes and shellfish
1992	Ashepoo, Combahee, and Edisto (ACE) River Basin (South Carolina)	136,000	salt marshes, brackish marshes, tidal flats, maritime forests, bird keys and banks, pine-mixed hardwoods	American alligator, loggerhead sea turtle, shortnose sturgeon, bald eagle, wood stork

1992	North Inlet and Winyah Bay (South Carolina)	12,300	salt marshes, intertidal oyster reefs, mudflats and sandbars, tidal creeks, shallow sounds, abandoned rice fields and canals	cordgrass, fiddler crab, American oyster, white shrimp, spot, alligator, white ibis
1993	Delaware (2 components)	8,600	salt marshes and open water, tidal shorelands and mud flats, freshwater wetlands and ponds, forests, farmlands and meadows	snowy egret and great blue heron, bald eagle, black duck, horseshoe crab and migratory shorebirds, blue crab and fiddler crab, American oyster
1998	Jacques Cousteau at Mullica River and Great Bay (New Jersey)	114,000	open water, pinelands, lowland forests, tidal and salt marshes, coastal dunes, barrier islands	piping plover, bald eagle, peregrine falcon, timber rattlesnake, eelgrass, pitch pine
1999	Kachemak Bay (Alaska)	365,000	open water, tidal marshes, upland forest glaciers, glacial streams	all species of Pacific salmon, halibut, bald eagle, clams, mussels, humpback and beluga whales, sea otters, Stellar's sea lion, black and brown bears, Sitka spruce
Anticipated 1999	Guana, Tolomato, and Matanzas (GTM) Rivers (Florida)	55,000	beach dunes, salt and fresh water marshes, cypress and hardwood swamps, shell mounds, xeric hammocks	Manatee, least tern, loggerhead, green and leather-back turtles, blue crabs, stone crabs, tarpon, snook, roseate spoonbill, bald cypress, sawgrass, sand cordgrass
Anticipated 1999	Grand Bay (Mississippi)	18,000	oyster reefs, sea grass, shallow water open bay, estuarine tidal marsh, wet pine savannah, coastal swamp, shell mounds	blue crab, brown and white shrimp, American alligator, osprey, brown and white pelicans, freshwater otter, muskrat, diamond-back terrapin, cordgrass and widgen grass

Source: Adapted from data from NOAA. National Estuarine Research Reserve System, 1999 (with thanks to Matthew Menashes).

The first appropriations for the program also became available in fiscal year 1974—but only $4 million of the $9 million authorized for the program. Because of the lead times required for states to identify and obtain purchase options on the watershed properties necessary to protect a prospective estuarine sanctuary, the $4 million appropriation was not exhausted until 1977 when a new appropriation of $1.5 million was received.

Similar to the CZM program as a whole, the estuarine sanctuaries program was designed as a state-federal cooperative effort. The coastal states or their political subdivisions are the owners and operational managers of these protected areas. The federal government's role as outlined in the CZMA was to set standards for their selection and operation, to assist states in the acquisition of the sites, and to ensure that the sites collectively form a national system representative of important biogeographic estuarine areas in the United States.

Following the designation of the first estuarine sanctuary in 1974, two additional sites were designated in Georgia and Hawaii in 1975 and 1976 respectively, and two more sites were added by 1979. During the 1980s, the annual appropriation for this program generally remained at the level of $3 million per year, about one-third of the $9 million that had been written into the authorizing legislation. Table 4.11 lists the sites that have been formally designated up to mid-1998, together with some relevant information about each site.

Despite the continued growth of the program, some confusion remained over its goals and objectives. Was the purpose of the designation to protect the sensitive resources of an estuarine area, to help complete a nationally representative system of estuarine sites, or to provide a site for applied research to benefit the coastal management programs of the coastal states? In 1985 Congress amended the CZMA of 1972 to change the name of the program to the National Estuarine Research Reserve System (NERRS) in order to highlight the research dimension of the program. Also, in the legislative changes, Congress more clearly defined the role of the federal government in the operation of the program, most notably in coordinating research and monitoring programs throughout the system and in encouraging educational programs benefiting the entire system.

In January 1993, NOAA asked an outside panel to review the operation of the program emphasizing the extent to which it was achieving the purposes contained in the authorizing legislation. The group's report, *The National Estuarine Research Reserve System: Building a Valuable Na-*

tional Asset, was submitted to NOAA in July 1993. It contained a number of recommendations for strengthening the national dimensions of the program and improving its coupling to coastal resource managers and their research needs (Knecht 1993). One recommendation called for NOAA to create a NERRS program office with a qualified director to lead the overall NERRS effort—a change that was subsequently implemented by NOAA. Up to that point, the NERRS program had been an integral part of a combined "sanctuaries and reserves" office headed by a single individual.

As of early 1999, a total of twenty-two research reserves have been established in nineteen states and Puerto Rico embracing a total area of 558,507 acres, with designations of five additional reserves expected by the year 2000 (see Figure 4.11).

Assessment of the Program

In terms of successes the program has seen steady growth over its twenty-four-year lifetime (since its first funding was appropriated) despite its very modest funding levels. Generally positive relations exist between the federal government and the state agencies administering the reserves, which is crucial to the ultimate success of a joint state-federal program such as this one. A nationally coordinated monitoring program

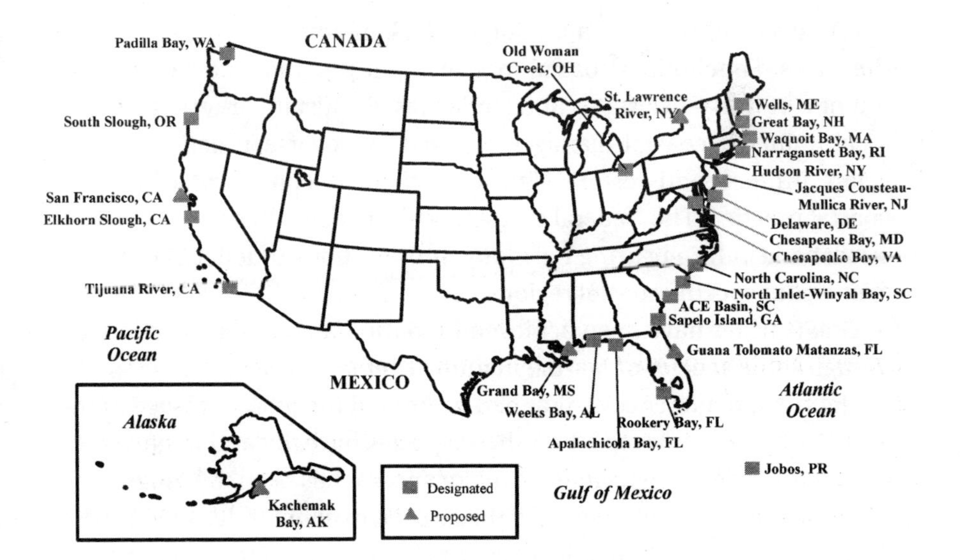

FIGURE 4.11.
Map of U.S. National Estuarine Research Reserve Sites.

Source: Adapted from NOAA, National Estuarine Research Reserve Program. 1999.

is now in place in all of the reserves, which should generate valuable comparative data that has been missing up to this point in the program. A recently inaugurated system of graduate fellowships at each reserve appears likely to significantly strengthen the research and educational dimension of the NERRS program.

Some problems continue to affect the program. Effective national coordination of the research program remains a goal, not an achievement; the coupling between the research needs to improve coastal zone management and the research program of the NERRS program remains weak. Positive linkages between estuarine research reserves and other types of marine protected areas in the United States remain relatively undeveloped. Federal support to the program remains at an inadequate level, given the very real needs and opportunities existing in the program. Regarding the last point, this situation could undergo improvement in 1999 if Congress responds favorably to President Clinton's $1 billion Lands Legacy Initiative, announced in January 1999, on preservation of open space.

Taking Stock: Some Successes, Some Failures, and No Overall Vision

Many successes have been attained in U.S. national ocean policy since the 1970s. Functioning coastal management programs now exist in all but one of thirty-five states and territories. Each state or territory has a focal point for coastal management that has overseen actions to develop, protect, and restore state coastal areas consistent with good coastal practice. The "federal consistency" provision has worked well as a mechanism for ensuring harmony in the actions of federal and state agencies in specific coastal regions.

Great strides have been achieved in marine mammal protection, although some species of marine mammals remain at risk. In general, the depletion of major species that had prompted the initial passage of the MMPA has been reversed, and these species have, by and large, recovered. In fisheries management, a major success has been the "Americanization" of the 200-mile zone and the replacement of foreign fishing with American fishers, giving U.S. commercial and recreational fishers broader opportunities for obtaining benefits from the EEZ.

Significant strides have also been made in the control of point sources of pollution to rivers and estuaries, and the number of people receiving appropriate wastewater treatment has more than doubled since 1973.

Under the National Estuary Program, twenty-eight estuaries have received special planning and management attention, and comprehensive plans for the reduction of pollution and the protection of key habitats in most of these estuaries have been completed.

The United States now has in place a network of marine sanctuaries and estuarine research reserve areas that, in the case of the sanctuaries, protect special conservation, recreation, ecological, historic, research, educational, or aesthetic values; and in the case of the estuarine reserves, serve as research and monitoring sites and as laboratories for coastal management.

Offshore oil and gas exploitation in the United States has developed significant domestic energy resources, generated sizable revenues for the federal government, and created jobs and contributed other revenues in coastal communities. The safety record of the offshore industry has been very good: No major accidents have taken place as a result of development in the OCS. The environmental requirements put into place following the 1978 OCSLA Amendments appear to have generally been successfully implemented. The industry has advanced new technological frontiers in recent years, such as the techniques, equipment, and procedures developed for operating in waters up to 10,000 feet deep in the Gulf of Mexico.

It appears that the worst failure experienced in U.S. ocean policy in recent years has been the serious decline in fisheries, even after the establishment of what appeared, in 1976, to be a very rationally designed system of regional fishery management councils. Perhaps the most difficult situation in ocean management is the need to be "hard-nosed" about fisheries conservation while knowing that drastic conservation measures will have substantial adverse impact on dependent fishing communities. Another major failure of policy has been the inability of the Department of Interior to expand the OCS program beyond the western Gulf of Mexico and central California into new frontier areas. There are many economic opportunities and returns to the federal treasury that have been foregone because of this failure.

Little evaluative data on the effectiveness of the CZM program has been available, which makes it difficult to demonstrate empirically the many facets and general success of this program. While estuarine management under the National Estuary Program has no doubt been advanced and improved, there is concern that the many management plans that have been developed under the NEP are not being effectively implemented. Although the Clean Water Act achieved notable successes with the control of point sources of pollution, the thorny task of controlling diffuse land-based or nonpoint sources of marine pollution will loom as the major issue in this area in the next decade. Regarding marine mammal protection, the expansion of some marine mammal populations beyond historic levels and without effective controls has created serious conflicts with fisher groups affected by the growth in marine mammal populations.

Implementation of all the programs covered in this chapter has been characterized by much conflict among various interests, users, government agencies, and levels of government. These conflicts have manifested themselves in coastal areas (sometimes incurring threats to public health and safety), in the courts, in the state capitols, in the agencies, and in the Congress. Such conflicts have often resulted in substantial delays and costs, at times in environmental damage, and at other times, in loss of potential economic and social opportunities. In each area there have been opportunities for examining the impacts of one use or activity on others, but each decision-making process has tended to give priority to the use or activity covered by its governing statute.

The U.S. did adopt a 200-mile exclusive economic zone and a territorial sea of 12 miles by presidential proclamations, but neither proclamation has been fully implemented. There has been no special exploration of the EEZ and no plan made for the mix of possible uses that could be encouraged in this resource-rich area roughly equal in size than the mainland United States. Similarly, the new 9 miles of U.S. territorial sea remain in federal hands. This area has not been incorporated into the formal coastal zones of the adjacent states and is not a part of their ongoing coastal planning and management programs.

While the Congress has played an important role in the oversight of each of these laws through periodic hearings, reauthorizations, and amendments, it has played this oversight role issue by issue and law by law, paying little, if any, attention to how well or how poorly the various parts, issues, and laws fit together. Interest groups, similarly, have tended to remain focused on particular issue areas, continually monitoring and mobilizing around their preferred marine policy cause. The last several years, however, have seen a great deal of interaction among interest groups around some issues (such as offshore oil), but most often in a competitive rather than a collaborative mode. In the last half of the 1990s, there has been more collaboration among various interest groups, at least on a bilateral basis, such as in the case of the coalitions formed on the marine mammal–fisheries issue.

Referring back to the recurrent tensions in U.S. ocean policy discussed in Chapter 1, the following situation appears to prevail at the end of the 1990s.

- Regarding its stance as a "coastal" or "maritime" state, the United States has increasingly emphasized its "coastal state" interests while the U.S. Navy has continued to focus on the "maritime state" interests. The Navy has been the main domestic advocate for ratification of the Law of the Sea Treaty in the U.S. Senate in order to protect the worldwide navigation interests of the United States.
- Regarding the tension of internationalism versus unilateralism, the United States appears to favor going it alone. While the international community has been extensively involved in the negotiation and implementation of a number of major new international agreements on oceans and coasts, the United States has often been a reluctant player in these fora. Its earlier position as a major leader in these international actions is now significantly eroded.
- As to the tension between federal and state control, there has been much discussion since the 1970s about federal devolution of authority to the states and to stakeholders. Ranging from the "New Federalism" rhetoric of the Reagan administration to current programs such as the National Estuary Program, which greatly emphasizes stakeholder involvement and partnerships, to fisheries where discussions of the virtues of community-level co-management of fisheries abound (see e.g., McCay 1997), the goals of increased state control and enhanced stakeholder and citizen participation are visible everywhere. However, this may be more rhetoric than reality. With regard to the implementation of ocean programs discussed in this chapter, little actual devolution of authority or of funds to the states has taken place. In fisheries, the ultimate decision-making authority remains with the secretary of commerce; in marine mammal management, although allowed by the MMPA, the federal government has not returned management authority for a marine mammal species to a state. In the OCSLAA, notwithstanding many comments by governors and coastal communities, the final decision-making authority for the conduct of the offshore oil program has remained squarely with the Department of Interior.

- With regard to the tension between favoring the environment or favoring development, these remain in opposition to each other on the U.S. domestic front, although there has been a fair amount of rhetoric and some efforts to bring these two perspectives together. Experiments with bringing together environment and development groups in several OCS regions (such as California and Alaska) have been at least partly successful. In fisheries management, although a rapprochement between environmental and fishing interests took place during the passage of the Sustainable Fisheries Act, it is not clear whether these coalitions will be maintained, especially as the environmental groups take independent steps to advocate fish conservation; for example, Seaweb's courting of important environmental reporters to highlight the plight of fisheries and use of public pressure to ban what it considers inappropriate practices, such as the harvesting and consumption of depleted swordfish.
- In terms of the tension between the roles of the government and those of the private sector in resource development, this time period definitely saw emphasis on the privatization of resource development in the United States. This was especially the case in fisheries management, where individual transferable quotas have become an important option and in the development of the aquaculture industry, a subject which we will address in the next chapter.

With this summary of the current state of oceans affairs, we next turn our attention to a review of major trends and challenges likely to affect national ocean policy in the next decade and to an analysis of the policy problems and opportunities that most likely will be posed.

FIVE
The Context of the Late 1990s
Challenges to U.S. Ocean Policy at the Dawn of the New Century

This chapter describes the changing context of the late 1990s and the way in which it is affecting U.S. ocean and coastal policy. Some changes are physical—the warming climate and the associated effects at the shoreline (accelerated sea-level rise and coastal erosion, increased storm frequency and, perhaps, intensity); some are social and demographic—the increasing flow of people and activities to coastal areas and the consequent changes in the coastal environment. Other changes are related to technology—the need for deeper navigation channels and harbors to accommodate larger, faster vessels and the need to respond to the challenges of new technology such as marine biotechnology; still others represent changes in public policy—a recognition that many environmental and resource problems are more effectively addressed by partnerships, both between levels of government and between the public and private sectors.

The Changed Context of the Late 1990s

More than ever, increasing numbers of people and interest groups are urgently concerned about the sharp declines in the abundance of fish stocks of high commercial and recreational interest. Groundfish off New England have sunk to levels one-third of what they had been only a decade ago. For example, landings of the region's "traditional" groundfish species (cod, haddock, and yellowtail flounder), fell from 72,000 tons in 1984 to 27,000 tons in 1993. The real value of these landings also decreased since 1984 by almost 50 percent, from $91 million to $46.8 million (NMFS 1996). At the same time, the growing numbers of red and brown tides coupled with the *Pfiesteria* scare of the summer of 1996 have sharply focused public attention on coastal water quality, especially in Mid-Atlantic estuaries bordered by intensive agricultural production (Garcia 1998).

The 1990s saw public interest in the oceans reach a new high. Globally, the decade saw two World Fairs dedicated to the sea—one in Genoa, Italy, in 1992 commemorating the 500th anniversary of Columbus's voyage to the new world, and another in Lisbon, Portugal, in 1998 celebrating the voyage of Vasco de Gama that had opened up the trade routes between Europe and the Far East 500 years earlier. Both fairs focused significant attention on ocean themes and involved, among other things, the construction of world-class aquaria. The 1998 Ocean Expo in Lisbon was only one event of the 1998 International Year of the Ocean designated by the United Nations General Assembly. It also was the year in which the Independent World Commission on the Oceans (IWCO 1998) issued its final report. The report of the commission's work (a three-year, multi-million-dollar effort headed by Dr. Mario Soares, former president of Portugal), *The Ocean, Our Future*, called for the establishment of a world ocean affairs observatory to independently monitor ocean governance activities, the convening of a periodic independent world ocean forum (emphasizing participation by NGOs), and the creation of an independent ocean guardian to take up grievances concerning international ocean agreements or misuse of the oceans and its resources (IWCO 1998).

Many nations, including the United States, had active Year of the Ocean (YOTO) programs that involved a wide range of events. In the United States, in late 1997 and early 1998, a dozen teams (largely from federal agencies) prepared twelve papers[1] describing important ocean and coastal issue areas and the programs now in existence to deal with them. The papers emphasized what was and was not working. This effort provided good background for a series of workshops organized by the Year of the Ocean steering group of the John H. Heinz III Center for Science, Economics, and the Environment in mid-1998. The report coming out of the Heinz Center work, *Our Ocean Future: Themes and Issues Concerning the Nation's Stake in the Oceans* (Heinz Center 1998), identified three obstacles that have blocked progress toward achieving greater benefits from the nation's ocean and coastal resources: (1) underinvestment in the physical and technological infrastructure necessary for efficient use of the oceans; (2) fragmentation in our national and international institutions and mechanisms for governing and managing ocean and coastal areas and resources; and (3) the devotion of insufficient effort to the development and application of the knowledge necessary for wise management (Heinz Center 1998). The U.S. Year of the Ocean effort also saw many local events related to marine affairs and ocean sciences. For example, the Consortium on Ocean Research and Education (CORE) sponsored a national "Ocean

Bowl" as part of the new National Ocean Partnership Program in which young people from around the nation tested their knowledge of the oceans and how they function (YOTO 1998k).

But from the perspective of national ocean policy, two developments in the United States dominated the 1998 Year of the Ocean. First was the passage of ocean policy legislation in both the Senate (in November 1997) and the House of Representatives (September 1998). Modeled after legislation that created the so-called Stratton Commission in the mid-1960s, this legislation authorized the appointment of a national Ocean Policy Commission to undertake a serious study of policy and to make recommendations for a coordinated, comprehensive, long-range national policy with respect to ocean and coastal activities. In the rush to adjournment in October 1998, the 105th Congress failed to act on a House-Senate compromise measure. It is expected that the legislation will be considered again by the 106th Congress, which took office in January 1999. In the second development, partially in response to congressional demands, a national ocean conference was held in Monterey, California, in June 1998, in which President Clinton, Vice President Gore, and several cabinet-level officials and congressional members participated. The president offered nine new initiatives, involving such issues as extending the moratorium on offshore drilling, building sustainable fisheries, constructing twenty-first-century seaports, protecting coral reefs, ratifying the 1982 Law of the Sea treaty, increasing ocean exploration, enhancing beach protection, addressing global warming, and increasing civilian access to military ocean data and technology (U.S. DOC et al. 1998). Although funding for the nine initiatives was valued at $224 million, it was not clear how much of this would come from new appropriations. The president also called on federal agencies to prepare a report on their ocean programs by June 1999, perhaps setting the stage for additional dialogue on national ocean policy.

Challenges to Be Faced in the New Century

It is expected that populations—both global and in the United States—will continue to concentrate in coastal areas. World megacities (cities larger than 8 million) that numbered twenty in 1990 will increase to thirty in 2010. Twenty of these thirty megacities will be coastal (Nicholls 1995). In the United States, coastal populations are expected to rise by 14 percent between 1990 and 2010, reaching a total of 127 million people in coastal areas (Culliton et al. 1990). By 2010, for example, two coastal states—California and Texas—will lead the nation in population, while

Florida's expected population of 16 million will rank fourth in the nation, up from tenth in 1960 (Culliton et al. 1990). Population density in coastal areas—in 1988 it was 341 persons per square mile, more than four times the U.S. average—is expected to increase more than 10 percent between 1988 and 2010 (Culliton et al. 1990).

Population pressures will most likely lead to increased user conflicts and competition for scarce ocean and coastal resources, result in loss of access to the oceans commons, and raise a variety of public health issues. Concomitantly, coastal ocean degradation in the form of declining water quality, coastal fisheries depletion, and destruction of important habitats is likely to continue. Controlling nonpoint (or land-based) sources of marine pollution—such as urban surface rain-caused runoff containing grease and oil, plastics, salt, and other substances; storm-water runoff; and agricultural runoff containing fertilizers, pesticides, and other chemicals used in farming practices—will pose one of the most significant challenges to decision makers since land-based sources such as these account for more than 75 percent of the pollutants entering the oceans (YOTO 1998, p. C-19). In addition, changes in the global climate, if they materialize, are expected to result in rising sea levels, increased damage by storms and floods, and changes in rainfall and freshwater flow to estuaries.

There is likely to be increased growth in coastal and marine tourism, as the growth in travel and tourism, the world's largest industry, continues to rise. In 1995, this industry employed 211.7 million people worldwide, produced 10.9 percent of world gross domestic product, invested $693.9 billion in new facilities/equipment, and contributed more than $637 billion to global tax revenues (World Travel and Tourism Council, nd). It is hoped that pressures for maintaining the health and attractiveness of coastal areas for tourism will provide the needed political will to support such programs as clean water protection and beach restoration and maintenance. Global trade, most of it by ship, will continue to grow in importance. The marine transportation industry will demand refurbished and modernized port facilities, including deeper channels, to accommodate the deeper draft of the larger and faster ships now coming on line. Given the decline of fisheries worldwide (in its 1996 report, the Food and Agricultural Organization of the United Nations concluded that 60 percent of commercial stocks are either overfished or fully harvested), the focus in this area will likely be on conservation and on rebuilding of stocks rather than on fisheries development.

Aquaculture currently accounts for about 25 percent of world food fish supplies, with China, India, Taiwan, and Thailand among the leaders in this field (YOTO 1998, p. C-28). Aquaculture is likely to grow as a substitute to wild fisheries, but ways will need to be found to avoid the environmental problems that have plagued aquaculture operations in some Asian and Latin American countries. Marine areas, too, will increasingly be used for the "bioprospecting" of novel marine organisms and marine organisms with unique properties (such as the heat-tolerant hyperthermophiles found in deep-ocean hydrothermal vents). Policy frameworks need to be developed to establish standards for allowing access to and exploitation of such resources, given the lack of any policy guidance at present. In offshore oil development, new challenges will be faced in at least two areas: (1) Environmentally sound methods of dismantling offshore oil platforms in depleted oil fields must be found. In the coming decade, approximately 4,000 platforms around the world will need decommissioning, including 1,000 in the Gulf of Mexico. (2) As industry develops oil resources in deeper and deeper areas of the Gulf of Mexico (Coy, McWilliams, and Rossant et al. 1997), new policy issues will arise related to marine safety, environmental impact, and relations with neighboring Mexico.

From an economic and political perspective, the globalization of the economy will continue and world economic and political interdependence will be even more apparent than it is today. Factors such as the emergence of regional economic blocs and growing international terrorism are likely to continue. The next several decades will see more, rather than fewer, demands for U.S. international leadership. Once the acknowledged leader in international ocean affairs, the United States now finds itself outside the ambit of some of the most important international agreements related to oceans ever concluded, particularly the Law of the Sea and the Biological Diversity Conventions, and will increasingly find it difficult to influence the course of actions decided by international bodies set up under these conventions. Regaining U.S. leadership in international ocean matters thus looms as one of the major ocean policy challenges in the next decade.

The main purpose of this chapter is to review the central challenges the United States is likely to face in the next decade in managing its ocean and coastal resources, to point to some of the major policy issues that will be raised, and to suggest some policy directions that U.S. decision makers should consider. We have elected to cover only the policy issues that we consider to be the most important. In this analysis, we have benefited from the insights garnered in the dozen papers on various aspects of national ocean policy prepared by the ocean-related federal agencies for the Year of the Ocean (the YOTO papers).

We first review the available data on the level of economic activity of various ocean sectors, noting the dearth of good comparative data on the contribution of ocean and coastal activities to the national economy. Next, we focus on the policy challenges in three interrelated issue areas: (1) navigation for defense and marine transportation; (2) ocean resource and space utilization; and (3) coastal management. For each, we highlight a set of policy priorities that should be considered by the Congress, the administration, and the coastal states.

The Overall Ocean and Coastal Wealth of the Nation: The Need for More Information

A first step toward more coherent ocean planning should be the analysis and assessment of the current ocean and coastal wealth of the nation—a detailed report of the level of economic activity generated by all major uses of the ocean and coastal zone, and an estimation of the other nonquantitative values—ecological, social, aesthetic, national security—of the oceans and coasts. This would be akin to efforts conducted by Pontecorvo and colleagues (Pontecorvo et al. 1980; Pontecorvo 1989) on the contribution of the ocean sector to the U.S. economy.

Similarly, for benefit-cost purposes, an estimate of the cost of managing the U.S. ocean and coastal zones—accounting for federal investments in major programs in fisheries and marine mammal management, oil and gas development, coastal management, beach management, erosion control, port infrastructure, and so on—is needed. Such an analysis would undoubtedly reveal underinvestments in some sectors and overinvestments in others. Unfortunately, this analysis would be nearly impossible to do at present because the data on these activities are scant and difficult to compare since all programs do not use the same measurement units and standards.

Table 5.1 summarizes the available data for the level of economic activity of major ocean and coastal uses, using the various estimates of economic activity found in the YOTO federal papers, which represent the best current estimates of the federal government. The notes to the table show that although all figures are intended to determine annual levels of economic activity, a variety of indices are used; thus, these data should be used with caution. In addition, the table does not address the ecological value of ocean and coastal areas—for example, their value as fish habitat, or the role of the oceans as a carbon sink in global warming. An effort to estimate such values globally may be found in Costanza et al. (1997).

TABLE 5.1 Level of Economic Activity of Various Ocean and Coastal Sectors.

SECTOR	ANNUAL LEVEL OF ECONOMIC ACTIVITY
Marine Transport	
Foreign trade	$590 billion[1]
Waterborne cargo	$78.6 billion[2]
Commercial Fisheries and Aquaculture	
Food harvested from the ocean	$38 billion[3]
Aquaculture	$0.8 billion[4]
Federal OCS Minerals	
Sales	$13–17 billion[5]
Federal revenue	$3 billion[6]
Tourism and Recreation	
Travel and tourism (total, all travel)	$746 billion[7]
Beach expenditures (seven states only)	$74 billion[8]
Recreational boating	$17.7 billion[9]
Saltwater fishing	$15 billion[10]

[1]YOTO 1998, p. A-4. In 1996, approximately $590 billion of goods were carried on the ocean and passed through U.S. ports.
[2]YOTO 1998, p. A-7. The demand for waterborne cargo initiates a chain of economic activity that contributes $78.6 billion to the GDP.
[3]YOTO 1998, p. E-3. Food harvested from the ocean generates approximately $38 billion in economic activity for the nation annually.
[4]YOTO 1998, p. C-28. The United States currently produces only about $800 million of the $33 billion annual value worldwide of aquaculture products, or less than 3 percent of the total.
[5]Federal Offshore Statistics, p. 72. Between 1986 and 1995, Federal OCS mineral sales would have averaged between $13 billion and $17 billion annually. Sales from state waters would add to this total. Total activity generated in the U.S. economy by OCS oil and gas production would be many times this range.
[6]Federal Offshore Statistics, p. 72. Revenue to federal government from offshore mineral leases has averaged about $3 billion annually from 1985 to 1995.
[7]YOTO 1998, p. F-2. In 1995, travel and tourism are estimated to have provided $746 billion to the U.S. gross domestic product (GDP), about 10 percent of U.S. output, making travel and tourism the second largest contributor to GDP just behind combined wholesale and retail trade.
[8]YOTO 1998, p. F-7. In seven states, beachgoers spent $74 billion with the most popular recreational activities being swimming, sunbathing, and walking in coastal areas.
[9]"Boating 1997" prepared by the Marketing Services Department of the National Marine Manufacturers Association. $19, 344,470,000 spent at retail during 1997 for new and used boats, motors and engines, accessories, safety equipment, fuel, insurance, docking, maintenance, launching, storage, repairs, and club memberships.
[10]YOTO 1998, p. F-6. Saltwater fishing generates expenditures of over $5 billion annually, a total economic output of $15 billion, total earnings (wages) of over $4 billion, and over 200,000 jobs.
Source: Prepared with the assistance of Rosemarie Hinkel.

The data suggest that a high level of economic activity, in particular, is associated with coastal tourism and recreation and with marine transportation and ports—two areas of policy that have not received much focused national ocean policy attention.

A new study announced by President Clinton at the June 1998 National Ocean Conference is intended to focus on the absence of good comparative data on ocean resources, uses, and activities. Under the oversight of the Department of Commerce, a group of policy analysts and economists is expected to undertake a comprehensive study of economic values represented by the U.S. ocean, with special attention on the EEZ (Kildow 1999).

Data are also difficult to obtain and compare on the level of federal investment in ocean and coastal programs. There is, at present, no federal budget "crosscut" on oceans and coasts, and it is difficult to construct one, given differences in the budget categories used by different agencies. In the late 1960s, the Marine Science Council performed a very useful role through its annual reports on federal agency expenditures on various ocean and coastal programs (Wenk 1998). The preparation of these reports ceased in 1971 when the council was terminated. At the National Ocean Conference, President Clinton requested that the federal agencies once again produce such a report.

Keeping in mind the difficulties of estimating the value of and federal investment in various ocean and coastal activities, we turn now to a discussion of current policy issues in three major clusters: (1) navigation for defense and marine transportation, (2) ocean resource and space utilization, and (3) coastal management.

Navigating the U.S. and World Waters

U.S. maritime security on the world's oceans depend on two conditions: freedom of navigation and safe domestic waterways. This section address them in turn.

The Changing Context of Maritime Security

Few events in the twentieth century have had as great an impact on American foreign policy as the end of the Cold War. The easing of tensions and the improvement in relations between the United States and the former Soviet Union have significantly changed the context of American national security. Naval power as an element of U.S. military strategy is also in flux. Naval forces have been reduced in size since the end of the Cold War, and the nature of potential conflicts that may involve U.S. forces is less clear. Although the probability of a "great power" war has markedly declined, the world has witnessed a proliferation of regional conflicts and the growing threat of international terrorism. The reductions of force size, coupled with the increasingly complicated political environment, have necessitated flexibility as the paramount characteristic of contemporary naval power. Not only must naval forces have the capacity for rapid mobilization, they must also be capable of conducting multiple missions (YOTO 1998, p. B-2).

Today's U.S. Navy must balance multiple objectives: ensuring the freedoms of navigation and overflight, power projection, deterrence, sea denial and operations other than war, information warfare, and intelligence, surveillance, and reconnaissance (YOTO 1998, pp. B-3–B-5). Rapid naval mobilization and flexibility are dependent on the freedoms of navigation and overflight. As noted by Department of Defense spokesman Bruce B. Davidson (Davidson 1995, p. 661), "To be effective, our armed forces must be where they are required and when they are required. Thus, they must be capable of moving within and between areas of operations in response to developing crises. . . . As forces move within and between areas of operations, they must pass through international straits . . . and may need to traverse the territorial sea of one or more coastal states . . ."

Power projection, or military presence on the high seas, is one method of maintaining international peace and stability, allowing U.S. foreign policy makers to retain the option to use naval power symbolically as a demonstration of diplomatic goodwill or as a means of deterrence. Offshore naval forces, because of their ability to arrive swiftly and remain on-site for extensive periods of time, serve as a powerful deterrent to prospective enemies (YOTO 1998, p. B-4).

Although the East versus West ideological confrontation of the Cold War has largely evaporated, a new set of tensions—"North" versus "South," related to the economic inequalities between developed and developing nations—has emerged. Concomitantly, there has been a resurgence of maritime interception operations in tense, but not openly hostile, situations. Besides "sea denial" missions, operations short of war—such as enforcing multilateral embargoes—have also been undertaken by coalition naval forces. In addition, naval forces take part in numerous small-scale contingencies, such as counterpiracy, drug interdiction, counterproliferation operations, migrant control and refugee operations, smuggling interdiction, peacekeeping operations, humanitarian emergency intervention, disaster relief, evacuation operations, and search-and-rescue mission support (YOTO 1998, pp. B-4–B-5).

Policy Challenges

The most salient marine policy priority for the navy is ensuring the freedoms of navigation and overflight. These freedoms are crucial on the high seas and particularly through straits. Therefore, the Law of the Sea and the comprehensive legal regime that it establishes regarding marine boundaries are of paramount importance to the United States (YOTO 1998, p. B-23; Galdorisi and Vienna 1997, pp. 147–148). As noted in a press statement on July 29, 1994, by Department of Defense Secretary Perry, "The Nation's security has depended upon our ability to conduct military operations over, under, and on the oceans. We support the Convention because it confirms traditional high seas freedoms of navigation and overflight, it details passage rights through international straits, and it reduces prospects for disagreements with coastal nations during operations" (quoted in Davidson 1995, p. 664). U.S. ratification of the Convention would also, according to the DOD, discourage the excessive maritime claims of other nations (Davidson 1995). For these reasons, U.S. ratification of the Convention is essential and long overdue.

While traditional ocean policy issues such as ensuring the freedom of navigation and power projection remain important priorities for the navy, it has also become engaged more fully in nontraditional issues and operations such as technology and information sharing and environmental protection. For example, the Navy, the Coast Guard, and the National Marine Fisheries Service have engaged in joint efforts to discover and curb illegal drift-net activity on the high seas. The navy has also adopted comprehensive measures to prevent injuries to and foster the rehabilitation of the endangered northern right whales, particularly in their critical habitats off the coasts of Georgia and Florida. The navy is also endeavoring to control and mitigate the effects of severe pollution events (for example, the navy is part of a team developing a Black Sea Regional Oil Spill Contingency Plan (YOTO 1998, pp. B-15–B-16). The navy has also set a goal of "Environmentally Sound Ships of the 21st Century" (ESS-21). These ships will generate only a minimum amount of new waste and have the capacity to treat or eliminate waste generated on board. Plastic-waste processors have been introduced on the majority of surface vessels in order to eliminate marine disposal of plastics. The navy has also taken steps to develop effluent minimization systems, as well as sewage and graywater treatment systems. Strict ballast water exchange procedures have also been instituted to preclude further introduction of exotic species into U.S. waters and elsewhere (pp. B-17–B-18).

Maritime Transportation and Port Infrastructure: Adapting to Technological Change in Environmentally Sustainable Ways

A vital element of the nation's economy, maritime transportation is frequently the most efficient means of transferring goods. However, the importance of marine transportation is often underestimated in comparison with air, rail, and highway transport. Economic activity generated by transportation of waterborne cargo contributes $78.6 billion to the nation's gross domestic product (GDP) and creates 15.9 million jobs nationwide (YOTO 1998, p. A-7). Waterway transportation has greater fuel efficiency than either land-based or air-based systems of transportation, and maritime transport of bulk goods is 35 percent less expensive than highway transport (p. A-3).

The United States is the leading trading nation in the world—almost one-fifth of international maritime trade is attributable to it (YOTO 1998, p. A-4). U.S. foreign trade, in 1995, was valued at over $600 billion (Heinz Center 1998). Marine transportation and port infrastructure are thus critical to the economic health of the nation. Maritime transportation and the associated port system, however, presently face several challenges.

Policy Challenges

1. *Faster and bigger ships will need enhanced port capacity.* Innovative technologies are enabling new ships to be larger, faster, and more efficient. Today's container vessels have a capacity of roughly 6,000 TEUs (20-foot-equivalent units). In light of ever-increasing trade and cargo volume, carriers have ordered vessels with capacities greater than 8,000 TEUs. It is expected that ships may have capacities of up to 13,000 TEUs by the year 2010 (YOTO 1998, p. A-8) and larger, faster, and more efficient ships will, of course, challenge port capacity. Even with the incorporation of technological developments such as state-of-the-art information systems that increase vessel traffic safety and cargo handling efficiency, ports suffer when ships cannot approach their terminals due to constraints imposed by navigational channel size. Many ports must engage in extensive dredging—more than has been customary to maintain normal port operations—to deepen channels and remain competitive (Brown 1998, p. 27).

The vice president of Sea-Land, a major U.S. and global liner and shipping company, for example, has warned that the Port of New York–New Jersey could lose 20 percent of its international cargo by 2015 unless its channels are deepened beyond 40 feet (Vulovic 1995).

Not all ports will be able to accommodate the larger ships, and yet ports are very independent and competitive entities at present. Because they may eventually need federal assistance to underwrite the changes needed to stay competitive (channel deepening, terminal modernization and expansion, etc.), there needs to be a national policy regarding the selection of "national ports" that are able to handle the new class of ships.

2. *Addressing the environmental effects of dredging.* Dredging is one of the key environmental issues associated with marine transportation. Perhaps the greatest controversy surrounding dredging pertains to the disposal of dredged materials, particularly when the sediments are contaminated. A variety of options exist for the disposal of dredge spoil, depending, in part, on the degree of contamination of the sediments. Ocean dumping remains the easiest and least expensive way to discard "clean" dredged material. Another option is aquatic contained disposal facilities such as containment islands (artificial islands), containment areas (facilities connected to land), and subaqueous pits (seafloor depressions). A third option, usually employed for contaminated materials, is upland disposal, which requires special measures to confine the dredged material. Substantial and deleterious delays in maintaining important navigational channels can and do occur when agreement cannot be reached on how to dispose of the material that will be dredged. In several notorious cases, delays of ten and twenty years were involved (Kagan 1994). It took seventeen years, for example, for the port of New York–New Jersey to receive the permits necessary to dredge a major channel (Brown 1999), and twenty years for channel deepening in Oakland Harbor (Kagan 1994). The problem was due, in part, to the number of different federal and state agencies involved in granting the permits necessary both for dredging and for the disposal of the dredged materials. If the dredging involves deepening existing channels or creating new channels, often some sort of mitigation is required. The federal Fish and Wildlife

Service standards for these activities may differ from the state fish and game department standards, and the federal EPA requirements can differ from those of state environmental agencies. The problem is made more difficult because few ports have long-term dredging plans with "pre-approved" disposal locations.

3. *Stemming the tide of exotic species introduction.* Another problem associated with increased vessel traffic is the introduction of exotic species. Nonindigenous plants and animals are often brought into new areas by means of a ship's ballast water (ballast tanks are customarily filled with local seawater to compensate for cargo that is off-loaded in order to maintain the ship's stability). Species introduced when this ballast water is discharged in a foreign port often have no natural predators in the new localities. Without predators, exotic species may outcompete and overwhelm indigenous populations (YOTO 1998, p. A-11). The well-known accidental introduction and subsequent proliferation of zebra mussels in the Great Lakes provides a striking example of the damage that can be caused by a highly invasive species introduced through ballast water.

   The National Research Council convened a committee under the auspices of the Marine Board to undertake a focused study of technologies for preventing and controlling the introduction of nonindigenous marine species by ships' ballast operations (NRC 1996). Recommendations included: (1) introduction of national voluntary guidelines to minimize risk of introductions until international standards now in negotiation are put in place; (2) sponsoring and encouraging further research and development for killing or removing organisms in ballast water; and (3) developing automated monitoring systems suitable for shipyard use (NRC 1996, pp. 89, 90).

   Marine transportation may also pose a threat to the environment in the form of biocides and other antifouling agents in marine paints used on vessels. While antifouling agents reduce drag and improve fuel efficiency, biocides used in these agents may harm nontarget species (YOTO 1998, p. A-12). Vessel groundings and spills represent yet another environmental concern raised by increased vessel traffic. Vessel groundings can result in physical damage to sensitive marine ecosystems such as coral reefs; furthermore, spills of oil or chemicals can seriously damage coastal habitats. Recovery time from events such as groundings or spills varies but can be very long (YOTO 1998, p. A-12).

The Heinz Center report *Our Ocean Future*, suggests that Congress create a demonstration program that provides planning funds to port regions. Port modernization partnerships and regional strategic planning would be supported by such funds (Heinz Center 1998, p. 24). The Year of the Ocean Papers also put forth a number of goals for the maritime transportation sector: an acknowledgment in legislation of the importance of maritime transportation; further development of federal ports and waterways; implementation of OPA 90 (Oil Pollution Act of 1990) pollution prevention and response initiatives; development of an inclusive management plan for dredged material; and, development of a "model port" concept (YOTO 1998, p. A-15).

Ocean Resource and Space Utilization Issues

This section addresses policy challenges in areas related to the utilization of ocean resources and of ocean space: fisheries, marine mammals, aquaculture, offshore oil and gas development, marine protected areas, marine biodiversity and marine biotechnology. In the past, these issues have been managed under separate decision processes governed by separate statutes. They are, however, intimately interrelated issues, as demonstrated by the many conflicts among and between various subsets of these activities.

Fisheries and Marine Mammals: Will the Legislative Fixes Work?

As discussed in Chapter 4, the Magnuson Act and the Marine Mammal Protection Act have been significantly amended in the last several years. The amendments to both acts are detailed and very complex and require extensive new work by the implementing agency.

Sustainable Fisheries Act Amendments The Sustainable Fisheries Act (SFA) reflects a precautionary approach to fisheries, an effort to apply domestically many of the principles that have been accepted internationally in the U.N. Agreement on Straddling and Highly Migratory Fishery Stocks and the FAO Code of Responsible Fishing (YOTO 1998, p. C-25). The SFA provides for (1) the establishment of guidelines to assist in the description and identification of "essential fish habitat" and impacts on that habitat, and to take measures that will further the conservation of that habitat; (2) to the extent practicable, the reduction of bycatch;

(3) stricter controls on the use of new fishing gear; (4) measures that will eliminate overfishing in domestic waters and identify management actions to rebuild those fisheries within ten years; and (5) the study and, if appropriate, implementation of a fishing capacity reduction program (p. C-25).

A major challenge in implementing the provisions requiring the description and identification of "essential fish habitat" and impacts on that habitat for each fishery will be to bring together NMFS efforts on this problem with at least two other relevant federal programs—the CZMA program under the National Ocean Service in NOAA and the National Estuary Program (in the NEP, in particular, much work has been done on causes of fisheries decline in estuaries). Cooperation from the Marine Sanctuaries Program will also be useful given the MPRSA experience with the establishment and operation of such measures as no-take zones for fisheries. Since much of the essential habitat for many fishes is found relatively close to shore and in state waters, the cooperation of state-level coastal management and fish management authorities will also be very important. Indeed, it would seem that the land and water use controls under existing state CZM programs represent one of the most effective ways of protecting essential fish habitat. A benefit from the ultimate incorporation of fish habitat management plans into state coastal management programs is that the "federal consistency" provisions will come into play, requiring that all federal activities be made consistent with such essential fish habitat plans.

Overall, the 1996 amendments, if fully implemented, may go a long way in turning around the fisheries crisis. The redefinition of OY in a more precautionary direction, in particular, and the requirements reducing bycatch, if effectively implemented and enforced, should help hold the line on further fisheries decline.

Policy Challenges

Some thorny problems remain and need attention:

1. *Reconsider the appropriateness of the regional council structure.* Although reasonable changes were made in the 1996 amendments to avoid further conflict-of-interest situations among council members, fundamental questions remain about the effectiveness of the council decision-making structure. When these regional arrangements were created in 1976, they were very innovative; they included participation by top fishery management authorities in each state, involved the relevant federal agencies

and the major stakeholder groups, and set up extensive scientific advisory mechanisms as well as industry advisory mechanisms. While recognizing that there are important regional differences in the performance of the councils (the Pacific and North Pacific councils are generally thought to have been most effective), overall the councils have been mired in controversy and have allowed the significant declines in fisheries to take place. Some argue that this is because the councils never had sufficient authority delegated to them (McCay 1997). Since the outset, the councils have not had power or authority over issues that profoundly affect fisheries, such as marine mammal management, management of habitats in estuaries and in state waters, and offshore oil development.

The regional fishery councils face an important challenge: Should they continue in their present form or be supplanted by some other form of regional, multiple-use, ocean management mechanism able to review, assess, and decide on the appropriate mix of uses, activities, and environmental protection efforts in particular ocean regions? This issue will need careful study and review, particularly of the features of the council system that have and have not worked well over the past twenty-three years. Some useful insights will likely be gleaned from an ongoing study of the U.S. fisheries management regime conducted by the Heinz Center, led by Professor Susan Hanna of Oregon State University (Hanna 1998).

2. *Review subsidies for impacted fishing communities.* Fisheries declines affect not only the fish but also the fishers. Many coastal fishing communities have been adversely impacted by the decline in fisheries, not only economically but socially as well. New England communities have received a $30 million emergency aid package from the New England Assistance Program to offset the effects of fisheries depletion (U.S. DOC, NMFS 1996). The United States is not alone in facing the economic and social impacts of decline in important fish stocks. In the late 1980s, Newfoundland saw its cod fisheries decline to the point that a total moratorium was put into place by the Canadian government. In the last decade, the Canadian federal government has attempted a wide variety of assistance programs, with very mixed results. To the extent that this experience is relevant to the New England case, we hope that it will be factored into U.S. local assistance programs for other affected fishing communities.

3. *Consider implications of privatization of fisheries.* For years, economists have been urging that market-based approaches be used in fisheries management as a way of increasing economic efficiency by reducing the cost of fishing operations. A number of countries, most notably New Zealand and Australia, have experimented with schemes involving individual transferable quotas (ITQs) or individual fishing quotas (IFQs) whereby individual fishers, based on their recent fishing history, are assigned percentages of the annual total allowable catch (TAC) in a given fishery. After the TAC is set, a fisher can then compute his or her permitted catch (by applying the percentage to the TAC) and then can take this fish in the most economically efficient manner. Fishers are free to sell their quotas (or portion thereof) at a market price that will fluctuate depending on the value attached to the percentage shares.

The U.S. is now employing ITQs in three fisheries—the Atlantic surf clam (quahog) fishery, the wreckfish fishery in the South Atlantic, and the Alaskan sablefish and halibut fishery (McCay et al. 1995). While it may be too soon to undertake a definitive assessment of the ITQ approach, some early indications suggest that problems may exist. Concentration tends to develop as large concerns buy out individual fishermen and aggregate large percentages of the TAC in a few hands. It has been reported that in New Zealand most of the quota in at least one fishery is now owned by a processing firm that, in turn, now employs the individual fishers and their boats to go out and catch the fish on its behalf (Donaghue forthcoming). Also, there are questions about the equity of "giving" some fishers a share of a public resource —in this case fish—at no cost when other users who want to use publicly owned ocean resources—such as oil and gas—have to pay for them.

The 1996 Sustainable Fisheries Act called for a National Research Council study on this issue and prohibited the further use of ITQs until October 1, 2000. The NRC report, *Sharing the Fish: Toward a National Policy on IFQs*, released in December 1998, urged that a "one-size-fits-all" approach to this issue not be taken, given the great variation that exists in individual fisheries, fishing communities, and fishing regions, and "to refrain from endorsing rigid blueprints at the expense of hard-won measures, carefully crafted to address unique local biologic and social conditions (p. 2)." The report highlighted a number of advantages of and concerns about the IFQ approach. Among the advantages, IFQs have been effective in dealing with overcapitalization; consumers have been able to purchase fresh fish during longer periods of the year; IFQs provide the opportunity to utilize better fishing and handling methods, reducing bycatch of nontargeted species and maintaining higher product quality; and gear conflicts may also be reduced by IFQs. Among the concerns expressed about IFQs were the fairness of the initial allocations; the effects of IFQs on processors; the increased costs for new fishers to gain entry; the consolidation of share quotas (and thus economic power); the effects of leasing; the confusion about the nature of the privilege involved; the elimination of vessels and reductions in crew; and the equity of gifting a public trust resource (NRC 1998). The committee concluded that IFQs should be allowed as an option in fisheries management "if a regional council finds them to be warranted by conditions within a particular fishery and appropriate measures are imposed to avoid potential adverse effects (p. 5)" and that, therefore, Congress should lift the moratorium on the development and implementation of IFQ programs. Among other, more detailed recommendations, the committee urged that the Magnuson-Stevens Act be amended to allow the public to capture some of the windfall gains and extract some of the fishery profits (rents) that are generated.

4. *Address overcapitalization.* Overcapitalization is a major problem in fisheries worldwide—in general, there are too many fishing boats for the amount of fish that should be caught. The FAO has estimated that there is 30 percent more capacity worldwide than is needed to take the present world catch (Heinz Center 1998). This situation is almost certainly the result of the fact that most of the world's fisheries are still open access. Overcapitalization not only causes economic inefficiencies but results in pressure to increase allowable catches above biologically safe levels. Clearly, government assistance programs to help coastal communities deal with drastically lower fish catches (lower because of reduced abundance or lower because of regulations imposed to allow stocks to recover) should have the effect of reducing the fishing effort going into particular fisheries. A number of possible negative consequences of the ITQ scheme were mentioned earlier. One of the positive aspects of ITQs is that they will tend to reduce capitalization in a given fishery as quota owners strive for increased efficiency.

The Marine Mammal Protection Act Amendments of 1994 The 1994 MMPA Amendments incorporated significant changes to many provisions of the act, including substantial revisions to the procedures governing incidental take of marine mammals during commercial fishing; establishment of take reduction plans by teams created for this purpose; consideration of effects on ecosystems of marine mammal management; reduction of fisheries service jurisdiction over the care and maintenance of captive marine mammals held for public display; provision for general authorization for noninjurious scientific research on marine mammals; and creation of a new permit category for photographing marine mammals (U.S. DOC, NMFS 1994b, p. 1).

The amendments required NMFS to undertake draft stock assessment reports for every population or stock of marine mammals in U.S. waters. Completed in July 1995, the assessments included information about each stock's range, an estimate of its minimum population and population growth rate, an estimate of its human-induced mortalities, a description of the commercial fisheries that have potential interaction with the stock, and an estimate of the stock's level of potential biological removal (PBR). PBR is defined as the maximum number of animals that may be removed from a population without affecting its ability to reach or maintain its optimum sustainable population (U.S. DOC, NMFS 1997, p. 1).

Under the new amendments, NMFS has categorized each fishery in the United States according to whether the likelihood of incidental mortality and serious injury to marine mammals is frequent (Category 1), occasional (Category 2), or remote (Category 3). NMFS also created take reduction teams (TRTs) to establish take reduction plans for Categories 1 and 2 fisheries. The primary aim of take reduction plans is to reduce the incidental take of marine mammals to insignificant levels approaching zero within five years of implementation (U.S. DOC, NMFS 1994b, p. 2).

The 1994 amendments also reasserted the zero mortality rate goal (ZMRG) of the 1972 act. The ZMRG, reducing incidental death and serious injury of marine mammals taken in the course of commercial fishing operations to insignificant levels approaching zero, was the original goal of the Marine Mammal Protection Act and is to be accomplished by May 2001 (U.S. DOC, NMFS 1994b, pp. 2 and Insert A).

As with the Sustainable Fisheries Act, the MMPA Amendments of 1994 raise questions about the potential implementation success and feasibility of the new provisions. The procedures governing the creation of take reduction plans by take reduction teams are very complex, and have placed extensive new burdens upon groups and agencies already struggling to meet preexisting mandates and tasks. As of June 1997, five take reduction teams had been established. Designing and agreeing on a take reduction plan is no easy task, and it remains to be seen whether the take reduction teams will be able to implement and enforce the new regulations that they develop. Furthermore, the ZMRG is perhaps more ambitious than it is practical. What exactly constitutes "insignificant levels approaching zero"? Suppose a fishery has an incidental take of marine mammals that does not qualify as an "insignificant level approaching zero," but does not present a danger to the particular population. If reducing its incidental take of marine mammals proves detrimental to the fishery, which would take precedence—the Marine Mammal Protection Act, or the Sustainable Fisheries Act?

Policy Challenges

We foresee three major challenges in the management of marine mammals in the years ahead:

1. *Full implementation of the 1994 MMPA Amendments.* As noted earlier, to complete well-crafted and scientifically based take reduction plans that are implementable, enforceable, and effective will be difficult to accomplish.
2. *Continue and complete the job of protecting marine mammals that remain at risk.* As noted in Chapter 4, a number of marine mammals remain at peril, including the West Indian manatee, the southern sea otter, the Hawaiian and Caribbean monk seals, the Guadalupe fur seal, the Steller sea lion, and a number of whale populations (northern right whale, bowhead whale, humpback whale, blue whale, finback whale, sei whale, and sperm whale and vaquitas). Plans for their recovery still need to be fully implemented.
3. *Develop a regime for management of marine mammals that are healthy and exceed their levels of optimum sustainable population.* For such populations such as West Coast seals and sea lions, the United States, as a society, will need to decide whether it is intent on managing marine mammals as other parts of marine ecosystems are being managed, or on keeping them wholly untouched as some type of national icon. As William Aron, former director of the National Marine Fisheries Service's Northwest and Alaska Fisheries Center put it in a 1988 article (which he wrote as a "private citizen and not as a government official," referring to

Americans' concerns with marine mammal protection), "If the concern is based largely on the need to correct abuses of the past and restore marine mammal populations to former abundance levels, continuing the current sweeping policy of virtually total protection for all species is no longer required. . . . If the issue is simply restoration, we must recognize that protection has worked and that many species and stocks are at high levels of abundance" (p. 107). Aron went on to suggest several kinds of action, such as ecosystem management of all marine species, employing a precautionary approach, and harvesting marine mammals by the most humane methods possible. Further, he advocated that an independent and scholarly entity such as the National Academy of Sciences document the current status of marine mammal populations and identify populations or species of continuing concern where rigid protection should be enforced, as well as those for which such protection is no longer required.

If, however, Aron notes, "the concern for marine mammals is based on the ethics or morality of their harvest, we probably should continue our present course." In so doing, however, "we should recognize that there is a difference between imposing a moral or ethical standard on U.S. citizens and imposing such standards on the international community" (1988, p. 107).

Can the Promise of Aquaculture Ever Be Realized?

The aquaculture industry has experienced swift growth throughout the world in recent years. A record for total aquaculture production was attained in 1994 with an output of 25 million metric tons (Heinz Center 1998). One-quarter of global food fish is provided by aquaculture; China, India, Taiwan, and Thailand are the leading contributors of aquacultured food fish (YOTO 1998, p. C-28). In the United States, the demand for seafood has steadily increased while harvests from capture fisheries have steadily decreased. Thus, the potential for aquaculture development in the United States is vast (NRC 1992c, p. 1). However, the United States presently lags far behind other nations in aquaculture output, producing approximately $800 million of the $33 billion total annual worldwide value of aquaculture products, less than 3 percent of worldwide aquaculture production (YOTO 1998, p. C-28). Moreover, to the degree that it has developed in the United States, the aquaculture industry has been largely dominated by farming of freshwater species; three-quarters of U.S. aquaculture output is comprised of freshwater organisms, especially catfish, crayfish, and rainbow trout (NRC 1992c, p. 23).

Constraints on Marine Aquaculture Development in the U.S.[2] The development of the marine aquaculture industry in the United States has been hampered by a number of factors. Marine aquaculture, an essentially agricultural enterprise, takes place in the special, essentially public realm of the coastal zone and ocean. Thus, it presents a number of concerns not normally raised in the conduct of agriculture or freshwater aquaculture (which generally occur on well-defined private lands)—such as the privatization of public waters and the conflicts with other uses of the ocean and coastal zone (e.g., navigation, commercial fishing, recreational fishing, municipal waste disposal, oil development, coastal residences, boating, and recreation). In addition, the marine aquaculture industry also faces problems in the siting of facilities (scarcity in the availability of suitable coastal sites) and the cumbersome permitting procedures from a myriad of federal, state, and local agencies.

Other factors that have hindered the development of the industry revolve around the potential environmental impacts of marine aquaculture operations. If not properly sited and continually monitored, marine aquaculture operations can cause serious environmental problems, such as contamination from waste effluents from the facilities; effects on nontarget species of the use of drugs to prevent disease in the fish; and possible genetic pollution from the interbreeding of the genetically altered marine aquaculture stocks with wild stocks of fish if escapement occurs. The industry also faces problems in the transportation and movement of cultured species, feeds, and products (such as problems related to the Lacey Act, which prohibits domestic and international trafficking in protected fish, wildlife, and plants interstate).

Politically, marine aquaculture is still an embryonic industry; hence it is weakly organized (although it is becoming better organized) and has relatively less clout that other, more traditional ocean interests (such as fishing). Concomitantly, there tends to be weak public recognition and support of the industry.

At the federal level, there has been little tangible encouragement to the industry in the form of incentives and the like, as the federal government has left the promotion and development of the industry to free market forces (Brennan 1997). This is in contrast to the experience in other nations with successful marine aquaculture industries, such as Norway and the United Kingdom, where the growth of these industries was given impetus through government research, development, and incentives.

Federal legislation on marine aquaculture has been primarily hortatory in character (e.g., encourage research and coordination). Thus, marine aquaculture, in contrast to most other uses of the ocean and coastal zone, lacks a solid base in the federal ocean regulatory framework. Moreover, federal legislation has involved low levels of funding and has generally worked to maintain the diffusion of responsibilities for aquaculture among various federal agencies, an arrangement that, as noted by Tiddens (1990), tends to create neglect in a bureaucratic system. The 1980 act declared that the development of the industry was in the national interest, required the secretaries of agriculture, commerce and interior to develop a National Aquaculture Development Plan, and established the Joint Subcommittee on Aquaculture (JSA) in the Federal Coordinating Council on Science, Engineering, and Technology to coordinate the work of twelve federal agencies involved with aquaculture research, promotion, and regulation. The chairmanship of the JSA was to rotate among the secretaries of agriculture, commerce, and interior. The act authorized a total funding level of $17 million in fiscal year 1981 (projected to grow to $29 million in 1993), but given the fiscal constraints characteristic of this time period, no funds were ever appropriated (NRC 1992c).

The National Aquaculture Development Plan of 1984, prepared by the JSA in response to the National Aquaculture Act, noted that crippling impediments still existed that prevented the growth of the industry. Subsequently, the 1985 National Aquaculture Improvement Act reauthorized the 1980 act and enacted two major amendments: (1) The USDA was designated as the lead federal agency with respect to the coordination and dissemination of national aquaculture information, and (2) two new studies were commissioned, on the effects of marine aquaculture on commercial fisheries and on the potential effects of the introduction of exotic species as a result of marine aquaculture operations. Funding was authorized for $1 million for a period of three years but was not appropriated (NRC 1992c). Notwithstanding these fiscal problems and lack of congressional support, the federal agencies most involved with marine aquaculture—the USDA, NOAA, FWS, and the National Science Foundation have continued to play active roles in aquaculture research and information efforts (NRC 1992c). New legislation enacted in 1998, the National Aquaculture Policy, Planning, and Development Act, extends the jurisdiction of the USDA and defines "private aquaculture as a form of agriculture" (P.L. pp. 105–185).

Notwithstanding problems at the federal level, it is actually at the state level that the majority of laws and regulations that specifically authorize, permit, or control aquaculture operations are actually found. Significant differences exist among the states on how they organize themselves for marine aquaculture (some states place the activity within departments of marine resources and others place it with agriculture departments). Some states, such as Hawaii, for example, have made significant investments in marine aquaculture, by creating a state aquaculture plan, designating a lead aquaculture agency, and creating pre-permitted marine aquaculture parks (De Voe and Mount 1989).

Policy Challenges

Marine aquaculture is an industry with great potential that remains relatively undeveloped in the United States. What steps are needed to move this industry forward? The 1992 National Research Council report *Marine Aquaculture: Opportunities for Growth,* puts forward several recommendations for settling the policy framework for marine aquaculture. Building on these and reflecting on intervening events, we offer the following recommendations.

1. *Incorporate marine aquaculture as a recognized use of the coastal zone under the Coastal Zone Management Act.* This would be a vehicle for giving marine aquaculture, a water-dependent industry, a basis of support and the prospect of being included in coastal zone management plans with their legal power of federal consistency. Some of the states, such as Massachusetts, have already developed aquaculture plans within the ambit of their state coastal zone management plans (Massachusetts CZM 1995).
2. *Develop a policy framework for leasing and management of marine aquaculture in offshore federal waters, where there is currently no well-delineated framework for management and control of such operations.* Marine aquaculture operations are taking place offshore in a number of countries around the world (especially Ireland, the United Kingdom, and Norway). Such an operation has been proposed by the American Norwegian Fish Farm, for a 47-square-mile offshore salmon farm located 37 miles off Cape Ann in Massachusetts—the project would require the exclusive use of an area approximately 50 square nautical miles in the EEZ (Rieser 1997). The proposal was challenged by litigation brought by the Conservation Law Foundation of New England, and is still awaiting approval.

In debates over the characteristics of a new offshore policy framework for aquaculture, consideration should also be given to how to ensure maximum benefits to the U.S. marine aquaculture industry, rather than to foreign companies (who have more experience and are attracted to the U.S. EEZ zone) for the use of public waters. One alternative would be to "Americanize" the EEZ for marine aquaculture purposes, much as the 200-mile zone was Americanized for fishing purposes. In the case of the marine aquaculture industry, this goal could be accomplished through a variety of means; for example, by requiring that marine aquaculture firms operating offshore be at least 51 percent domestically owned (NRC 1992c, p. 87).[3]

3. *Resolve competition between federal agencies, especially the Departments of Commerce and Agriculture, over who shall be the lead for aquaculture development.* While the Departments of Commerce, Agriculture, and Interior and other affected federal agencies have worked cooperatively for years under the aegis of the JSA, recent congressional efforts to designate the USDA as the aquaculture lead have led to significant competition among the agencies. NOAA, in particular, has objected to this proposal and has created an internal aquaculture task force to plan for a more effective future industry. There are two sides to the "lead agency" question. The USDA definitely has knowledge and experience regarding farming operations and has an extensive system of farm agents that have worked with farmers over the years in the development of the agriculture industry. NOAA, however, has the experience and legal responsibilities regarding marine resources, better understands environmental impacts that may ensue, and knows how to operate in a situation where many other users have legal rights to use offshore waters.

4. *In partnership with the states, provide incentives and support to the most promising elements of the industry, and work to streamline the permitting process for marine aquaculture.* Government research, development, and incentives have helped the marine aquaculture industry develop in other countries. Similar efforts are needed in the United States, if a domestic industry is to take hold and compete on an international level.

   Approaches to streamlining the permit process for marine aquaculture operations (typically cited by industry representatives as a major impediment to the development of the industry) might include such measures as promoting the inclusion of marine aquaculture in state planning; developing model guidelines for permitting; promoting joint intergovernmental review of projects; promoting the development of state plans for marine aquaculture and the development of marine aquaculture parks (following the Hawaii example). The federal government can also bring the states together to share common problems and approaches, and to exchange technical information.

Offshore Oil and Gas: Achieving Predictable and Orderly Development and Benefiting Affected Communities

In the past several years, the Minerals Management Service, in cooperation with its OCS Policy Committee and Scientific Committee, has made efforts to set a new direction for the OCS program. In 1993, the OCS Policy Committee prepared a report, *Moving Beyond Conflict to Consensus*, which stated that the "OCS program in the past several years has been regressing rather than progressing and it is now at a crossroads" (U.S. DOI, MMS 1993, p. 49). The committee recommended a number of major changes in the OCS program: The enactment of impact assistance and revenue sharing; the establishment of regional task forces representing all OCS program stakeholders to focus on attaining consensus on OCS leasing decisions to obviate the need for moratoria; resolving issues associated with lease cancellation and buybacks; and provision of incentives to industry, especially royalty relief for OCS development in deep water (which is significantly more expensive). A study on environmental studies in OCS areas under moratoria (conducted jointly by the OCS Policy Committee and the OCS Scientific Committee) concluded that the MMS should be conducting environmental studies (including studies of natural gas resources, natural resources, socioeconomic factors, etc.) in moratoria areas in order to be prepared, as it were, in the event that these areas are opened up in the future (U.S. DOI MMS 1997).

Unfortunately, these efforts at consensus building and conflict avoidance were cut short by the extension of the OCS moratoria announced by President Clinton at the National Ocean Conference in Monterey, California, in June 1998. In what was perceived by some as a partisan political move (designed to attract California votes and contributions in an election year), the president announced a ten-year continuation and expansion of the OCS moratoria. President Clinton withdrew from oil and gas leasing through June 30, 2012, all the areas already under the previous moratorium established by President Bush, plus parts of the eastern Gulf of Mexico (15 miles off Alabama and more than 100 miles off Florida) (ENN 1998). The president also withdrew indefinitely all areas of the OCS currently designated as National Marine Sanctuaries (going a lot further than President Bush, who had not withdrawn any areas in sanctuaries from OCS leasing). The size of the areas under these moratoria exceed that previously established by President Bush (Bush's withdrawal applied to a smaller portion of the eastern Gulf of Mexico [about 20 million acres withdrawn compared to 70 million acres in President Clinton's directive], and did not apply to the North Aleutian Basin in Alaska or to the Mid- and South Atlantic Planning Areas [MMS 1998]).

The reaction of industry to the moratoria announcement was swift. In a letter to the organizers of the Monterey Conference, Paul L. Kelly, President of Rowan Industries (an offshore oil support company) and former chair of the OCS Policy Committee, noted that there was nothing in the background studies leading up to the event (e.g., the YOTO papers) that would justify or substantiate the president's action, and stated that the president was sacrificing long-term economic and strategic interests of the United States in favor of political short-term interests to help reelect Senator Barbara Boxer and other California congressional members, and that offshore drilling was not the culprit when it comes to pollution of the seas (Kelly 1998b). Clinton's announcement may well cloud future prospects for cooperation between the industry and other ocean and coastal interests. The offshore oil industry had already expressed some reservations about the projected National Ocean Policy Commission in March 1998 testimony on the Oceans Act of 1998, claiming that the past such commission, the Stratton Commission, after all, had resulted in the passage of the Coastal Zone Management Act, which had been detrimental to the industry (Kelly 1998a).

Policy Challenges

In our view, OCS moratoria are not the appropriate way to run the nation's OCS program. A more balanced and lasting solution to the OCS moratoria approach should entail: (1) amendment of the OCSLAA to incorporate a greater role for the coastal states, (2) revenue sharing and impact assistance for the coastal states, and (3) some kind of regional mechanism for making decisions about OCS development. The political forces may not, however, be well aligned for OCSLAA reform—the environmental interest groups have generally been satisfied with the moratoria and don't want to see new offshore development. The industry, busy with precedent-setting new developments in deepwater areas in the Gulf of Mexico and with joint ventures abroad, may not be receptive to opening up a possible Pandora's box of OCSLAA reform.

Great advances in technology have made deepwater drilling an option for many oil companies. "Geosteering" drilling technology has dramatically reduced exploration costs for companies such as Exxon, which reports an 85 percent decline in exploration costs over the last decade. Improvements in floating rigs have greatly contributed to the accessibility of deep waters. Computer-controlled thrusters now allow drilling ships and floating rigs to remain in place regardless of inclement weather. Such technological advances have led to lower costs, and companies are now exploring development of smaller fields (less than 80 million barrels) that were once considered unprofitable. The British-Borneo Petroleum Syndicate, for example, reports that fields with as few as 30 million barrels can be profitably drained by the company. Furthermore, tests with air-injection techniques have doubled the amount of oil extracted from a particular field. Tension leg platforms (TLPs), floating platforms "tethered" to the seafloor, are making production possible at depths exceeding 5,000 feet. These innovative technologies and their concomitant lower costs have made deepwater drilling much more attractive to industry (Coy, McWilliams, and Rossant 1997).

1. *Amendment to the OCSLAA to give coastal states a greater voice in OCS decisions.* When the OCSLAA was enacted in 1978, the Congress added Section 19 in an attempt to create a formal role in the leasing process for the governors in coastal states. The section reads, in part:

> The Secretary shall accept recommendations of the Governor and may accept recommendations of the executive of the opportunity for consultation, that they provided for a reasonable balance between the national interest and the well-being of the citizens of the affected State. For purposes of this subsection, a determination of the national interest shall be based on the desirability of obtaining oil and gas supplies in a balanced manner and on the findings, purposes, and policies of this subchapter. (U.S.C. & 1345[c])

As discussed in Chapter 4, when Secretary Watt largely ignored the coastal state concerns, it appeared that Section 19 was never implemented as intended by Congress, since the secretary of interior ultimately made all the decisions. Van de Kamp and Saurenman (1990) have proposed a reasonable amendment of Section 19: The Department of the Interior must defer to the recommendations of coastal state governors, and Interior may reject these recommendations only if the secretary makes a factual determination that they are arbitrary and capricious. The secretary's decision to reject a governor's recommendation would also need to be supported by substantial evidence in the record. This approach would give coastal governors a much greater voice in leasing decisions while still allowing the Department of the Interior leeway to reject unreasonable recommendations.

2. *Enact revenue sharing and coastal impact assistance.* The only existing sharing of OCS oil and gas revenues with the coastal states comes from the 1985 settlement involving Section 8 (g) (of the OCSLAA) funds that now provide each state with 27 percent of the OCS revenues derived from tracts immediately adjacent to the state's submerged lands (generally the 3- to 6-mile zone). But there is no program for revenue sharing for tracts beyond this area where most of the federal offshore activities are located (CIAWG 1997). As noted by Kitsos (1994b), bills on OCS revenue sharing were passed in the House in 1982, 1983, and 1984, but were opposed by the administration and defeated in the Senate. In 1997, however, a subgroup of the OCS Policy Committee prepared a report on the revenue-sharing issue, and recommended that 27 percent of new OCS revenues be shared with the coastal states and affected coastal counties. Eligibility to receive revenues would be extended to all coastal states. Coastal counties and local governments identified as affected by OCS activity would also be eligible to receive funds and would receive payments directly instead of having them passed through the state. A formula giving weighted consideration to OCS production (50 percent), shoreline miles (25 percent), and population (25 percent) would be utilized to determine the amount of funds for which each state and territory is eligible (OCS Policy Committee 1997). Such a revenue-sharing proposal is currently being considered by the Congress: S.2256, introduced by Senator Landrieu on October 7, 1998, would

establish in the Treasury an Outer Continental Shelf Impact Assistance Fund in which the secretary would deposit 27 percent of the revenues from each leased tract or portion of a leased tract. It also includes formulas to determine payment to the states.

3. *Develop and implement a regional mechanism for making OCS decisions.* The great diversity of conditions facing OCS development in various regions of the country calls for regional approaches to deciding the conduct of OCS operations and mitigation measures that may be appropriate. The Minerals Management Service has had success with convening regional task forces in Alaska and in California to address outstanding policy issues. These regional task forces, however, are the product of administrative initiative, which can be wholly overturned by a subsequent administration. Therefore, such a regional approach, in our view, should become embodied in legislation (in amendments to the OCSLAA) or as part of an omnibus oceans act. The regional approach could be devoted only to OCS issues or, alternatively, be part of a wider regional multiple-use decision-making mechanism.

Marine Protected Areas, Biodiversity, and Biotechnology

As noted by the 1998 YOTO report on ocean living resources (pp. C-31–32), as part of a larger integrated area management scheme, marine protected areas can provide one of the most effective mechanisms for conserving marine living resources and the habitats on which they depend. They can:

- Be a management tool—a refugia, in effect—for protection of areas that are repositories of especially rich marine biodiversity.
- Protect unique and/or ecologically significant resources.
- Provide a living laboratory against which to test the effectiveness of management measures.
- Provide potential future benefits from marine biotechnology development.

As mentioned earlier, the Marine Sanctuaries Program and the National Estuarine Research Reserves Program have grown to a network of fourteen marine sanctuaries and twenty-two NERRS sites. These two protected-area programs, however, have not developed specific targets and timetables on their ultimate size and rate of growth.

Policy Challenges

1. *The policy challenges that lie ahead in this area are primarily to develop a vision for the future of these programs and to consider what proportion and specific areas of the EEZ will need special protection because of the presence of special ecological, aesthetic, cultural, or recreational assets, or for the protection of biodiversity, and/or for the orderly exploitation of marine biotechnology.*

There has been growing global concern with the protection of the world's biodiversity in both terrestrial and marine areas. As noted by Thorne-Miller (1999), ". . . it is certain that the loss of biodiversity will be acutely felt by all, as the failure of biologically impoverished ecosystems to adapt to further changes in the global environment results in the loss of one life support system after another and the inability of the natural world to meet the needs of humans and other species that remain. As E. O. Wilson . . . says in his masterpiece of insight, *Biophilia,* 'This is the folly our descendants are least likely to forgive us' . . ." (p. 169). The Convention on Biological Diversity (CBD), adopted at the Earth Summit in Rio de Janeiro in June 1992 and entering into force in December 1993, is aimed at the conservation of biological diversity, the sustainable use of its components, and the equitable sharing of the benefits arising from the utilization of genetic resources. One of its major provisions is the requirement that nations develop national strategies, plans, or programs for the conservation and sustainable use of biodiversity and to integrate that conservation and use into other policies and programs.

The United States has signed but not yet ratified the CBD. But, to our knowledge, it has done little to develop a national plan for marine biodiversity mapping and conservation. This, in our view, is an important policy challenge in the years ahead: to determine what areas of the U.S. EEZ warrant special attention and protection, most likely through the Marine Sanctuaries Program, to maintain the ocean's biodiversity.

Marine biotechnology—the use of living organisms (or parts of organisms) to make or modify products, to improve plants or animals, or to develop microorganisms for specific purposes—is expected to grow into a multibillion-dollar industry in the next century, exploiting the potential biotechnology applications that may be derived from various types of marine organisms. These include human health and biomedical applications, seafood supply enhancement, environmental remediation, and marine resource management and monitoring (Cicin-Sain et al. 1996). Internationally, the regime for governing access to marine resources and organisms under the jurisdiction of coastal nations for marine biotechnology purposes (both for samples and experimental research and for harvesting and production purposes) is in the process of redefinition. In a nutshell, access to marine zones for this purpose was governed, if at all, under the marine scientific research provisions of the Law of the Sea Convention. Now, access to genetic resources in ocean areas under national jurisdiction must also take into account the requirements of the 1992 Convention on Biological Diversity, which, as noted earlier, calls for the fair and equitable sharing of the benefits arising out of the utilization of genetic resources (Cicin-Sain and Knecht 1993). A related policy question is, Which regime should govern access to the unique marine organisms found in and around seafloor vents associated with mid-ocean spreading centers (Knecht, Cicin-Sain, and Jang in preparation).

2. *There is a need for the United States to develop an appropriate regime for the exploitation of marine resources for marine biotechnology purposes in its own EEZ as well as for it to take a lead role in clarifying the international regime on this issue.* As noted by Roger McManus, president of the Center for Marine Conservation, in his 1998 testimony on the Oceans Act,

> . . . we must lead by example and establish national legislation and standards for utilizing marine wildlife important for commercial ventures such as the pharmaceutical industry. Each day seems to bring to light new and important medicinal uses of microorganisms and rare plant and animal species, even as these species are rapidly disappearing. For example, hyperthermophilic bacteria found in temperatures of up to 104 degrees Celsius in the Gulf of California have aided scientists in the process of DNA amplification, fueling a growing biotechnology industry that already reaps millions of dollars in profits. Deep sea hyperthermophiles are an untapped resource and their unchecked and unregulated exploration and harvesting are currently taking place at unprecedented rates. Blood from horseshoe crabs helps hos-

> pitals test for toxins in pharmaceutical products that cause septic shock, which accounted for one-fifth of all hospital deaths. Unless we develop safety standards for the use and protection of these resources, we risk losing them and their vital pharmaceutical and economic benefits forever. (McManus 1998, p.12)

A Regional Approach to Ocean Resources and Space Utilization?

Is a regional mechanism needed to bring state and federal interests together to manage interrelated ocean resources and space? Iudicello, Burns, and Oliver (1996), for example, refer to an emerging living resource management regime related to marine mammal and fisheries management that grew out of the coalitions formed to negotiate the 1994 MMPA Amendments and the 1996 passage of the Sustainable Fisheries Act. Fisheries and marine mammals, though, interact frequently with the other spatial and resource uses of the ocean—with the offshore oil and gas industry, with aquaculture operations, with marine protected area management. On the part of the OCS regime, too, there has been a move toward a regional approach and toward involving state and environmental group stakeholders. Thus, it makes sense to think about the formation of some type of regional mechanism for multiple-use ocean management to plan for, and decide among multiple uses and resources in particular ocean regions. Such a mechanism was called for in the recent National Research Council report *Striking a Balance: Improving Stewardship of Marine Areas* (1998). Such a regional mechanism would have to be structured very carefully, most likely involving the highest levels of decision making at the state level (the governors) and the policy-level federal officials from the region, and perhaps involving an interstate compact. In addition, this mechanism would have to be overseen by some type of national ocean council in order to ensure that national and regional interests are properly balanced. We will return to this question in more detail in Chapter 7.

Issues in Coastal Management

Three major clusters of interrelated policy issues involve both sea and land areas of the coastal zone—coastal tourism and recreation, clean water, and coastal hazards. We examine each in the following sections, and also discuss strengthening the capacity of the coastal states to address such coastal issues as well as the ocean issues covered in the previous section.

Coastal Tourism and Recreation: The Driver of Coastal Development

While there is general recognition that coastal tourism and recreation are important in the coastal zone, we believe that their impact is systematically undervalued both economically and as the most important driver of coastal development in many U.S. coastal areas. In California alone, it is estimated that coastal tourism is the largest "ocean industry," contributing $9.9 billion to the California economy, compared to $6 billion for ports, $860 million for offshore oil and gas, and $550 million for fisheries and mariculture combined (Wilson and Wheeler 1997). Travel and tourism are estimated to have provided $746 billion to the U.S. domestic product, about 10 percent of U.S. output, making travel and tourism the second largest contributor to the GDP, just behind combined wholesale and retail trade (Houston 1995). Although there are no precise estimates of the magnitude of coastal travel and tourism in the United States, studies have shown that beaches are America's leading tourist destination, ahead of national parks and historic sites. Approximately 180 million people visit the coast for recreational purposes, with 85 percent of tourist-related revenues generated by coastal states (Houston 1996, p. 3).

The following examples highlight the very high value of coastal travel and tourism in the United States (YOTO 1998, p. F-5). A 1996 EPA study on the benefits of water quality improvement in terms of the numbers of people involved and the economic value of the activities in which they partake, found that saltwater fishing generates expenditures of over $5 billion annually and more than 200,000 jobs. Over 77 million Americans participate annually in recreational boating, with the total number of recreational boats by the year 2000 estimated to be 20 million. Over 80 million Americans participate in outdoor (nonpool) swimming, and in seven states, beachgoers spent $74 billion. Finally, birdwatching—a great deal of which occurs in coastal regions—generates around $18 billion annually.

Given these figures, it is significant to note that there is no federal agency with a mandate to manage coastal travel and tourism, and that there is no overall national policy in place to plan for, and achieve, sustainable tourism in the United States. In addition, although recognized as a highly valuable revenue earner, promotion and marketing of travel and tourism in the United States lags well behind that of other countries; the United States ranks thirty-first in international tourist market advertising, with Spain, for example, spending ten times more in advertising than this country (Houston 1996, p. 3).

A major reason for the lack of a formal program at the national level is that travel and tourism is viewed as a sector that requires relatively little formal management and is primarily a private sector endeavor. The benefits of tourism on coastal areas are great, yet its adverse effects are often not immediately visible, which leads to a sort of "management apathy." Also, most aspects of coastal travel and tourism that need managing are already dealt with at one governmental level or another, but in separate programs and run by different agencies rather than as a coordinated, interconnected whole.

The YOTO paper on coastal tourism and recreation (1998, p. F-9)[4] notes that sustainable development of coastal tourism depends on a number of factors, including the following:

- Good coastal management practices, especially related to location of infrastructure and provision of public access.
- Clean air and water, and healthy ecosystems.
- Maintenance of a safe and secure recreational environment, specifically relating to management of hazards, and provision of adequate levels of safety for boaters, swimmers, and other recreational users.
- Beach restoration, including beach nourishment and other efforts that maintain and enhance the recreational and amenity values of beaches.
- Sound policies for coastal wildlife and habitat protection.

Healthy and sustainable coastal tourism requires attractive, safe, and functional recreational beaches, clean coastal waters, and healthy coastal ecosystems producing abundant fish and wildlife. In most parts of the burgeoning U.S. coastal zone, these factors do not exist by chance. Most recreational beaches have to be maintained with occasional replenishment of sand lost to storms and erosion. Clean and healthy coastal waters are the result of effective programs of pollution control—of municipal sewage treatments, of septic tanks, of agricultural runoff, and of many other point and nonpoint sources. Coastal fish and wildlife depend on the existence of healthy ecosystems; wetlands have to be protected and, where already degraded, restored. Failure in any of these areas can seriously affect tourism. A failed sewage treatment plant can close a beach to swimming—in 1996, there were nearly 3,000 such closings or advisories (Heinz Center 1998) at U.S. beaches. New Jersey reportedly lost $800 million in tourism revenues following reports that medical wastes had washed up on some of its beaches (NRC 1995).

While there are already programs in place dealing with each of these areas, no agency or mechanism exists to coordinate them toward the overall goal of sustainable tourism development. Federal programs most relevant to coastal travel and tourism include the following:

- *Coastal management and planning* is administered by NOAA's Office of Ocean and Coastal Resource Management (OCRM) and includes programs in thirty-four states and territories. Three management practices under the Coastal Zone Management program are particularly important in the context of sustainable tourism development: provisions for the management of coastal development; provisions to improve public access to the shoreline; and provisions to protect and, where necessary, to restore coastal environments.
- *Management of clean water and healthy ecosystems* is especially important. A number of federal agencies and programs are involved with water quality, including the Clean Water Act (e.g., the National Estuary Program) administered by the EPA; protection of the marine environment from oil spills, covered by the Oil Pollution Act of 1990 and administered by the U.S. Coast Guard; and NOAA's work with states under the CZMA to deal with nonpoint-source water pollution.
- *Management of the impacts of coastal hazards, including flood and erosion protection,* and the use of siting methods such as setback lines is dealt with under both the FEMA National Flood Insurance Program and the Coastal Zone Management Program. Also important here is safety and accident prevention for visitors involved in coastal recreation—the U.S. Coast Guard is the principal federal agency responsible for user safety and accident prevention. Beach restoration and nourishment programs are managed at the federal level through the Army Corps of Engineers. Increasingly, however, it is local communities, sometimes with state assistance, that are having to undertake such restoration programs.

Given the very large contribution to the economy associated with coastal tourism and recreation, it would seem that special policy and pragmatic coordination efforts are needed among the federal, state, and local agencies responsible for the activities mentioned above.

Policy Challenges

1. *Federal policies and programs essential for sustainable tourism development are interrelated and should be treated as such.* Consideration should be given to the creation of a standing interagency group devoted to coastal tourism among the various federal agencies with programs in this area. State and local government representatives should also be included.

2. *Little guidance is available to states and communities for sustainable tourism development in coastal areas.* The federal government could play a role in providing guidelines to communities and states (standards, codes of conduct, manuals, etc.) to assist in their efforts to manage coastal tourism and recreation sustainably.

3. *There is little systematic collection of data and information on the magnitude, nature, and economic and social impacts of tourism in the coastal zone.* This needs to be changed to provide greater information on issues, trends, and value of tourism at all levels in the United States. The availability of this kind of information will help attract the appropriate level of attention to this issue.

4. *U.S. recreational beaches are in great demand by both U.S. citizens and foreign tourists, yet no national program of beach standards exists.* The EPA is launching a beach action plan dealing primarily with water quality (U.S. EPA 1999). While this represents a good first step, we think that a national program on beach standards should be broader in scope. The European Blue Flag program, now in place at about 1,000 beaches in different nations of the European community, provides a good model. The flag can only be flown at beaches that meet pre-set standards in water quality, safety (lifeguards, first aid, storm planning), beach management (erosion control, replenishment, cleanup), and environmental information and education (information on fish and wildlife, beach dynamics, tides, currents, etc.). The program has been encouraged by the European Union and individual governments, but the actual operation (judging beaches against the standards) is performed by

nongovernmental committees set up in each nation. The United States could benefit from a program similar to this one.

Clean Water: Control of Land-Based Sources of Marine Pollution

More than any other ocean or coastal issue, the attainment of clean water in the United States has suffered from the policy fragmentation brought about because a number of different federal agencies each have a piece of the problem and no agency has overall responsibility for this important goal. The seeds for this problem were sown as early as 1972 when the federal Coastal Zone Management Act was worded to exclude water quality concerns since they were already being handled under the Clean Water Act. The problem was compounded by the creation of separate non-point-source pollution control programs in the EPA (the Section 319 program) and in NOAA (the 6217 program). The third federal participant is the Department of Agriculture and its National Resources Conservation Service (formerly the Soil Conservation Service). In each case, the federal agencies have their state counterparts (again, different state agencies are involved), and in the case of Agriculture, there are local organizations as well (the conservation districts). More effective programs to clean up the nation's water bodies must address this institutional problem. Ways must be found to cause these federal agencies to coordinate their programs more effectively and to have them motivate their subnational counterparts and their constituents to do the same (Hinrichsen 1998).

But achieving clean water will take more than improved coordination among government agencies. A brief review of the goals of one of the principal interest groups in this area—the Clean Water Network—illustrates the kinds of problems that remain to be solved:

- Elimination of the use and release of toxic pollutants.
- Prevention of polluted runoff (urban, stormwater, and agricultural).
- Elimination of toxic releases (from industry) into sewage treatment plants.
- Stopping of raw sewage discharges (as a result of combined sewer overflows [CSOs]).
- Strengthening the protection of wetlands.
- Protection of groundwater.
- Reduction in pollution associated with confined animal and poultry feedlots.

Given the existing institutional and legislative realities, these problems raise a host of policy challenges.

Policy Challenges

1. *Passage of the Clean Water Act Reauthorization.* Reauthorization of the Clean Water Act has been pending since 1993, which has had a number of negative consequences. Agencies funded under the program and constituencies benefiting from their programs are unable to do the kind of long-term planning that is beneficial in this field. In addition, the lack of reauthorization legislation leaves considerable uncertainty with respect to public policy. For example, will wetland laws have to be tightened or liberalized? Will money be available to fund significant efforts in non-point-source control? While a certain measure of success appears to have been obtained in securing funding for the administration's Clean Water Action Plan as part of the fiscal year 1999 appropriations process, this is still no substitute for passage of a five-year reauthorization measure that should include language making controls mandatory in situations where voluntary approaches to the management of non-point-source pollution associated with agricultural runoff have not proven effective.
2. *Putting in place an effective program to manage non-point-source pollution entering coastal waters.* The present scheme for managing non-point-source pollution is not working well for a number of reasons—some related to the complexities in the relationship between the EPA 319 program and the NOAA 6217 program, some related to lack of funding, and some related to a lack of political will to backstop voluntary measures with enforceable ones, especially in terms of agricultural runoff. A clearly articulated national policy is needed to clarify the nation's goals in this area and to support it with a workable administrative structure.
3. *Securing substantial reductions (one-half or more in the next decade) in (a) acreage of waters closed to shellfishing, (b) fish consumption advisories, and (c) beach advisories and closings.* Of the more than 6 million acres of shellfish-growing waters currently closed to shellfishing, most have been closed since the 1970s (EPA 1998). But the situation can be reversed with proper attention. In 1997, for example, the Navesink River in New Jersey was reopened to shellfishing after years of collaborative effort by state and local agencies and groups, and now generates an estimated $10

million annually for the local economy (EPA 1998). The nation needs to set a firm policy, including targets and timetables, to finally solve this problem. In 1996, 2,193 public advisories restricting the consumption of locally caught fish were in effect (EPA 1998). Such advisories now apply to 15 percent of the nation's lakes, 100 percent of the Great Lakes, and a large portion of the nation's coastal waters. Clearly, higher priority needs to be given to a problem that has been with us since the mid-1970s. Similarly, conditions that lead to advisories or closings of the nation's recreational beaches need to be reduced (and eventually eliminated). Nationwide beach water quality standards should be put into place as soon as possible and, subsequently, reductions in the number of closings or advisories targetted. Setting such goals and publicizing annual progress toward meeting them will no doubt increase public interest in and support for clean water programs generally.

4. *Capitalizing on the investment in the National Estuary Program.* As discussed in Chapter 4, the National Estuary Program is overseeing the formulation of comprehensive conservation and management plans (CCMPs) for the twenty-eight estuaries of national significance that make up the program. By and large, these CCMPs, each developed with the benefit of at least $5 million for research and other support, present forward-looking, carefully laid-out action programs to revitalize and restore the nation's most important estuaries and the resources that inhabit them. Yet progress has been limited, due in part to lack of funding for implementation. While some federal funding is going into public education, outreach, and (more) studies, significant additional funding earmarked solely and clearly for implementation of the CCMPs is urgently needed.

Natural Hazards in the Coastal Zone: Emerging Issues

As more and more people choose to live and recreate along the nation's shorelines, more people and more property are put at risk from periodical coastal storms and other natural disasters. Shorelines are the most attractive and the most vulnerable part of U.S. geography. Many people, when vacationing on the coast, want to be close enough to the ocean to hear the waves breaking. They do not realize (or they realize and don't care) that the 3-foot waves they hear today could, under the right conditions, grow to 20 feet or more very quickly.

A succession of hurricanes in recent years—Hugo in 1989, Andrew and Iniki in 1992, and later Opal, Marilyn, and Fran—has raised public concern about coastal hazards and the threats that they pose. Losses due to natural disasters have been increasing in the United States and worldwide. In 1970, losses were estimated at between $4 billion and $5 billion per year; they are now estimated at about $50 billion per year (YOTO 1998). The rise is undoubtedly due, in part, to increased exposure along the nation's coasts as populations and development grow. Of the $50 billion, 80 percent of the losses are meteorologic in origin; the remainder involve earthquakes and volcanoes.

At the federal level, two agencies are primarily concerned with coastal hazards—NOAA's National Weather Service (NWS) and the Federal Emergency Management Agency (FEMA). NWS issues predictions and warnings of hurricanes and the expected landfalls; it also calculates storm surge heights and flooding potential. FEMA has programs that deal with preemergency planning, assistance during the disaster itself, and postdisaster relief and reconstruction. Recently, FEMA has placed increased attention on a national mitigation strategy that involves actions that can be taken before disasters such as hurricanes to reduce the likelihood of and/or the severity of damage and associated losses.

FEMA also operates the National Flood Insurance Program (NFIP) which insures coastal (and riverine) property against flood damage if the community in which the property is located has adopted federal flood standards (for example, building above the level of the 100-year flood or some variation of that elevation). Unfortunately, because of the political opposition of coastal land developers and others, Congress has not modified the program in such a way that coastal property owners pay insurance premiums consistent with the risk inherent in their property and shoreside locations. Thus, the taxpayers, in general, continue to subsidize government flood insurance, which, in turn, probably leads to more concentration in the coastal zone than would otherwise be the case.

An emerging issue, global climate change, could exacerbate the coastal storm threat. Scientific opinion, for the most part, seems to support the notion that increasing emissions of greenhouse gases (most notably carbon dioxide, but also methane, CFC, and others) are beginning to trap heat in the earth's atmosphere and increase the earth's temperature. Although observational proof positive of this hypothesis is still lacking, sufficient evidence is now in hand to convince most, if not all, scientists that the theories are basically sound and that temperature rises have begun. Predictions have also been made of a 40- to 60-centimeter (16–24 inches) rise in sea level over the next century, as glaciers and ice shelves melt and warmer ocean water expands. Warmer sea-surface temperatures could provide energy for more intense and more frequent hurricanes, though the theorists seem less certain of this aspect of their work. Nonetheless, it would be prudent to build these possibilities into long-term coastal planning.

Much of the emphasis in current mitigation programs is on building structures in the coastal zone that can withstand the forces present during hurricanes conditions—for example, better ties between the structure and its foundation, better ties fastening the roof to the structure, stronger windows, and better roofs. A recent study found that relatively little of what was learned during hurricane disasters was fed back into the land use planning processes of the state coastal management programs (Knecht, Cicin-Sain, and Kempton 1999). In our view, more attention needs to be paid to land use planning as a potentially effective mitigation measure.

Policy Challenges

A number of policy challenges need to be addressed in the area of coastal hazards:

1. *Reform of the National Flood Insurance Program.* Removal of remaining subsidies for federal flood insurance is a high priority. Coastal property owners should be aware of and bear the full costs of insuring for the risks inherent in their coastal location.

2. *Inclusion of land use planning measures in mitigation programs.* The best hope for lowering risks to both life and property in coastal areas is through land use planning and management. Lessons learned in disasters should inform changes in zoning and land use regulations (setbacks, etc.). Unfortunately, this tends to run counter to the mood of generosity and "aid for rebuilding" that generally accompanies disasters of all kinds.

3. *Establishment of a long-term policy of retreat from hazardous areas of the coastal zone.* The federal and state governments would do well to adopt a policy of encouraging "retreat" over time from the most hazardous areas of the coastal zone. As structures

in high-hazard areas reach the end of their useful lives, programs should be in place to promote the relocation of the activity to a safer site, most likely inland. Engineering works built to strengthen or armor a stretch of shoreline in order to protect adjacent structures eventually fail and have to either be rebuilt on an ever-grander scale or be abandoned. Either way, the beach and any amenity values present are lost. Ultimately, given the dynamics of barrier islands in particular, Americans need to be shown that their occupation of any specific part of the sandy shore is temporary and they may build or not build accordingly. Perhaps movable or "disposable" structures of limited lifetime would be best suited for the coast.

Strengthening the CZM Programs of the Coastal States

The completion of the national network of state/territorial CZM programs provides an appropriate time to reflect on the policy challenges that lie ahead. Ninety-nine percent of the U.S. shoreline will be under the aegis of the CZM program when the states of Indiana and Minnesota join the program in 1999. Only Illinois, with its 59-mile shoreline (fully incorporated), has chosen to remain outside the program. Also, the twenty-seven years since the original CZM legislation was enacted have seen many changes in coastal and ocean affairs in the United States. Are the act and its mechanisms still appropriate to today's and tomorrow's problems?

The CZM legislation has undergone significant amendment and updating four times since its enactment in 1972. The fact that four of the five states outside the program in the 1980s have chosen to join in the last several years suggests that the program continues to be seen as relevant and important to the coastal states.

Policy Challenges

1. *State CZM programs should move in the direction of setting measurable, on-the-ground goals, and develop performance monitoring programs to systematically measure progress toward meeting such goals.* To the extent possible, the goals should refer to on-the-ground changes: How many acres of wetlands are newly under protection? How much more public access has been provided to public beaches? How much more beach is under a higher level of management (e.g., under a regular sand replenishment schedule)?

2. *Development of an effective policy coordination mechanism among the core coastal management programs of the country—CZM, NEP, Coastal Barrier Program, National Flood Insurance Program, the beach maintenance programs of the Army Corps of Engineers, the 404 wetlands program, and the non-point-source pollution programs of EPA, NOAA, and others as appropriate.* Given the multiplicity of coastal programs, such a mechanism is obviously needed. Certainly NOAA's coordinating responsibilities under CZMA would be greatly facilitated by such a body. When formed, it could be a subunit of a national ocean council.

3. *Clarification of and support for an expanded state role in the newly expanded territorial sea.* As mentioned earlier and discussed more fully in Chapter 7, there is a need to "domesticate" the new 9-mile band of territorial sea that the United States acquired in 1988. The adjacent coastal states are in the best position to do the required planning for the area. They also can assure a seamless connection to the planning and management regime they have formulated for state waters, which are also the first 3 miles of the U.S. territorial sea.

Addressing the Policy Challenges

In the preceding pages, we have laid out what we foresee will be major domestic policy challenges in many of the areas comprising national ocean policy—maritime security, marine transportation and port infrastructure, fisheries, marine mammal protection, marine aquaculture, offshore oil and gas development, marine protected areas, marine biodiversity, marine biotechnology, coastal tourism and recreation, clean water, coastal hazards, and coastal management. Table 5.2 summarizes the major policy challenges that will be facing the United States in each of these areas in the new century.

Properly addressing these challenges will require many decisions about what to protect and what to develop, where, when, and how, and for how long, in the American ocean. These are choices that will involve cross-cutting planning, analyses, and wise decision making. And yet we are ill equipped to undertake such choices under our existing ocean governance system, which still treats each ocean use separately and not as

TABLE 5.2 Major Domestic Policy Challenges.

POLICY AREAS AND CENTRAL POLICY ISSUE(S)	PRIORITY POLICY CHALLENGES
Policy Cluster #1: Navigating the U.S. and World Waters	
Policy Area: MARITIME SECURITY	
Central policy issue: Ensure freedom of navigation and overflight, which are essential to military mobility.	• Ratify the Law of the Sea Convention.
Policy Area: MARINE TRANSPORTATION AND PORT INFRACTRUCTURE	
Central policy issue: Modernize U.S. port infrastructure, in an environmentally sensitive manner, in response to new technological developments.	• Adapt port capacity to faster and bigger ships. • Address the environmental effects of dredging. • Stem the tide of exotic species introductions.
Policy Cluster #2: Ocean Resource and Space Utilization	
Policy Area: FISHERIES	
Central policy issue: Reduce the decline in fisheries and reduce bycatch.	• Fully implement the 1996 Amendments (Sustainable Fisheries Act), especially the provisions to eliminate overfishing, rebuild fisheries within ten years and designate essential fish habitats. • Reconsider the appropriateness of the regional council structure. • Address problems of overcapitalization and the appropriateness of using management schemes such as individual transferable quotas.

continued

TABLE 5.2 (*continued*)

POLICY AREAS AND CENTRAL POLICY ISSUE(S)	PRIORITY POLICY CHALLENGES
Policy Area: MARINE MAMMAL PROTECTION	
Central policy issue: Complete the job of protecting marine mammals that remain at risk, while considering appropriate management measures for healthy marine mammal populations	• Fully implement the 1994 MMPA Amendments, especially the timely development and implementation of plans to reduce take of marine mammals in commercial fisheries operations. • Continue and complete the job of protecting marine mammals that remain at risk. • Develop a regime for management of marine mammals that are healthy and exceed their levels of optimum sustainable population.
Policy Area: MARINE AQUACULTURE	
Central policy issue: Develop the industry in an environmentally sustainable manner with the aim to reduce negative trade deficits in fish products, meet domestic consumer supply, and create industry jobs.	• Incorporate marine aquaculture as a recognized use of the coastal zone under the CZMA. • Develop a policy framework for leasing and management of marine aquaculture in federal waters where there is currently no well-delineated framework for management and control of such operations. • Resolve competition between federal agencies, especially the Departments of Agriculture and Commerce, over who shall be the lead for marine aquaculture development. • In partnership with the states, provide incentives and support to the most promising elements of the industry and streamline the permitting process.

continued

TABLE 5.2 *(continued)*

POLICY AREAS AND CENTRAL POLICY ISSUE(S)	PRIORITY POLICY CHALLENGES
Policy Area: OFFSHORE OIL AND GAS DEVELOPMENT	
Central policy issue:	
Restart the OCS program in offshore areas outside the western Gulf of Mexico in an environmentally sound manner and in partnership with and giving benefits to the coastal states.	• Amend the OCSLAA to give coastal states a greater voice in OCS decisions. • Adopt a program of revenue sharing and coastal impact assistance to the coastal states. • Develop and implement a regional mechanism for OCS decisions.
Policy Area: MARINE PROTECTED AREAS, MARINE BIODIVERSITY, AND MARINE BIOTECHNOLOGY	
Central policy issue(s):	
Develop a vision and a specific plan detailing what proportion and specific areas of the EEZ will need special protection because of the presence of ecological, aesthetic, cultural, or recreational assets; for the protection of marine biodiversity; and for the orderly and equitable exploitation of marine biotechnology.	• Assess the nation's marine biodiversity resources. • Develop a specific plan for expansion of the marine sanctuaries program. • Develop an appropriate regime for the exploitation of marine resources for marine biotechnology purposes in the U.S. EEZ as well as take a lead role in clarifying the international regime on this issue.

continued

TABLE 5.2 *(continued)*

POLICY AREAS AND CENTRAL POLICY ISSUE(S)	PRIORITY POLICY CHALLENGES
Policy Area: RATIONALIZING MANAGEMENT OF INTERRELATED OCEAN RESOURCES AND OCEAN SPACE	
Central policy issue Since ocean resources and uses—fisheries, marine mamals, offshore oil and gas development, marine protected areas, marine biodiversity, and marine biotechnology—affect one another and often occur in the same ocean area, ways must be found to rationalize decision making about protection and exploitation of these resources and uses.	• Consider the formation of some type of regional mechanism for multiple-use ocean management to plan for, and decide among multiple uses and resources in particular ocean regions, with oversight from a national ocean council to ensure that national and regional interests are properly balanced.
Policy Cluster #3: Issues in Coastal Management	
Policy Area: COASTAL TOURISM AND RECREATION	
Central policy issue(s): Travel and tourism (the world's largest industry) and accompanying recreation are the most important drivers of coastal development in most U.S. coastal areas, and yet their impacts (both positive and negative) are systematically underestimated and addressed in a piecemeal fashion by federal and state governments.	• There is a need to recognize that federal policies and programs essential for sustainable tourism development (such as proper siting of facilities and clean water programs) are interrelated and should be treated as such. • Little guidance is currently available to states and communities for sustainable tourism development in coastal areas. The federal government should play a role in providing guidance to communities and states (standards, codes of conduct, manuals, etc.) to assist in their efforts to manage coastal tourism and recreation sustainably.

continued

TABLE 5.2 *(continued)*

POLICY AREAS AND CENTRAL POLICY ISSUE(S)	PRIORITY POLICY CHALLENGES
Policy Area: COASTAL TOURISM AND RECREATION *(continued)*	• There needs to be more systematic collection of data and information on the magnitude, nature, and economic and social impacts of tourism in the coastal zone. • As recreational beaches in the United States are in great demand both by U.S. citizens and by foreign visitors, national programs to establish and maintain beach quality standards and for periodic nourishment of beaches are needed.
Policy Area: CLEAN WATER: CONTROL OF LAND-BASED SOURCES OF MARINE POLLUTION	
Central policy issue(s): The United States has made significant achievements in the control of point sources of marine pollution, but the current most difficult challenge in this area is the control of nonpoint sources of pollution, which account for over 70 percent of marine pollution worldwide.	• The Clean Water Act needs to be reauthorized and strengthened. • An effective program to manage non-point-source pollution must be put into place. • Efforts are needed to substantially reduce, through more effective water pollution control programs, the number of ocean areas closed to shellfishing and those affected by fish consumption advisories and by beach advisories and closings.

continued

TABLE 5.2 *(continued)*

POLICY AREAS AND CENTRAL POLICY ISSUE(S)	PRIORITY POLICY CHALLENGES
Policy Area: CONTROLLING COASTAL HAZARDS *Central policy issue(s):* The impact of natural hazard events in the coastal zone (such as hurricanes, coastal storms, etc.) has been increasing and is likely to continue to increase, also in relation to new factors such as sea-level rise associated with global warming. More efforts need to be made to tie coastal management siting practices (where to allow housing and other facilitites to locate) with coastal hazard prevention and mitigation measures.	• Reform the National Flood Insurance Program to remove incentives for locating housing and infrastructure in hazard-prone locations. • Include land use planning measures (such as the enforcement of setback lines) in hazard mitigation programs. • Establish a long-term policy of retreat from hazardous areas of the coastal zone, particularly in view of potential sea-level rise.
Policy Area: STRENGTHENING THE CZM PROGRAMS OF THE COASTAL STATES *Central policy issue:* State coastal management programs, now present in all but one of the U.S. coastal states and territories, are the major vehicle at the state level for managing coastal areas, both on the land side and regarding state waters. As such, these programs are in need of constant support and strengthening.	• CZM programs need to move in the direction of setting measurable on-the-ground goals, and performance monitoring to systematically measure progress toward meeting such goals. • A policy coordination mechanism needs to be developed among the core federal coastal management programs, such as CZM, NEP, Coastal Barriers Program, National Flood Insurance Program, beach maintenance programs of the Army Corps of Engineers, the 404 wetlands program, the non-point-source pollution programs, etc., possibly as a subunit of a national ocean council. • The unclear legal status of the extended territorial sea—the 3- to 12-mile zone—needs to be clarified, to produce, most likely, an expanded state role in the management of this zone.

part of an interconnected whole. Addressing these challenges will require reform of the U.S. ocean governance regime, a topic to which we will turn in the final chapter, Chapter 7.

Before addressing the need for structural reform of how the United States. is organized to deal with its ocean, we turn, in Chapter 6, to one final policy challenge we consider a priority—regaining U.S. leadership on international ocean policy matters.

SIX

The United States and The World

Regaining Leadership in International Ocean and Coastal Affairs

The United States has been a leader in international ocean affairs for much of the time since World War II and was a guiding force in the first U.N. Conference on the Law of the Sea (Geneva 1958); in the second U.N. Conference on the Law of the Sea (Geneva 1960); in the Seabed Committee deliberations in the U.N. General Assembly between 1968 and 1972 that led up to the third U.N. Law of the Sea conference; and in six of the eight years of negotiations in that third conference (between 1974 and 1980). On the environmental front, the United States was a leader in the development of international agreements dealing with ocean dumping (the London Convention of 1972), with vessel source pollution (MARPOL 1973/78), and in a host of related oil pollution and safety of life at sea agreements during the period 1950–1980. The nation was also a leader at the U.N. Conference on the Human Environment held in Stockholm in 1972 and in implementing parts of the Action Programme coming out of that conference. It played a positive role helping to broker the 1985 Vienna Protocol on ozone-depleting substances and the follow-up 1987 Montreal Protocol that actually set targets and timetables for the reduction in both the production and the use of these harmful substances. The active leadership role played by the United States in forging these agreements helped ensure that the resulting instruments were technically sound, economically viable, and consistent with U.S. national interests.

However, with a few exceptions, in the 1980s the United States ceased to be an active and progressive leader and became a more conservative, skeptical follower. The status of U.S. participation in current agreements relating to international ocean affairs—and the 1994 Agreement on Part XI of the 1982 LOS Convention, the Agreement for the Implementation of the Provisions of the LOS Relating to the Conservation and Management of Straddling Fish Stocks and Highly Migratory Fish Stocks, Agenda 21, the Framework Convention on Climate Change and the Kyoto Protocol, the Convention on Biological Diversity, the Jakarta Mandate and the Biosafety Protocol, and the Global Programme of Action on land-based sources of marine pollution—is shown in Table 6.1.

Some of the changes in the U.S. stance on international ocean affairs were brought about by the conservative administration of Ronald Reagan (1981–1989). President Reagan and his appointees sought to incorporate their political philosophy into ongoing negotiations such as those involving the Law of the Sea agreement. The Congress also grew more conservative during this period, as in 1980 the Republicans took control of the Senate and, with it, the chairmanship of the Foreign Relations Committee. The late 1980s and the early 1990s saw the growth in Congress of an anti–United Nations sentiment among a number of vocal senators and congressmen. Even noncontroversial U.N. programs such as those under the World Heritage Convention, where natural or cultural sites of universal value are declared to be part of the global heritage and "listed" as such, were attacked as examples of a threat to private property in the United States.

This chapter discusses in some detail three important international conventions related to the oceans that were being negotiated or implemented in the 1990s—the Convention on Law of the Sea, the Framework Convention on Climate Change, and the Convention on Biological Diversity—to illustrate the manner in which the United States is currently conducting its foreign policy on oceans and coasts. Following these accounts is a review of the role of the United States in the implementation of several "soft law" international agreements related to oceans and coasts, including Agenda 21, the Programme of Action for the Control of Land-Based Sources of Marine Pollution, and the International Coral Reef Initiative.[1,2] We conclude with some suggestions on ways in which the nation could regain its leadership role in international ocean and coastal affairs.

Ratifying the Law of the Sea Convention

The 1982 United Nations Convention on the Law of the Sea (UNCLOS, hereafter referred to as "the Convention" or "LOS") entered into force on 16 November 1994, one year after the deposit of the sixtieth instrument of ratification. As of July 1998, there were 127 parties to the Convention, including virtually all OECD countries and the vast majority

TABLE 6.1 United States Participation in Major Recent International Agreements Related to Oceans and Coasts.

INTERNATIONAL AGREEMENT	OPEN FOR SIGNATURE	ENTERED INTO FORCE	DATE OF U.S SIGNATURE	DATE OF CONGRESSIONAL HEARINGS	DATE OF U.S RATIFICATION
United Nations Convention on the Law of the Sea	Montego Bay, Jamaica, December 10, 1982 [a]	November 16, 1994 [a]	July 29, 1994 [a]	August 11, 1994, none since [b]	Not yet ratified [a]
Agreement on Part XI of the LOS Convention	New York, USA, July 29, 1994 [a]	July 28, 1996 [a]	July 29, 1994 [a]	August 11, 1994, none since [b]	Not yet ratified [a]
Agreement for the Implementation of the Provisions of the LOS Relating to the Conservation and Management of Straddling Fish Stocks	New York, USA, December 4, 1995 [a]	Not yet entered into force [a]	December 4, 1995 [a]	June 20, 1996 [c]	August 21, 1996 [a]
United Nations Conference on Environment and Development					
Agenda 21	Earth Summit, Rio de Janeiro, Brazil, June 14, 1992 [d]	Not applicable	June 14, 1992 [d]	February 26, 27, July 21, 28 1992 [d]	Not applicable
Framework Convention on Climate Change	Earth Summit, Rio de Janeiro, Brazil, June 4, 1992 [e]	March 21, 1994 [e]	June 12, 1992 [e]	September 18, 1992 [f]	October 15, 1992 [e]
Kyoto Protocol	Kyoto, Japan, December 11, 1997 [e]	Not yet entered into force [e]	December 11, 1997 [e]	September 30, 1997 [g]	Not yet ratified [e]
Convention on Biological Diversity	Earth Summit, Rio de Janeiro, Brazil, June 5, 1992 [h]	December 29, 1993 [h]	June 4, 1993 [h]	April 12, 1994 [i]	Not yet ratified [h]

continued

TABLE 6.1 *continued*

INTERNATIONAL AGREEMENT	OPEN FOR SIGNATURE	ENTERED INTO FORCE	DATE OF U.S SIGNATURE	DATE OF CONGRESSIONAL HEARINGS	DATE OF U.S RATIFICATION
Jakarta Mandate	Jakarta, Indonesia, November 15, 1995 [h]	Not applicable	Not applicable	None	Not applicable
Biosafety Protocol	Under development [h]	Not applicable	Not applicable	None	Not applicable
Global Plan of Action for the Protection of the Marine Environment from Land-Based Activities	Adopted Washington, D.C. November 3, 1995 [j]	Not applicable	Not applicable	None	Not applicable

[a] United Nations Division for Ocean Affairs and the Law of the Sea (http://www.un.org/Depts/los).

[b] United States Congressional Records. 1995. Senate Committee on Foreign Relations (S. Hrg. 103-737) 95 *CIS S* 3812.

[c] United States Congressional Records. 1996. Senate Committee on Foreign Relations (S. Hrg. 104-537) 96 *CIS S* 38115.

[d] United States Congressional Records. 1992. House Committee on Foreign Affairs, Hearing. 93 *CIS H* 38129.

[e] United States Information Agency Climate Change section (http://usiahq.usis.usemb.se/topical/global/environ/envcl.htm)

[f] United States Congressional Records. 1993. Senate Committee on Foreign Relations (S. Hrg. 102-970) 93 *CIS S* 3817.

[g] United States Congressional Records. 1998. Senate Committee on Energy and Natural Resources (S. Hrg. 105-331) 98 *CIS S* 31121.

[h] United Nations Secretariat of the Convention on Biological Diversity (http://www.biodiv.org)

[i] United States Congressional Records. 1994. Senate Committee on Foreign Relations (S. Hrg. 103-684) 94 *CIS S* 38124.

[j] U.S. federal government, 1998. *Year of the Ocean Discussion Paper J: A Survey of International Agreements*, pp. J-7.

Source: Prepared with the assistance of Nigel Bradly and Ampai Harakunarak.

of the world's maritime nations (UN Division for Ocean Affairs and the Law of the Sea website [http;//www.un.org/depts/los/los94st.htm]). The Convention is seen as the legal guide for the world's oceans and covers virtually all ocean space and its uses, including vessel navigation and overflight, resource exploration and exploitation, conservation and pollution, fishing, and shipping. The 320 articles and 9 annexes provide a framework for behavior by states in the oceans of the world, defining maritime zones; laying down rules for creating sea boundaries; assigning legal rights, duties, and responsibilities of states; and creating a dispute resolution process (Cicin-Sain and Leccese 1995).

While most of the 1982 Convention met with approval from most states, the provisions of Part XI, related to deep-seabed mining, were seen as a significant obstacle to the universal acceptance of the Convention. The main objections, raised primarily by the United States, related to the procedures for granting mining licenses, production limits for seabed minerals, the financial rules of contracts; decision making in the Council of the Seabed Authority; and mandatory transfer of technology (Cicin-Sain and Leccese 1995, p. 5).

In response to what was seen as a "deeply flawed" Convention (because of Part XI and the U.S. "no" vote on the adoption of the LOS Convention in 1982), President Reagan in 1983 outlined the official policy position of the United States in an Ocean Policy Statement. The basis of the statement was that "the United States was prepared to accept and act in accordance with the balance of interests relating to traditional uses of the oceans. . . . The United States would recognize the rights of other states in the waters off their own coasts, as reflected in the LOS Convention, so long as the rights and freedoms of the United States and others under international law were recognized by these coastal states" (Malone 1990, p. 2). It was clear then, that at this stage of the LOS, the fundamental flaws in Part XI were seen by the United States as sufficient reason not to ratify—a position that was also followed by the Bush administration. Key senators were also negative on the LOS. Senator Helms, chair of the Senate Foreign Relations Committee, described the LOS as "a hardy perennial weed that keeps coming back and back, no matter how hard you try to pull it out by its roots. And it is a weed that remains obnoxious." Helms further described the treaty as "dangerous," and proponents in favor of resuscitating it were dismissed as using "the same old discredited arguments" (U.S. Senate 1990).

Given that difficulties associated with Part XI effectively prevented the ratification of the Convention by a number of the major maritime nations, the U.N. secretary general undertook consultations with interested parties over a period of four years, with a view to modifying its provisions to address the objections of the developed maritime nations (Cicin-Sain and Leccese 1995, p. 5). The outcome of these negotiations was the 1994 Agreement Relating to the Implementation of Part XI of the United Nations Convention on the Law of the Sea of 10 December 1982. The agreement restructured the deep-seabed mining regime along free-market principles and met the U.S. goal of guaranteed access by U.S. firms to deep-seabed mineral resources. It also eliminated the mandatory transfer of technology and removed controls on production. The agreement effectively removed the obstacles that had led many industrialized nations to withhold ratification of the original Convention, opening the way for near universal acceptance of the Law of the Sea Convention (Center for Oceans Law and Policy 1997, p. 28). One of the major benefits of the agreement from the U.S. perspective was that it guaranteed U.S. membership on the International Seabed Council by virtue of being the state with the largest economy at the date of entry into force of the Convention.

The deep-seabed mining agreement paved the way for the Clinton administration's support of the LOS Convention. In a letter of submittal to President Clinton in September 1994, Secretary of State Warren Christopher recommended that U.S. interests would best be served by becoming a party to the Convention and the agreement. The letter stated that all interested federal agencies and departments unanimously concurred with this point of view, and described the primary benefits of the Convention to the United States:

- The Convention advances U.S. interests as a global maritime power, preserving the right of the military to use the world's oceans to meet national security requirements, and of commercial vessels to carry seagoing cargoes.
- The Convention advances U.S. interests as a coastal state by providing for a 200-nautical-mile EEZ, and securing rights over the full extent of the continental shelf. This is consistent with U.S. oil and gas leasing practices, and domestic and international fisheries management practices.
- The Convention provides for, and promotes improvement in, the health of the oceans by addressing vessel source pollution, pollution from seabed activities, ocean dumping, and land-based sources of marine pollution.

- It sets out criteria and procedures for access to the oceans for scientific research.
- It provides dispute resolution mechanisms to enhance compliance with its various provisions.
- The 1994 agreement changes the deep-seabed mining regime of the original Convention, providing a stable and acceptable framework for future mining (U.S. DOS 1994).

As Professor John Norton Moore (former principal U.S. negotiator on the LOS Convention) notes, there are substantial disadvantages to not ratifying the Convention (Center for Oceans Law and Policy 1997). That is, if the United States wants to have a serious role in shaping the Convention through its implementing institutions, it must ratify sooner rather than later. Given the worldwide adherence to and support of the Convention, it is certain to affect U.S. interests regardless of whether the United States becomes a party. The Convention offers the United States the opportunity to exercise much-needed leadership with regard to many of the pressures that are being exerted on the world's oceans, and to protect the many U.S. ocean interests at stake (e.g., fisheries, oil and gas, deep-seabed mining) (Center for Oceans Law and Policy 1997; U.S. DOC et al., 1998).

As the major problems with Part XI of the LOS have now been resolved to the satisfaction of the United States, it would seem logical that ratification should follow. Yet the date when provisional membership in the International Seabed Authority ended if ratification had not occurred—November 1, 1998—passed without U.S. ratification thus forgoing the guaranteed U.S. seat on the authority. Ratification has not occurred because the political stance of the 104th and 105th Congresses was and is unfavorable, and, most prominently, Senator Jesse Helms, chair of the Senate Foreign Relations Committee, remains in opposition.

It is also worth noting that many of the international agreements related to oceans and coasts that have been negotiated since the LOS in 1982 have incorporated the Convention as the basis for many of their provisions. Failure of the United States to ratify the Convention makes little sense given the significant role the LOS agreement has played in subsequent international agreements, including Agenda 21 (especially Chapter 17 on Oceans and Coasts), the U.N. Straddling Fish Stocks and Highly Migratory Species Agreement (1995), the Global Programme of Action for Protection of the Marine Environment from Land-based Activities (1995), and the 1996 Protocol to the London Convention.

Implementation of UNCED

U.N. Agreement on Straddling and Highly Migratory Fish Stocks

One of the tougher ocean issues faced in the preparations for UNCED and the Rio meeting itself was the issue of straddling and highly migratory fish stocks. While the 1982 LOS Convention deals with these issues in a general way, more detailed guidance was needed to deal with the growing conflicts in several parts of the world. In particular, problems were being experienced in the northwest Atlantic between Canada and fishing nations of the European Union, most notably Spain, and in the Bering Sea "donut hole" between Russia and the United States. Although no agreement could be reached as a part of the UNCED process, it was agreed to hold a post-UNCED conference on the matter under the auspices of the United Nations. That conference took place between April 1993 and August 1995 and resulted in "The Agreement for the Implementation of the Provisions of the United Nations Convention on the Law of the Sea of 10 December 1982, Related to the Conservation and Management of Straddling Fish Stocks and Highly Migratory Fish Stocks" (the Agreement).

The Agreement has been hailed as a "groundbreaking step toward reducing overfishing and corollary damage to marine ecosystems" and as representing a "sea change" in the management of the world's marine fisheries (Speer and Chasis 1995, pp. 74, 71). In one of the most important provisions, it holds that conservation and management measures established for the high seas (i.e., beyond EEZ or national fishing zone boundaries), and those adopted for areas under national jurisdiction should be compatible, and that coastal states and states fishing in the high seas shall have a duty to cooperate in the management of straddling stocks and highly migratory species (Cicin-Sain 1996). The Agreement calls for the creation and/or strengthening of regional and subregional fishery management bodies to be governed in a transparent manner and with appropriate involvement of NGOs.

The Agreement also breaks new ground in several other areas. First, it clearly incorporates the Precautionary Principle into its framework. Article VI states, among other things, that states should be more cautious when information is uncertain, unreliable, or inadequate. The absence of scientific information shall not be used as a reason for postponing or failing to take conservation or management measures. Second, the Agreement incorporates strong compliance and enforcement procedures. For example, in high-seas areas covered by regional agreement, a state party may board and inspect fishing vessels carrying the flag of another party to the Agreement; if there is evidence of illegal fishing found, the Agreement authorizes the inspecting state to detain the vessel and bring it into port.

Unfortunately, as of now—three and one-half years after the conclusion of the Agreement—it has not yet come into force, since only nineteen of the required thirty nations have ratified. The United States was one of the early ratifiers, ratifying the agreement on August 21, 1996.

Framework Convention on Climate Change

One of the important actions taken in Rio in June 1992 was the opening for signature of the Framework Convention on Climate Change. For a dozen years, scientists had been increasingly concerned that human-made emissions associated primarily with industrialization were modifying the earth's climate. More specifically, they feared that increasing emissions of carbon dioxide, methane, and other so-called greenhouse gases were causing an increase in the earth's temperature and other changes in climatic conditions.

This prospect has several implications that are relevant to U.S. ocean policy. First, in a physical sense, the oceans are closely interconnected with the atmosphere; ocean circulation directly affects the climate and weather in many locations. Further, the ocean represents an immense sink for greenhouse gases and as such is a dominant factor in setting the global balance of such gases at any one time. Like the forests, another important absorber of carbon dioxide, the oceans help slow the rate of increase of greenhouse gases in the atmosphere and hence represent a factor that helps buffer the earth's climate against civilization-induced changes.

In addition, one of the principal effects of the warming of the atmosphere is expected to be an accelerated rise of sea level caused both by increased melting of glaciers and ice caps and, especially, by expansion of the volume of seawater due to heating effects. It is also likely that increased water temperature will modify ocean currents and these in turn will affect local and regional climates and fishing conditions. Possibly, increases in the frequency and intensity of tropical hurricanes and typhoons could follow warmer sea-surface temperatures at the lower latitudes. The possibility of sea-level rises of up to 50 centimeters or more in the next century and the prospect of increased storm activity represents a major challenge for coastal and ocean policy in the years ahead.

As mentioned earlier, the Framework Convention on Climate

Change (FCCC) was opened for signature at the Earth Summit in Rio in 1992 after two years of negotiation. The Convention provides a broad policy framework for nations to follow with the overall goal of "stabilization of greenhouse gas concentrations in the atmosphere at a level that would prevent dangerous anthropogenic interference with the climatic system."

The policy response to scientific concerns began in 1988 with the establishment by the World Meteorological Organization (WMO) and the United Nations Environment Programme (UNEP) of the Intergovernmental Panel on Climate Change (IPCC). The role of the IPCC was to assess the scientific information available on climate change, its impacts, and possible policy responses (Soroos 1997b). This was provided in a working report of the IPCC in 1990, *Policy Makers' Summary of the Scientific Assessment of Climate Change,* which subsequently informed the negotiations for the FCCC.

The second World Climate Conference, held in Geneva in 1990, began formal negotiations on the FCCC. The negotiations leading up to the FCCC were complicated by the fact that some countries, notably the United States, opposed the original intention to impose binding limits on greenhouse gas emissions. Although most countries declared their intentions to freeze or reduce greenhouse gas emissions by either 2000 or 2005, the United States stood firm against any binding limits until such a time as there was more definitive scientific evidence (Soroos 1997b). The perspectives of developing nations on the emerging climate change regime differed a great deal from the more industrialized nations, with the majority of developing nations unwilling to limit their use of fossil fuels at the expense of economic development until the industrialized nations made serious commitments. At the other extreme, however, were small island developing states that believe their nations to be most susceptible to the effects of sea-level rise (Soroos 1997b).

The final outcome embodied in the FCCC in 1992 was that developed countries were recognized as being largely responsible for the buildup of greenhouse gases in the atmosphere. They are therefore supposed to "aim" to return their emissions to 1990 levels by the year 2000. The lack of mandated reductions by a specified date was primarily due to the U.S. opposition to specific targets by specific dates (Soroos 1997a).

There was, however, a significant shift in attitude from the Bush to the Clinton administrations. While President Bush had advocated delaying commitment to specific reductions pending increased scientific knowledge, President Clinton and Vice President Gore, in the fall of 1993, declared the U.S. intention to return greenhouse gas emissions to 1990 levels by the year 2000. The Clinton plan included fifty new or expanded "cost effective domestic actions" aimed at reductions—most of which were based on voluntary partnerships between government and industry to increase energy efficiency (Dernbach 1997).

Even though the attitude shift from Bush to Clinton was significant, the Clinton plan significantly underestimated the reductions required to achieve its aim by the year 2000 (Dernbach 1997). In addition, Congress failed to provide sufficient funding for programs that were needed to implement the plan, and voluntary partnerships were unable to make significant headway against the counterincentives provided by current law and practice.

The first meeting of the Conference of Parties for the Climate Change Convention was held in Berlin in 1995; the two main issues were adequacy of commitments in the FCCC, and whether to sanction joint implementation. "Joint implementation" provides a mechanism for industrialized countries to receive credit for emission reductions by financing emission reductions in developing countries for a lower cost —this is an approach the United States strongly supported in Berlin (Lanchberry 1997). A "mandate" calling for industrialized nations to agree to binding targets and timetables at COP2 was also adopted.

The second meeting of COP was held in Kyoto, Japan, in December 1997 and was intended to be the forum at which specific, binding figures for greenhouse gas reductions were to be agreed to (Kopp, Morgenstern, and Toman, 1998a, p. 1). The Kyoto protocol calls for industrialized nations (Annex I countries in the FCCC) to reduce average national emissions over the period 2008–2012 to approximately 5 percent below 1990 levels. The United States pledged to reduce its emissions by 7 percent, slightly less than the European Union countries, and slightly more than Japan. According to Kopp *et al.* (1998b, p. 1), both the proposed target and timetable from Kyoto will impose significant costs on the U.S. economy—the limit agreed to by the United States means a reduction of around one-third of what the Department of Energy estimates carbon dioxide emissions will be by the turn of the century. The DOE considers that if the United States is to effectively implement the goals it set at Kyoto, there needs to be a better understanding of the benefits and costs of the protocol in both the Senate and the general population (p. 2).

Kopp *et al.* (1998a, p. 1) identify a number of unresolved issues that must be addressed before implementation of the Kyoto protocol can occur. These include the following:

- The rules and institutions that will govern international trading of greenhouse gas emissions among Annex I countries must be better established.
- The rules and institutions governing joint implementation must be developed in detail.
- The criteria used to judge compliance, and any penalties for noncompliance, must be clearly articulated.
- A binding agreement on the part of the major developing countries to limit their emissions at some specified point in the future must be obtained. The lack of any early commitment by developing countries aggravates short-term concerns in the United States.
- To make longer-term objectives more credible, moderate but specific near-term goals should be set for Annex I countries, and these countries should be able to use early emissions reductions to meet longer-term requirements.

The third COP meeting was held in Buenos Aires in November 1998. After "tumultuous, marathon talks," 160 nations agreed to deadlines and an action plan to implement the Kyoto protocol. They set rules for enforcing the Kyoto agreement by late 2000, including tough measures to guard against cheating and penalties for countries that fail to comply (Warrick 1998b).

The Clinton administration formally signed the accord on November 12, 1998, but U.S. officials insisted that they will not submit the pact for Senate ratification until improvements in the agreement are made, including new commitments by the largest developing nations. Critics argued that the treaty would unfairly punish U.S. industries and would cripple the economy by raising energy prices. Congressional observers at the talks had urged President Clinton not to sign at all—a move that would have killed the protocol. After U.S. signature, however, they vowed to defeat the protocol in the Senate. As noted by Senator Chuck Hagel (R-Nebr.), a congressional observer at the talks, "This is not going to pass the Senate—it's not going to come close. Obviously the President knows that. He's doing something very dishonest by signing the treaty and telling Americans it's good for them by not having the courage to debate it and try to get a vote on it" (Warrick 1998a, p. A26).

While the U.S. formal position on climate change at the highest levels of government has been subject to intense administration-congressional confrontation, and has, at times, been quite out of step with the stance of other industrialized nations, it should be noted that at lower bureaucratic levels and in more informal ways, U.S. agencies have been supportive of consideration of climate change issues, particularly those regarding impacts on coastal zones, such as sea-level rise and increased frequency of storms. The United States has been an active participant in the work of the Intergovernmental Panel on Climate Change, especially in the work of the IPCC subgroup on coastal zone management (aimed at exploring the threat of sea-level rise and recommending adaptive strategies) (Carey and Mieremet 1992). Entities such as NOAA and the Country Studies Program (a U.S. interagency effort to assist less developed nations in preparing assessments on vulnerability to climate change) have worked extensively with other nations in the development of vulnerability assessments using a common methodology, and in the preparation of country plans for adaptation to sea-level rise in the context of integrated coastal management.

Convention on Biological Diversity

Coastal and ocean areas are among the most biologically diverse places in the world. In fact, a greater number of different species are found in the ocean environment than in the terrestrial environment. In some respects, coral reefs are the marine equivalent of rain forests in the diversity of species that they contain. Further, the deep ocean with its unusual environments (high pressures, low temperatures, low light levels, high-temperature seafloor vents, etc.) are home to a myriad of marine organisms, many of which are adapted to these unique environments and are of great interest to the field of marine biotechnology.

Hence, ensuring that ocean and coastal biodiversity is adequately protected and that access to such resources is appropriately managed are important ocean policy objectives, especially for nations concerned with assuring equitable access to these potentially valuable resources.

The international regime for managing global biodiversity is centered around the United Nations Convention on Biological Diversity (CBD), the second convention opened for signature at the Earth Summit in Rio in June 1992. Biodiversity refers not only to diversity of species (i.e., the number of species in a given ecosystem) but also to genetic variation within species. Biologists had become very concerned at the increasing rate at which species, in both the animal and the plant kingdoms, were going extinct.

The lead-up to the CBD began in November 1988, when the UNEP convened an Ad Hoc Working Group of Experts on Biodiversity. Formal negotiations for a legally binding Convention began in 1990 with the Ad Hoc Working Group of Legal and Technical Experts, later the Informal Negotiating Committee (INC)—a group that met eight times until the Convention was adopted in Nairobi in May 1992 (Elliot 1998, p. 76). Overall, the most controversial aspects of the CBD regime related to ownership of genetic resources, intellectual property rights, and the distribution of benefits of genetic exploitation (Elliot 1998, p. 77). Throughout the negotiation period, these issues firmly divided the "northern" and "southern" nations. The northern nations are generally more technologically capable of exploiting genetic resources but have far less biodiversity of their own; the southern nations, in general, are more biologically diverse but tend to have far lower capabilities to exploit their own resources.

The divisions occurred over the question of whether genetic resources should be a common property resource (this was the northern view), or if they were more appropriately viewed as sovereign national resources (the southern view). The attitude of southern nations was that genetic biodiversity is a sovereign national resource, to be utilized according to their own development and environmental priorities. They further argued that if the North wanted to continue to utilize the biodiversity of the South, equitable sharing of benefits and increased access to biotechnology for the South must first occur. This view was countered by the northern nations, notably the United States, who responded by claiming the need for legal protection (to be provided through the CBD) to safeguard the huge investment northern nations were making in the biotechnology field. In the end, the southern nations won, with the CBD containing specific provisions safeguarding their point of view, although not always as definitively as they had wanted (Elliot 1998, p. 76).

The Convention on Biological Diversity was opened for signature in Rio de Janeiro in June 1992, and was signed by more than 150 countries, coming into force on 29 December 1993. The overall objectives of the CBD are the conservation of biological diversity, the sustainable use of its components, and the fair and equitable sharing of the benefits arising from the utilization of genetic resources (Article 1). The objectives are to be met through a variety of means, including the development of national-level strategies, plans, and programs (Article 6); identification and monitoring (Article 7); in-situ and ex-situ conservation; and impact assessment (Articles 8, 9, 14).

From the coastal perspective, an important dimension of the biodiversity regime was formulated during the second Conference of

Parties in Jakarta, Indonesia, in November 1995. The Jakarta mandate is a program of action for implementing the Convention with respect to coastal and marine biodiversity. It identifies five areas of importance, and requires action by nations having coastal and marine areas. These are: (1) Institute integrated coastal area management; (2) establish and maintain marine protected areas; (3) ensure sustainable use of fisheries and other marine living resources; (4) ensure that mariculture is sustainable; and prevent introduction of, and control or eradicate, harmful alien species (de Fountabert, Downes, and Argady 1996, p. 4).

Notably, the United States was one of the most significant nonsignatories of the Biodiversity Convention at the Earth Summit—on the grounds that its provisions went beyond legitimate biodiversity protection goals, and that the Convention would unduly restrict the biotechnology industry. President Bush felt that the Convention did not adequately protect industry's intellectual property rights, and was of the opinion that the issues of access to, and transfer of, technology were poorly dealt with in the CBD (Rogers 1993, p. 189). Rogers cites President Bush's speech (in person) to the Earth Summit: "We come to Rio prepared to continue America's unparalleled efforts to protect species and habitat . . . [o]ur efforts to protect biodiversity itself will exceed—will exceed—the requirements of the Treaty (sic). But that proposed agreement threatens to retard biotechnology and undermine the protection of ideas . . . [a]nd it is never easy . . . to stand alone on a principle, but sometimes leadership requires that you do. And now is such a time" (p. 174).

The U.S. position, however, was reversed by President Clinton in his first major speech on the environment (on Earth Day, April 22, 1993), when he announced that the United States would sign the CBD (139 *Congressional Record* S 4795, Volume 139, No. 52). In the president's letter of transmittal to the Senate, he overturned the policy of the Bush administration, stating "[t]he Administration . . . supports the concept that benefits stemming from the use of genetic resources should flow back to those nations that act to conserve biological diversity and provide access to genetic resources. . . . We look forward to continued cooperation in conserving biological diversity and in promoting the sustainable use of its components. . . . Prompt ratification will demonstrate the United States' commitment to the conservation and sustainable use of biological diversity and will encourage other countries to do likewise" (139 *Congressional Record* S 16572, Volume 139, No. 162—Part 2).

As with the Law of the Sea Convention, however, the Convention on Biodiversity remains unratified, despite having been signed in

1993 by President Clinton. As with the LOS and Climate Change agreements, there is strong congressional opposition to ratification. Senator Sarbanes, speaking to the 104th Congress in October 1995, likened the refusal of Senator Helms, chair of the Foreign Relations Committee, to sign the Convention on Biodiversity (and a number of other key international agreements such as LOS) to holding the Senate "hostage" (U.S. Congress 1995).

Regardless of the Senate's unwillingness to ratify, the United States has acted on several aspects of the Convention and on the Jakarta mandate. According to Secretary of State Madeline Albright, in an Earth Day speech in April 1998, "the Administration believes we can implement the Biodiversity Convention in a way that protects our commercial interests while enabling those who protect biodiversity to share in the benefits . . . [s]o I hope the Senate will use common sense and approve the Biodiversity Convention as soon as possible" (Albright 1998, p. 3).

Much as with the Law of the Sea Convention, being outside the ambit of the Biodiversity Convention is already tangibly hurting U.S. interests. The Conference of the Parties (the Convention's decision-making body) have already met four times to further operationalize and implement the terms of the Convention. Particularly troublesome has been the absence of U.S. participation in the negotiation and expected agreement (in 2000) of a new global regime on biosafety under the CBD.

Agenda 21

Agenda 21 is the major action plan coming out of the deliberations of the four UNCED preparatory meetings and the Rio meeting itself. It spells out a course of action in forty areas that, if followed, will lead to a more sustainable world in the twenty-first century. It includes not only the full range of environment and resource issues—from oceans and coasts (Chapter 17, the longest chapter) to freshwater resources, radioactive waste, forests, agricultural practices and the like—but also chapters on such crosscutting issues as poverty, population, financing, and the role of major groups (women, indigenous people), business, labor, and local communities).

Chapter 17—the Oceans and Coasts Chapter Chapter 17 of Agenda 21 emphasizes throughout that new approaches to marine and coastal area management will be needed, approaches that are "integrated in content and precautionary and anticipatory in ambit." The introduction to Chapter 17 highlights that the LOS Convention provides "the international basis upon which to pursue the protection and sustainable development of the marine and coastal environment and its resources." Seven major program areas are included in Chapter 17: integrated management and sustainable development of coastal areas, including EEZ; marine environmental protection; sustainable use and conservation of living marine resources of the high seas; sustainable use and conservation of living marine resources in areas under national jurisdiction; addressing critical uncertainties for the management of the marine environment and climate change; strengthening international and regional cooperation and coordination; and sustainable development of small islands.

The first section, on integrated management and sustainable development of coastal and marine areas, underlies all other sections of Chapter 17. Coastal nations commit themselves to "integrated management and sustainable development of coastal areas and the marine environment under their national jurisdiction." The text stresses the need to reach integration (e.g., identify existing and projected uses and their interactions); promote compatibility and balance of uses; apply preventive and precautionary approaches, including prior assessment and impact studies; and ensure full public participation.

The Chapter calls for integrated policy and decision-making processes and institutions: "Each coastal State should consider establishing, or where necessary strengthening, appropriate coordinating mechanisms . . . for integrated management and sustainable development of coastal and marine areas, at both the local and national levels." It provides a series of suggested actions such coordinating institutions should consider undertaking—for example, preparation of coastal/marine use plans (including profiles of coastal ecosystems and of user groups); environmental impact assessment and monitoring; contingency planning for both human-induced and natural disasters; improvement of coastal human settlements (particularly in terms of drinking water and sewage disposal); conservation and restoration of critical habitats; and integration of sectoral programs (such as fishing and tourism) into an integrated framework. Also called for is cooperation among states in the preparation of national guidelines for integrated coastal management and in undertaking measures to maintain biodiversity and productivity of marine species and habitats under national jurisdiction.

The framework of integrated coastal management (ICM) has also come to be embraced by a number of the other major agreements emanating from the Earth Summit: the Framework Convention on Climate Change (under whose ambit much work has been done to address the effects of sea-level rise resulting from climate change through an ICM framework); the Convention on Biological Diversity and especially the Jakarta mandate; the Programme of Action for the Protection of the Marine Environment from Land-Based Activities; the Global Conference on Sustainable Development of Small Island Developing States; and the International Coral Reef Initiative (Cicin-Sain, Knecht, and Fisk 1995).

There has been much activity at both the international level (by U.N. entities, international NGOs, international donors such as the World Bank, Asian Development Bank, and individual nation aid agencies such as the Swedish SIDA, the Canadian CIDA, etc.) and at the national level to further define and operationalize the concept of integrated coastal and ocean management, and many efforts to put it into practice have taken place around the world. The growth of these activities is detailed in Cicin-Sain, Knecht, and Fisk (1995). They include the development of international guidelines on ICM (by the World Bank [1993], the World Coast Conference (1994); UNEP (1995); IUCN (1993); and OECD (1991, 1997); efforts at capacity building in ICM through the conduct of training courses and the like by U.N. entities such as the Intergovernmental Oceanographic Commission, the U.N. Division of Ocean Affairs and Law of the Sea, the Food and Agriculture Organization, the U.N. Environment Programme, and the U.N. Development Programme; the World Bank, the International Ocean Institute, and many others. At the national level, many nations have begun new programs in integrated coastal management, often, in the case of developing countries, with the assistance of international donors. A 1996 survey of twenty-nine nations, both developed and developing, (Cicin-Sain and Knecht 1998) showed that in 50 percent of the nations surveyed, some type of national coordination mechanism for managing oceans and coasts had been created (such as establishment of an interagency or interministerial commission, a special coordinating commission or committee, the naming of a "lead agency," or by establishing such a mechanism in the prime minister's office). Of the countries noting that they had set up a national-level coordination mechanism for ICM, 58 percent reported that the creation of an ICM coordination mechanism was highly or somewhat related to Agenda 21. Canada, Australia, and Korea are examples of three nations that have been very influenced by the Chapter 17 prescriptions and have all created new national institutions and processes for comprehensive national ocean and coastal policy (as discussed futher in Chapter 7).

In contrast to other nations—for example, Sweden, Denmark, Canada, and Australia—that have played very important roles in guiding and funding the implementation of the ICM concept in many developing countries, and have incorporated many of the prescriptions of Chapter 17 into their own domestic institutions, the United States has not been a very active player in this area. This is somewhat ironic since the United States was the first nation to enact, a quarter of a century ago, a far-reaching body of law relating to coastal and ocean management. Now, when the opportunity to share the extensive experience gained in this area and to assist the diffusion of good coastal management practices around the world is so readily at hand, it is unfortunate that the United States is not fully rising to this challenge. Its involvement in promoting good coastal management practice internationally has been somewhat modest, consisting mainly of assistance on ICM to half a dozen countries by the AID, through the University of Rhode Island's Coastal Resources Center and work undertaken by NOAA, especially by the NOAA/National Ocean Service/International Programs Office, newly created in 1998. The United States has also not adopted domestically any of the Chapter 17 prescriptions regarding the organization of ocean affairs. While it has created and operates a President's Council for Sustainable Development, the council has not addressed ocean and coastal matters.

The U.N. Commission on Sustainable Development Although national governments and intergovernmental organizations are expected to be the major actors in implementing Agenda 21, a new institution, the Commission on Sustainable Development (CSD), was suggested as a way to regularly monitor the progress being made in accomplishing the goals of this ambitious program. The CSD was established by a United Nations General Assembly resolution of December 1992 and it was formally brought into being in February 1993 under the U.N. Economic and Social Council (ECOSOC). The CSD is charged with ensuring the effective follow-up of the decisions of the UNCED conference, including Agenda 21. A small staff has been located administratively in the new department of Policy Coordination and Sustainable Development in the office of the U.N. secretary general.

Since 1993, annual meetings of the CSD have been held at the United Nations in New York to follow progress in specific issue areas. The oceans and coasts issue area was reviewed in the 1996 meeting where progress was noted with respect to the Straddling Stocks Agreement, land-based activities affecting the marine environment, and issues of concern to small island developing states, and were reviewed again at the 1999 CSD session. Unfortunately, the work of CSD has been hindered by its small staff, a relatively unaggressive style, and what appears to be the reluctance of a number of nations to provide reports of their Rio follow-up actions. A major meeting to assess progress in implementing Rio decisions after five years (the Earth Summit +5) was held by the U.N. General Assembly in 1997 with mixed results. Progress was reported in some areas (e.g., reform of the global environment fund procedures), but other areas such as new and additional financial assistance to developing countries, were faltering. While many nations have taken the prescriptions contained in Agenda 21 very seriously, especially regarding oceans and coasts (see for example, the survey of nations conducted by Cicin-Sain and Knecht 1998, and Hong and Lee 1995), there is little evidence that the United States has done so, nor has the U.S. government sought to strengthen the role of CSD.

Global Programme of Action on the Protection of the Marine Environment from Land-Based Activities

Land-based activities account for up to 80 percent of the pollution of the marine environment, but similar to the case of straddling stocks, the controversies surrounding the issue of land-based sources of marine pollution were too large and diverse to be dealt with at UNCED. Again, however, UNCED called for a post-UNCED meeting to be arranged by the United Nations Environment Programme (UNEP). The conference was convened by UNEP and hosted by the United States in Washington, D.C. from 23 October to 3 November 1995. A Global Programme of Action (GPA) aimed at preventing the degradation of the marine environment from land-based activities was adopted by the meeting, with UNEP designated as the secretariat. The GPA is designed to assist national governments in taking actions that will lead to the prevention, reduction, and control and/or elimination of the degradation of the marine environment from land-based activities. It identifies nine marine source categories that have the potential for degrading the marine environment: sewage, persistent organic pollutants, radioactivity, heavy metals, oils, nutrients, sediment mobilization, litter/plastics, and physical alterations and destruction of habitats. The GPA specifies national, regional, and international actions that can be taken to control these sources (Cicin-Sain 1996).

UNEP prepared a GPA implementation plan in 1997 that resulted in a GPA coordination office being set up in the Netherlands (as host country) and a GPA clearing house that UNEP is organizing. As a first step, UNEP is exploring the strengthening of its regional seas program as the principal regional element of the GPA.

The United States has been generally supportive of this program, hosting the initial Washington conference in 1995, and promoting efforts in the Pacific at the regional scale. Except for a possible formal agreement relating to persistent organic pollutants, the United States does not support a new, binding agreement in this area, but instead believes that a softer approach such as is contained in the GPA is more appropriate at this time.

International Coral Reef Initiative (ICRI)

This initiative had its roots in a U.S. decision in 1994 to focus on the plight of coral reef ecosystems in both U.S. and international waters and to encourage the use of integrated coastal management as an appropriate tool to deal with this problem. Building on the mandates contained in Chapter 17 of Agenda 21 (Mieremet 1995) and with leadership and financial support from the U.S. State Department and NOAA, a program was developed and additional partners were recruited to the effort. By 1995, six additional nations were involved—the United Kingdom, France, the Philippines, Sweden, Australia, and Jamaica. A 1995 workshop in the Philippines issued a "call to action" that established the set of objectives and regional strategies. Preferring to minimize formal administrative structures, the ICRI partners (nations, intergovernmental organizations, donor organizations, and NGOs) have established a set of largely informal arrangements coordinated by a rotating secretariat that was located at the Great Barrier Reef Marine Park Authority in 1997–1998. Active programs now include a Global Coral Reef Monitoring Network and a Reef Check Program. In addition, the ICRI was instrumental in having 1997 designated as the Year of the Coral Reef. The ICRI seems to be having considerable success in raising public awareness of present threats to coral reef health and sustainability and in many of its regional training and monitoring efforts. Nonetheless, despite having U.S. support and that of other bilateral donors and intergovernmental organizations (such as IOC, UNEP, and IUCN), the program continues to search for a more secure source of long-term funding.

Suggestions for Regaining U. S. Leadership

The United States is paying a price for its failure to lead in these three areas of international environmental policy—Law of the Sea, climate change, and protection of biodiversity. In the LOS area, the United States is almost certain to lose the opportunity to have its appointees named to the new institutions (the tribunal, the Continental Shelf Commission, and the organs of the International Sea Bed Authority), even though it certainly will eventually accede to this Convention. Similarly, in the biodiversity area, it is losing the opportunity to help shape the biosafety protocol, a protocol that could be very important to the U.S. biotechnology industry and its competitiveness. In climate change, the reluctant involvement of the United States and the strong division between the Congress and the administration will almost surely blunt U.S. leadership in the coming negotiations related to the implementation of the Kyoto protocol. What can the U. S. do as a nation to move back toward a leadership role? We believe such movement must begin in the State Department.

National ocean policy in its global context must be seen as more than simply LOS policy. The protection of navigation and overflight is surely important to national interests, but so is the protection of biodiversity, reversing ozone depletion, constructively addressing global warming, and aggressively dealing with land-based sources of marine pollution. A first step could involve the replacement of the existing State Department–led interagency committee dealing with LOS and putting in its place a broader interagency committee dealing with the full range of international ocean issues, including, of course, LOS-related concerns. This group could logically be a subgroup of the national ocean council that is recommended in Chapter 7. Until this happens, the international aspects of U.S. national ocean policy will continue to be handled entirely within the limited LOS framework.

In our opinion, national ocean policy needs to be conducted within a global framework consistent with national interests. The United States needs to be actively engaged on both the domestic front and the international front simultaneously if it is to fully protect and enhance the interests of the American people in both the U.S. ocean and the global ocean. A global ocean regime that develops without active and positive U.S. involvement will not necessarily protect U.S. interests. Global political realities do not support the view that the United States can unilaterally protect its interest through its superpower status alone. For example, unilateral approaches have not brought success vis-à-vis Iraq,

Yugoslavia, or Somalia. Even if, in principle, U.S. military or economic force alone could solve a particular problem, the costs might be unacceptably high. For example, efforts by the United States to unilaterally apply economic sanctions in an effort to change global environmental policy regarding porpoises and sea turtles killed in commercial fishing operations ran afoul of the World Trade Organization and GATT agreements in international trade.

In the last decade and a half, the United States has increasingly turned to the use of unilateral economic sanctions as a way to achieve its environmental and resource management goals. It has imposed sanctions on nations whose fishers do not apply U.S. standards in catching tuna and shrimp. It has also sanctioned nations that violated U.S. policies on the taking of whales, even though the whales in question (minke whales) are not endangered and the taking is not in violation of the terms of the International Convention for the Regulation of Whaling, the agreement covering the actions of the International Whaling Commission.

Our prescription for regaining U.S. leadership in international ocean policy is simple: The United States needs to back away from the unilateral use of power, economic or other, to accomplish its ends and begin again to lead in fashioning multilateral approaches to global problems. The United States has been a leader in creating and supporting the International Coral Reef Initiative, a successful multilateral post-UNCED endeavor. It has also been a leader in guiding the Antarctic treaty regime from its beginning in 1959. That kind of sustained, long-term, positive leadership is needed now in connection with the completion of the new regimes for the Law of the Sea, climate change, and biodiversity and the creation of a new regime for land-based sources of marine pollution.

All four of these agreements are now at their most difficult stage—implementation. And effective implementation, obviously, is the key to their success. We feel strongly that the implementation of these agreements should take place in a coordinated fashion, one that minimizes the burden placed on individual nations, which, after all, will be the level where the actual work takes place. Common approaches and frameworks such as integrated coastal management can go a long way toward helping nations respond positively to the obligations contained in these agreements, provided that the secretariats responsible for each of the agreements cooperate in such an approach. The United States and others (the European Union, for example) can play a pivotal role in bringing about this kind of positive interaction and cooperation.

Another area where U.S. leadership is urgently required involves the need for mechanisms to improve the information flow between the myriad of organizations, governmental and nongovernmental, that now populate the global ocean field. In 1998, a nongovernmental group—the Independent World Commission for the Oceans—after a three-year study, offered several recommendations in this regard. We strongly support one of their suggestions, the concept of a periodic global ocean forum that could bring together governments, nongovernmental organizations, and intergovernmental organizations to review the status of the world's oceans, how well various management programs are performing, how they can be strengthened, and new and emerging problems.

We turn now to the concluding chapter, Chapter 7, in which we outline a set of structural reforms which, if made in the U.S. system of ocean governance, will aid in addressing the multifaceted domestic policy challenges outlined in Chapter 5 and the international ocean policy issues set out in this chapter.

SEVEN

Today

Toward a New System of National Ocean Governance

The preceding chapters have reviewed the evolution of the present system used for governing the nation's ocean resources and space and have described the kinds of conflicts that are endemic to the existing approach. This chapter addresses the question that is central to the purpose of this book: What kind of a system of ocean governance will best meet the needs of this country as it enters the twenty-first century? In order to answer that question, one must first look at the basic problems with the existing system as well as the improvements that are needed to correct those problems.

Basic Problems with the Existing System

Fundamental Structural Problems

The rational management of ocean and coastal resources is presently an elusive and difficult matter for three fundamental reasons. The first and second reasons (jurisdictional splits and single-purpose management) deal with the present system of ocean governance; the third (the system's inherent complexity) relates to the nature of the system itself.

The Jurisdictional Split among Levels of Government Coastal and ocean areas are governed by three separate bands of jurisdiction—local governments generally control shoreland and shoreline use; state governments have jurisdiction in the belt of ocean from the tidemark out to the 3-mile limit; and the federal government has jurisdiction from 3 to 200 miles. Two major problems are posed by these jurisdictional splits:

- Many of the most important ocean activities traverse or impact all three jurisdictions, adding complexity to the planning and management of these activities, given the absence of effective mechanisms to coordinate the actions of all levels of government.

- The benefits and costs of ocean resource exploitation frequently fall disproportionately on different jurisdictions, exacerbating interjurisdictional frictions.

The Sector-by-Sector Approach in the Management of Different Ocean Resources and Uses Within the two offshore jurisdictions (federal and state), each resource or use typically falls under the jurisdiction of a different agency operating under a different legislative framework. Major problems posed by this single-purpose approach include the following:

- Few opportunities exist for examining the ramifications that decisions in one ocean sector (such as oil development) have for other ocean sectors (such as fisheries). While most of the laws do call for examination of the consequences of a proposed action on other ocean uses, these reviews take place within a specialized context that tends to be biased toward a particular outcome, either protection or development, depending on the particular law in question.
- Few opportunities exist for rational and long-range planning for the protection, enhancement, and use of ocean resources in specific regions.
- Because resources are managed on a use-by-use basis, few opportunities exist for the interested public to debate overall priorities and goals for a particular resource or region or to contribute to making trade-off decisions among different sets of values expressed by user groups.
- Conflicts among different ocean sectors, including those among different users and different government agencies, are difficult to solve through public means because no agency or other authoritative source has jurisdiction over such conflicts. These marine conflicts can be costly in many ways; they can result in extensive delays, threaten public order and safety, threaten the long-term well-being of marine resources, and involve excessive duplication and waste on the part of government.

Narrowly Based and Adversarial Decision Making Decision making about ocean and coastal resources is often driven by narrowly based interest groups representing particular resources or commercial concerns. Although there is widespread public concern about the health of

America's oceans, most decisions about ocean use are prompted by special interests wishing to preserve or promote their own particular aspect or use of the marine environment, without much consideration for the effects on other aspects or uses, or for broader societal considerations. In some cases, policy has oscillated between unmitigated development thrusts followed by the adoption of wholly preservationist approaches. This either-or view of development and conservation is costly and prevents the United States from achieving sustainable development of its ocean to maximize multifaceted benefits for the American public.

Short-Term Thinking Another characteristic of the U.S. system that exacerbates problems is the short time frame in which policy makers need to show results to maintain support for their programs, versus the need for longer-term planning in ocean management (Juda and Burroughs 1990). The result is that "multiple short-term programs may be adopted in place of long-term programs. Yet successive short-term programs may not be the best way to achieve long-term results" (Juda and Burroughs 1990, p. 32).

Complexity of the Ocean System The highly dynamic nature of the ocean, its mobility and fluidity, combined with the presence of complex and interdependent ocean ecosystems, make detailed prediction of the impacts of an ocean use activity exceptionally difficult, especially given the current rather rudimentary understanding of ocean processes and behavior. The complexity of the ocean system poses the following management problems:

- The full consequences of ocean exploitation activities are often poorly understood and therefore difficult to predict with any certainty.
- Uncertainty generally exists about the fate and effects of discharges of various types and about the nature and severity of impacts.

These basic problems have resulted in a national ocean management system that does not serve the nation well and will serve it even less well as ocean use and competition increase in the future. The present system fails for several reasons:

1. It attempts to superimpose a rigid jurisdictional framework with fixed boundaries onto a highly fluid and dynamic environment.

2. Ocean use decision making is fragmented and compartmentalized, yet ocean resources and activities themselves are interactive and often interdependent.

3. No overarching statements of national policy or priorities exist to guide or harmonize the ocean programs of federal agencies or to deal with the ocean use conflicts between them.

4. There is no organized or coherent way for the federal government to deal with the coastal states on ocean planning and use issues.

Fundamental Institutional and Policy-Related Problems

In addition to the basic structural problems, equally serious and fundamental problems exist in the institutional and policy aspects of the present system. Two related problems have the largest impact: (1) the lack of policies or principles to guide decision making with respect to activities in federal waters (3 to 200 miles offshore—the 3- to 12-mile portion of the territorial sea, and the EEZ), and (2) the relatively broad discretion contained in some ocean use legislation, for example, the Outer Continental Shelf Lands Act Amendments of 1978.

In the absence of any overarching national policies with regard to ocean conservation and use, and to the extent that discretion exists in relevant ocean legislation, federal ocean managers are left to set the policy goals of their programs according to their own desires or the desires of their agencies. In this situation, not surprisingly, the goals of the administration then in power are pursued by the federal ocean agency rather than by any hortatory policies that may have been written into the legislation.

These problems are exacerbated, of course, by the fragmentation and compartmentalization that exists in today's ocean management system. Since each federal agency is both judge and jury regarding its own ocean use decision making (except for the occasional litigious environment group), and since there tends to be little coordination among agencies, outcomes are often inconsistent with good ocean management practice.

Growing Recognition of Ocean Governance Problems

A number of studies, reports, workshops, and other efforts in the past decade have examined these problems and have put forward various suggestions for improvement. The Ocean Governance Study Group (OGSG), organized in 1991, brought together ocean policy academics to focus on the theme of national ocean policy reform. The OGSG has sponsored a number of conferences and reports laying out directions for improvement (e.g., Cicin-Sain and Knecht 1992; Caron and Scheiber 1993; Cicin-Sain and Denno 1994; Cicin-Sain and Leccese 1995; Knecht, Cicin-Sain, and Foster 1998; Scheiber 1999; Cicin-Sain, Knecht and Foster 1999 [the latter three in association with NOAA National Ocean Service's series of Dialogues on National Ocean Policy]).

In 1993, the Marine Board of the National Research Council held a workshop on the EEZ, bringing together the major national-level ocean interest groups and selected congressional and agency staff members to determine to what extent there was consensus on the need for changes or reforms in the existing framework. The workshop found extensive agreement among participants on the need for a more coherent national ocean policy strategy, including the need for an EEZ plan. To develop this further, the Marine Board, with the support of a number of government ocean agencies, undertook an eighteen-month study of the governance of marine areas. The recommendations of the study—*Striking a Balance: Improving Stewardship of Marine Areas*—called for, among other things, the establishment of a national marine council to establish and monitor national ocean policy and of regional marine councils to oversee developments at the regional and local level (NRC 1997).

In 1998, the Heinz Center for Science, Economics, and the Environment held three major workshops bringing together participants from all sectors (ocean interest groups, government agencies, academics, ocean user groups, etc.) around three themes: conservation and management of the coastal environment, protection and restoration of fisheries and other living marine resources, and advancement and application of ocean science and technology (Heinz Center1998). Among the recommendations coming out of the Heinz effort was the clear call for a more integrated approach to the management of ocean and coastal resources. Galdorisi, writing from the perspective of the Defense Department and emphasizing the importance of the Law of the Sea Convention to the United States, also highlights the urgent need for a review of national ocean policy (Galdorisi and Vienna 1997). In the next sections we present our own ideas for ocean governance reform, having profited by the many good ideas in the efforts we have mentioned.

A Conceptual Framework For Multiple-Use Ocean Governance

If one were to design a multiple-use ocean management system from scratch, in a situation where no previous experience with management existed (e.g. in some developing nations), the tasks would be relatively straightforward. In broad terms, this might be a likely set of steps:

1. Designate a responsible agency or an interagency group to conduct an inventory of national ocean resources and public needs and desires.

2. Develop a national strategy for sustainable development of the ocean under the nation's control and delineate a set of coherent ocean regions in need of management.

3. Examine, through a participatory process with stakeholders, existing and potential ocean resources and uses in these regions.

4. Create an institutional process for authoritative decisions about ocean uses in these regions, setting forth priorities among uses and adopting rules for managing all uses, by using some combination of national and regional/local institutions.

5. Establish a system for implementing, monitoring, and enforcing the decisions reached.

6. Regularly assess the performance of the management program and provide for "midcourse" corrections based on such assessments.

In the United States, given the presence of many established ocean laws, interest groups, and institutions, and given the importance of intergovernmental factors, it is a much more difficult challenge to design a multiple-use approach that takes into account and generally accepts as "given" the existing array of ocean-related laws, institutions, experiences, interest groups, and public values. It would be virtually impossible, in terms of political and administrative feasibility, to design and implement an "ideal" multiple-use ocean management system in the United States.

The following section explores what steps the United States could take to "move toward" a more integrated multiple-use ocean management regime, accepting the fact that an extensive body of ocean law and policy is already in place. Although, of course, this body of law and policy is subject to amendment and to some reformulation, given the politically potent constituencies behind much of the existing legislation, realistically its basic characteristics must be accepted as given.

Table 7.1 depicts two situations. The column on the left summarizes the current United States approach to ocean management, which, as discussed earlier, is largely sectoral. The column on the right summarizes the major features of a multiple-use ocean management framework. One can visualize a hypothetical middle column that would depict a "middle ground" toward which the United States could move by adapting a number of features of a multiple-use regime to the existing situation.

In the next section, we discuss possible improvements that could be made in each of seven categories: structural basis of the governance system, national guidance, value and ethical components, conflict resolution capacity, planning capacity, inter-agency integration, and intergovernmental integration.

In considering means whereby the United States could move toward more integrated multiple-use management, the following caveats should be kept in mind (Cicin-Sain and Knecht 1998).

1. *Not every interaction between ocean users is problematic.* Attention should be focused only on addressing those conflicts that pose serious problems; not everything needs to be integrated (for a list of marine conflict conditions that require resolution, see Cicin-Sain, 1992).

2. *"Ocean governance" and "ocean management" do not necessarily replace "marine resources management."* "Ocean governance" and "ocean management" connote an areal and multiple-use focus, while "marine resources management" focuses on specific resources, and their control, development, and allocation. Individuals and institutions concerned with ocean management focus on the policy coordination and harmonization functions, while individuals and institutions concerned with the management of specific resources continue specialized management of these resources (Cicin-Sain et al. 1990). Clearly, both orientations are needed.

3. *The costs of policy integration must be kept in check.* As Underdahl (1980) points out in his concise statement on "integrated marine policy," policy integration can be both difficult and costly, and

TABLE 7.1 Features of National Ocean Governance Systems.

FEATURES	CURRENT U.S. OCEAN MANAGEMENT FRAMEWORK	"IDEAL" MULTIPLE-USE OCEAN MANAGEMENT FRAMEWORK
1. Structural Basis	Sector-based Single uses managed separately Different regimes in state waters (0 to 3 miles) and in federal waters (3 to 200 miles)	Area-based approach encompassing multiple uses in the 0- to 200-mile zone
2. National Guidance/Goals and Principles	No overall guidance on use of U.S. waters (0 to 200 miles) No national consensus on use of U.S. waters Guiding principles lacking	Development of a national policy that establishes goals, objectives, priorities, and lays down basic principles and criteria that provide guidance for the formulation of a strategy for the sustainable development of the ocean and its resources
3. Value and Ethical Components	Stewardship responsibilities not well defined	Code of stewardship values and ethics developed and implemented
4. Conflict Resolution Capacity	Little capacity for resolution of ocean use conflicts Often no public mechanisms for resolving multiple-use conflicts Ad hoc approaches to conflict management prevail	Capacity to understand multiple-use conflicts, their characteristics, costs and benefits, and consequences Capacity to establish priorities among uses Decision-making mechanisms present to make authoritative decisions regarding ocean uses and to resolve conflicts Use of zoning and other techniques to separate ocean uses
5. Planning Capacity	Lack of planning capacity to plan for future mix of uses	Proactive, anticipatory planning approach Development of national and regional/local ocean use planning process
6. Level of Interagency Integration	Lack of integration/harmonization of federal ocean agencies Frequent conflicts prevail	Achievement of *horizontal integration* through such means as naming of a federal ocean council, interagency coordinating committees, memoranda of understanding, naming of lead agency
7. Level of Intergovernmental Integration	Lack of integration/harmonization among agencies at different levels of government (federal, state, local) Frequent conflicts prevail among federal, state, and local governments	Achievement of *vertical integration* through such means as establishment of joint state-federal management of specific areas; strengthening existing harmonizing mechanisms; sharing of revenues

Source: Adapted from Cicin-Sain 1994.

should, therefore, not be attempted unless the marginal gain is sufficient to be worth the cost of the effort.

Needed Ocean Governance Improvements

Following the categories found in Table 7.1, we offer this discussion of strategies that the United States could adopt to improve ocean governance.

Structural Basis

The present system of U.S. ocean governance is largely structured around single-sector management, with the exception of three programs that are structured around area management: coastal zone management, the National Estuary Program, and the Marine Sanctuaries Program. A major advantage of an area-based approach over a single-sector approach is that it allows governing authorities to better address the effects of one ocean use or resource on other uses, resources, and the environment. A second advantage is that it can be made consistent with management on an ecosystem basis. In recent years, it has become apparent that resource management can be better informed and more successful if the size and configuration of the key ecosystems are taken into account in setting management boundaries.

To move toward a more multipurpose and area-based regime, the United States should consider: (1) developing better methods of linking the existing sectoral programs to one another and to the area-based programs, building, inpart, on the relatively successful record of the federal consistency provisions of the federal Coastal Zone Management Act of 1972 as amended; (2) developing good connections among the area-based programs, for example, by ensuring that the comprehensive management plans (CCMPs) developed under the National Estuary Program become a formal part of the relevant state coastal management programs, thereby giving them legal standing as well as providing an additional mechanism for implementation funding; and (3) building any additional increments to area-based programs within the existing governance scheme.

Guiding Principles

Each of the major ocean laws currently in effect is guided by a set of principles. The Magnuson Act, for example, calls for fisheries to be managed "throughout their range" and "using the best scientific data available." While such principles provide an adequate measuring stick for evaluating the particular law in question (e.g., the Magnuson Act), they offer little guide for wise management of the entire ocean area under the jurisdiction of the United States or for the management of ocean activities that are not yet governed by federal law (such as marine aquaculture). No overall national guidance or overall national ocean goals exist—the only goals and policies that exist are found in individual laws dealing with single resources or single uses.

In the absence of overarching national ocean goals, there is no rational basis for solving conflicts or assigning priorities for the expenditure of limited resources. Conflicts are apt to be resolved "late" and, to the extent that the judicial system ultimately becomes involved, within a narrow, legalistic framework. The broader public interest is rarely served in such circumstances. Some conflicts, of course, never reach the courts, being resolved (or at least temporized) as a part of the eternal power struggles between government agencies, their constituencies, and their backers in the Congress.

A code of stewardship ethics needs to be developed for the U.S. ocean to provide guidance to both government officials and the involved private sector, above and beyond the guidance offered in specific statutes. The code of stewardship ethics should, of course, build on recent international advances in this area, as reflected in both the Law of the Sea Convention and the 1992 Earth Summit agreements (especially Agenda 21 and the Rio Declaration of Principles) coming out of the United Nations Conference on Environment and Development.

A first cut at a set of principles for the U.S. ocean has been offered by Van Dyke(1992a, 1992b), who lists nine such principles: the precautionary principle; the need for environmental assessments; protecting of rare and fragile ecosystems and endangered and threatened species; giving priority to living, over nonliving marine resources in cases of conflicts; using the public trust doctrine to protect the interests of the whole community and the interests of intergenerational equity; utilizing ocean resources in a sustainable development mode; governing in partnership with states, territories, and commonwealths; paying special attention to the historically based claims of indigenous peoples to ocean space and ocean resources; and accepting the responsibility of developed countries to assist poorer developing countries in undertaking the responsibilities outlined in these principles.

Other principles worthy of consideration include: use and orderly development of the ocean zone to benefit the American public; use and orderly development of the ocean zone to create and maintain jobs, to benefit U.S. industry, and to enhance U.S. economic competitiveness;

ensuring a good return to the public from the use of commonly held ocean and coastal resources; and promotion of efficiency and effectiveness in government operations (reduction of duplication and overlap).

No doubt there will be much discussion and debate over the "right" mix of principles to adopt. The list should be comprehensive and yet only as long as it needs to be to address all relevant aspects of ocean governance: stewardship toward ocean resources and space, U.S. ocean publics, the international community, and future generations. Adoption of such a code of stewardship ethics for the U.S. ocean will require legislative action that could be done in conjunction with suggested legislative action on other governance reforms discussed in the following sections.

The principles discussed here pertain mainly to the issue of stewardship and ethics. Additional principles related to the special nature of oceans and coasts and their biophysical character are also needed to help develop a workable ocean management system (Cicin-Sain and Knecht 1998). Principles of this type are discussed in connection with our suggestions related to an EEZ governance strategy.

National Guidance

All the existing ocean laws have some set of explicit goals and guiding statements of intent, for example, achieving optimum yield of fisheries under the Magnuson Act. The problem with the amalgam of goals from existing federal ocean programs is that they often don't aggregate well together, since they were originally crafted separately and without regard for other existing policy goals. For example, maximizing fisheries in a given area may be impossible to accomplish at the same time as maximizing oil development in the same area. In some existing programs (e.g. marine sanctuaries), national goals have not yet been fully developed. In emerging areas of ocean use activity, such as marine aquaculture or minerals exploitation, there has been little articulation of national goals and targets. And no mechanism exists for anticipating and encouraging new uses in the future through such means as research, technology development, or industry incentives. In short, developing national goals, objectives, and priorities, with specific targets and timetables is needed to harmonize existing sectoral policy goals and articulate compatible goals for uses where little policy guidance exists and/or for emerging uses. Furthermore, development of national guidance and evaluation of progress toward national goals must be done in conjunction with states and localities, given the great diversity in regional circumstances that characterizes different coastal regions of the

United States. This must also be a process that continually adapts and changes over time, in response to changing national and international developments.

How can such national guidance be developed? It is difficult to imagine that this task could be accomplished by any one of the component units of our existing system of federal ocean governance (given the absence of explicit mandates on this question) or through an interagency effort. While important in achieving exchange of information and sometimes in resolving conflicts and in adapting agency goals to one another, interagency committees (of agencies at similar hierarchical levels), generally lack the capacity to take a broad perspective and to assess goals and objectives across the federal government and into the future. Indeed, experience suggests that many interagency committees are seen as primarily defensive mechanisms. That is, agencies attend meetings of this type to make certain that their missions, resources, and programmatic bases are protected and, if the opportunity arises, expanded. Advisory committees, such as the former NACOA (National Advisory Committee for Oceans and Atmosphere) (advisory to NOAA), could logically be thought of as playing a national guidance role, but such committees are often limited in what they can accomplish because they only have advisory powers and are most often attached to a particular line agency and not to a higher level, such as the executive office of the president.

While we are reluctant to recommend adding a new institution in a situation already characterized by institutional complexity, we think that consideration should be given to the creation, through legislative action, of a national ocean council. Ideally, such a council would be connected to the highest levels of government, as the Marine Science Council of the late 1960s was connected to the vice president's office. If the President's Council on Sustainable Development were to achieve continuing oversight responsibilities, a national ocean council could conceivably be part of such an entity. One of the principal functions of such an entity, made up of the directors (or their policy-level designees) of all of the ocean- and coastal-related agencies, would be to provide the kind of broad national guidance on the overall goals for the nation's oceans and coasts and to articulate the principles that should be used to guide decision making throughout the ocean governance system. As one of our principal recommendations, the concept of a national ocean council is discussed in greater detail later in the chapter.

Building Better Capacity for Conflict Resolution and for Planning

The efforts to provide national guidance outlined above would go a long way toward enhancing capacity for conflict resolution and for planning at the federal level, because national guidance activities are aimed precisely at these questions—reconciling conflicts among federal laws and programs, proactive planning for possible new uses, and identification of problems and opportunities.

It is at the local level, however, that the ocean resources are found and that the actual uses and conflicts occur; it is at this level that the most urgent needs exist for ocean use planning and conflict resolution. As Bailey (1994) cogently puts it, "Ocean planning and management, like all politics, is local. Overall policies and management programs must eventually work in rock-by-rock, cove-by-cove, reef-by-reef situations where the abstract world of policies and planning meets the real world of birds, fish, SCUBA divers, fishermen, tourists and local residents" (p. 38). At the local level of real-life interactions among stakeholders and agencies, it is the state and local governments who must (and generally do) take a leading role in problem resolution and in proactive planning for state waters and, in some instances, working with the federal government for federal waters. Much could be done, however, to better assist the states and the localities in these endeavors and to develop better state-federal partnerships in both state and federal waters. These options are discussed later, in the section on intergovernmental integration.

Achieving Interagency Policy Integration

There are currently no regularized mechanisms for periodically bringing the representatives of federal ocean programs together, much less harmonizing agency policies. The Department of State does periodically convene an interagency group, but this effort is mainly oriented toward coordinating the national position on internationally driven ocean developments, especially those related to Law of the Sea issues or foreign policy. Interagency efforts on specialized ocean use issues such as the dredging of navigational channels or storm hazards as they affect the nation's beaches have taken place, but to our knowledge, no ongoing interagency effort currently exists that cuts across the broad range of national ocean issues. Informal groups such as the Ocean Principals (the leaders of the principal ocean agencies) have existed in the past, but their level of activity and impact has varied over time.

Following the Earth Summit, there has been much discussion of "integrated coastal management" (encompassing land, nearshore areas, and EEZs). In a recent book (Cicin-Sain and Knecht 1998), we attempted to explain the meaning of integrated management, and noted that this concept should be thought of as a continuum rather than as an absolute goal. Figure 7.1 illustrates this concept.

The current U.S. situation with regard to oceans and coasts is fragmented: Individual government entities pursue their largely single-purpose mandates, with coordination efforts occurring mainly in the context of decisions about specific development projects (e.g., through the environmental impact assessment process, the federal consistency review process, the endangered species consultation process, etc.). The challenge for the federal agencies, in our view, is to move (or be moved) to a state we define as "harmonization": Independent entities continue to operate their own programs but coordinate and synchronize their actions, guided by a set of national policies and criteria (these policies and criteria are generally established at a higher bureaucratic level or by the legislative branch).

"Moving toward integrated management" thus does not necessarily imply full integration, in the sense of government reorganization and creation of a larger bureaucratic entity. Instead, it can mean taking any of a range of measures to better mesh agencies' actions with one another. This requires a regularized mechanism for interagency coordination and, one hopes for harmonization. This can take place by means of administrative or legislative action, such as the creation of a national ocean council. One could envision such a council (a "national ocean council") operating at two levels: (1) a high political level whereby agency heads would meet periodically (such as two or three times a year) to set overall goals and policies, and (2) a working group level that would bring together, on a more frequent basis, staff from the different federal agencies to work out implementation details. Such a council of ocean agencies could also establish special task forces to address "problem clusters" of

FIGURE 7.1 Continuum of Policy Integration.

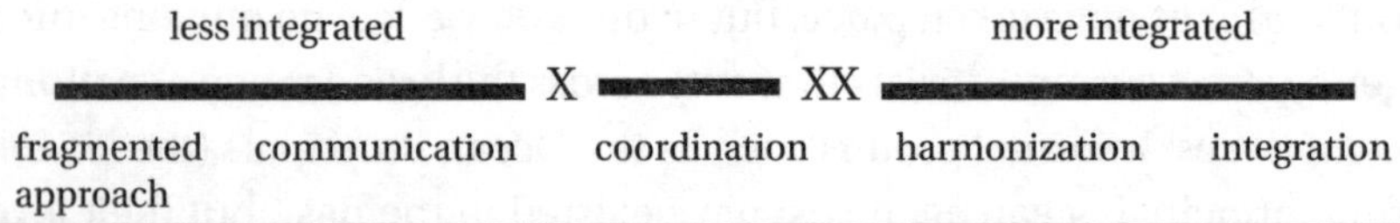

less integrated — more integrated

X — XX

fragmented approach — communication — coordination — harmonization — integration

Source: Adapted from Cicin-Sain 1994.

ocean issues that have proven to be particularly problematic or conflictual. Possible examples of such problem clusters include: marine mammal–fisheries conflicts, conflicts related to the outer continental shelf oil and gas program, conflicts associated with port and navigational channel dredging, and problems related to beach erosion and replenishment, and coastal hazards.

As we discuss later in this chapter, to make such an effort work considerable care would need to be exercised in the initial establishment of the interagency council—regarding its charge and scope of activities, how it is staffed and operated and directed, and to whom its reports and recommendations are directed. Agencies, too, must be given positive incentives for collaboration with other agencies, such as creating possibilities for funding from special sources aimed at achieving interagency cooperative activity.

Intergovernmental Integration

A clear delimitation of state and federal jurisdictions in the ocean exists in federal statutes, but in practice these distinctions have been somewhat blurred in recent years. Through the national Marine Sanctuaries Program, for example, the federal government has, with the approval of the state, designated protected areas in state waters. The states have moved to fill ocean policy voids, for example, through the crafting of ocean plans that include federal waters (Hershman 1996).

In a number of cases (e.g., Oregon, Hawaii), states have developed comprehensive plans to guide ocean use activities, to resolve conflicts, and to anticipate new uses. In a number of cases, too, states have come together in regional groupings to begin to provide a regional perspective on ocean use and protection issues. Among the regions that are the furthest along are the Gulf of Maine region (with the Gulf of Maine Council, a state-initiated regional grouping involving three U.S. states and two Canadian provinces); the Pacific Coast states (through the regional-level work of the Western Governors' Association and the Western Legislative Conference); the Pacific islands region (through the Pacific Basin Development Council); and the Gulf of Mexico region (through the federally initiated Gulf of Mexico Program) (Cicin-Sain 1995).

Such activities are taking place both in individual states and in groupings of states, sometimes with the full support and blessing of federal agencies, sometimes not. It should be noted, too, that interest in and capacity for ocean governance is not evenly spread out among the nation's states and regions. In some states and regions, there is little interest in ocean governance questions, generally reflecting relatively sparse ocean resources, lack of economic interests in the adjacent ocean, and other such factors.

Given the significant level of activity in ocean governance that is already taking place in a number of states and regions, the role of the federal government should be to encourage and facilitate such efforts, while ensuring that they are consistent with national interests and policies and are properly coordinated with the federal ocean programs.

Since the early 1970s, the United States has had the benefit of a reasonably well-functioning intergovernmental integration mechanism in the form of the federal coastal zone management program. As an ocean coordinating device, its main deficiency has been that state CZM programs have been relatively weak on the water side of the coastal zone. With some notable exceptions, state CZM programs have tended to concentrate on shoreland use issues and less so on water uses in the adjacent territorial sea. A related issue involves the question of jurisdiction over the 9-mile-broad territorial sea newly created when, by the presidential proclamation in 1988, the United States expanded its territorial sea from 3 to 12 miles in width. Although it can be readily argued that this 9 miles of new U.S. territory, like virtually all other new acquired territory in U.S. history, should become incorporated into the states of the union for purposes of domestic governance, the federal government has shown no interest in such a move. It is difficult to see that an indefinite perpetuation of this governance anomaly will be in the long-term interests of either the nation as a whole or the individual coastal states. Clearly, it would make sense to have the initial 3 miles of the territorial sea and the new 9 miles planned and managed in a seamless and coherent manner. New governance arrangements ought to provide incentives for developments to move in this direction (Forman, Jarman and Van Dyke 1992). We take up the question of ultimate jurisdiction over the new 9-mile belt later in this chapter.

A possible approach for achieving improved intergovernmental integration of ocean planning might include the following:

1. Providing federal grants (with state match) for the development of ocean plans to be crafted by individual states or (preferably) by regional groupings of states.

2. Fostering an ocean plan approval process that incorporates a partnership between state and federal levels.

3. Working to achieve proper integration between these efforts and the existing area-based, federally supported ocean and coastal programs (coastal zone management, estuary planning and management, marine sanctuaries) as well as with the single-sector programs (e.g., fisheries, oil and gas development).

4. Providing support for these ocean plans to have legal standing through their incorporation into the state coastal management process, thus invoking the powers of federal consistency review (some adjustments to the state coastal management process would have to be made if regional entities are carrying out the ocean use planning effort).

It would be best to test the approach described above on a voluntary basis, allowing coastal states and regions with particular interests and capacity to participate in such a program upon petition and on a pilot basis.

Attaining Efficiency and Effectiveness in Government Operations vis-à-vis the Ocean

Most of the options discussed above should work to achieve greater efficiency in government operation and less duplication and overlap among government programs. Some of the options outlined will, however, require new expenditures of funds, particularly the proposal for the creation of a national ocean council and the federal grants to states and regions for ocean use planning. It is always difficult to finance new ventures, yet the expected benefits of more efficient government operation and the encouragement of appropriate economic development (of both existing and emerging industries) in the 200-mile ocean zone should outweigh these costs over time. These ocean governance improvements could be financed through a combination of revenue sharing from government revenues earned from leasing and operation of oil and gas development on the outer continental shelf, higher fees for ocean users (including the recreational groups that traditionally resist such fees), and harbor maintenance funds. While the mere mention of a new tax is political dynamite, it can be argued that a good number of our coastal and ocean problems are related to the relentless population pressures on the coastal areas of the nation. A tiny tax on real estate transfers in the nation's coastal counties would raise a significant amount of money to address many of these problems, including those related to coastal and ocean governance improvement.

Operationalizing Suggestions for Improvement

In this section, we offer three concrete suggestions for changes in national ocean policy that we believe are of vital importance, both because of their inherent merit and because together they could be the foundation stones upon which the United States can build a more robust national ocean policy process suitable for the beginning of the next century. The first involves the urgent need to formulate a national plan for the conservation and sustainable use of the 200-mile EEZ. The second concerns the equally urgent need for a formal, continuing policy coordination mechanism at the federal level—a national ocean council. Our third suggestion is also structural in nature and involves the coastal states and territories—the twenty-three states bordering U.S. oceans, the seven states bordering the Great Lakes, and five ocean territories/commonwealths—and the need to develop a set of equitable and resilient partnership arrangements with them in the management arena. The coastal states and their coastal communities have strong economic, environmental, and cultural interests in the seas adjacent to their shorelines and therefore must be an integral part of any enduring ocean governance regime.

A National Strategy for the EEZ

President Reagan's proclamation establishing a 200-mile EEZ claimed "sovereign rights" over all marine resources found therein, with the exception of the highly migratory tuna. In contrast to existing marine laws, which are largely sectoral and single-purpose in nature and have varying geographical jurisdictions, the EEZ proclamation declared U.S. jurisdiction over a single geographical area, including all living and nonliving marine resources therein (management of tunas was also ultimately included in the 1990 MFCMA amendments). Theoretically at least, the creation of a large area of uniform jurisdiction encompassing all resources offers the possibility of developing a multiple-use approach to marine resources management and of redressing some of the coordination problems inherent in the sectoral, use-by-use approach.

The declaration of a United States EEZ introduced two novel elements into the existing ocean governance framework: the notion of a common area covering all resources, and the concept of "sovereign rights." For the first time, the United States declared formal jurisdiction over a huge, common ocean area, encompassing close to 3 million square nautical miles. And, also for the first time, a type of jurisdiction encompassing all marine resources was created, thus, in principle, establishing a common multipurpose zone.

The second new aspect inherent in the EEZ was the notion of sovereign rights over all resources. The concept of "sovereign rights" was used in the 1982 Law of the Sea Convention to describe the nature of the rights that a coastal nation acquires over the resources in its 200-mile EEZ. While the full implications of the declaration of sovereign rights may be unclear, what is clear is that the notion of "sovereign rights" implies a higher level of government authority over, and greater responsibility for, marine resources than existed before. Prior to the EEZ declaration, the authority of the United States over marine resources was more limited. It included "exclusive fishery management authority" in the fishery conservation zone (under the Magnuson Act) "jurisdiction and control" over the resources of the continental shelf (under the OCSLAA), and "jurisdiction over the taking or importation" of marine mammals (under the MMPA).

Speculation as to the exact meaning of sovereign rights is best left to the lawyers, but it appears that the EEZ notion of sovereign rights brings with it the idea of more responsibility for the common property resources found in the ocean—an increased role of public or common stewardship. The notion of sovereign rights may thus lend new meaning to the concept of "common property," which has come to be cast in largely negative terms—in the international as well as the national context.

Internationally, "common property" traditionally has meant freedom from restraint, which in fact turned out to be the freedom to overexploit the resources of the world's oceans on the part of those nations with the technological capability to do so. The representation of the oceans as the "common heritage of mankind," which was advanced during the LOS negotiations and is now built into the 1982 Convention, could be interpreted as an effort to cast the common property concept in a more positive light—viewing ocean resources as a common heritage implies international responsibility for conserving as well as wisely utilizing that heritage. This idea, however, has not yet been fully operationalized at the international level, in part, because there may not be sufficient commonality of interests, traditions, and ties among nations to produce the necessary level of "communal" involvement and responsibility. In short, the lack of "community" at the international level has thus far prevented the firm rooting of concepts of common responsibility, in spite of the fact that the deep seabed has been declared the common heritage of humankind by the United Nations General Assembly and procedures now exist for operationalizing the concept in the 1982 LOS Convention, which entered into force in 1994.

At the nation-state level, on the other hand, there exists a sufficient level of commonality—the common traditions and ties that bind the nation together as a unit—that could be relied upon to create a more structured level of societal involvement vis-á-vis the oceans. In the United States, however, the notion of common property, as applied to the oceans, has come to be cast in largely negative terms—as the freedom for all to go and take, causing detrimental consequences such as resource depletion and economic inefficiency (Crutchfield 1961). As discussed elsewhere (Cicin-Sain 1980) and by others (e.g. McCay and Acheson 1987), we do not believe that "tragedy" is necessarily inherent in the "commons" (Hardin 1968). The notion of sovereign rights contained in the EEZ could be used, instead, to avoid depletion of the commons by building a higher level of societal responsibility over ocean resources.

As pointed out earlier, we see a parallel between the evolution of governance of a large new ocean area like the EEZ and the manner in which governance came to the western part of our country during the last century. At first, the needs for government in the western lands were minimal—perhaps a place to pick up such maps and survey information as existed at the time and an office to register land claims. Of course, the sheriff was needed early, as, eventually, were judges and a court system. As settlements took root, services such as roads, water systems, and schools were needed and general purpose governments (towns and villages) were created to meet these needs. Likewise, as activities in the U.S. ocean zone increase and, with them, interactions and conflicts, the need for something closer to an area-based, general purpose governance system will become clear.

With sovereign rights go "sovereign responsibilities." When the United States claimed sovereign rights over the EEZ and its resources, the nation accepted certain obligations associated with its assertion of this higher level of control, obligations that can only be met if it puts an active stewardship program in place. Americans have a right to expect ocean resources claimed by the government to be managed and conserved in a sustainable manner for this and succeeding generations. And this is not a passive undertaking in an ocean zone already under pressure from a variety of economic activities in many areas.

We believe that "a national strategy for the sustainable development of the U.S. ocean" is needed. We use the term U.S. ocean to mean both the U.S. territorial sea (from the shoreline to 12 nautical miles offshore) and the U.S. EEZ (from 12 nautical miles to 200 nautical miles from the shoreline). The coastal states currently have jurisdiction only over the first 3 miles of the U.S. territorial sea, with the remaining 9 miles not incorporated into state boundaries and constituting a vacuum in domestic governance. Although the first 3 miles of the territorial sea are included within state boundaries, and by statute are included in state coastal zones for planning and management purposes, the remaining 9 miles of the territorial sea and the entire 188 miles of the EEZ are not now a part of any general planning or management framework. Hence, until the federal government can see its way clear to extend coastal state boundaries to include this 9 miles of new U.S. territory, this band of "orphan ocean" should, we feel, be treated as part of the U.S. EEZ for planning and management purposes, and we do so in the following proposal.

A National Strategy for Sustainable Development of the U.S. Ocean

Although the United States has the largest and richest ocean zone of any nation, it has no overall national policy for its sustainable development and use. Development of a national strategy is essential if the American people are to receive full benefit from their oceans and coasts while the interests and choices of future generations are protected as well. Adoption and implementation of a national strategy for sustainable development of the U.S. ocean will also enable the United States to, once again, lead in ocean management internationally and by way of example, share to other nations its technological and managerial know-how regarding oceans and coasts.

Given the great diversity that exists in different regions of the coastal ocean in the United States (e.g., distribution of offshore resources, physical factors, and socioeconomic and political-administrative conditions), the national strategy will need to be tailored to take into account important differences in regional circumstances and interests. To succeed, such tailoring must be done in close partnership with the affected state and local governments.

The national strategy for sustainable development of the U.S. ocean should contain the following (Cicin-Sain 1994):

1. A set of goals and specific targets for sustainable development, with an accounting of projected benefits for current and future generations.

2. A set of principles to govern use of the U. S. ocean, with special

emphasis on operationalizing the concepts of "stewardship" and "intergenerational equity."

3. Priorities for action and mechanisms for establishing priorities.

The national strategy for the U.S. ocean should also contain plans to accomplish the following:

1. A decade-long ocean exploration program to inventory the resources, both living and nonliving, that are contained in the seafloor, the water column, vent systems, sea mounts, and the other domains of this vast area.

2. An economic assessment of the value of these resources and possible timetables and scenarios for sustainable development and/or use of appropriate subsets of these resources.

3. An EEZ-wide plan for the systematic designation of the special places in the U.S. ocean meriting protected-area status under an expanded national Marine Sanctuaries Program.

4. Options for a management regime for the U.S. ocean that incorporates the interested coastal states and territories as full partners with the federal government, giving specific attention to the possibility of developing regionally oriented regimes consistent with an ecosystem approach.

5. A timetable and means of implementation of the various elements of this strategy. In terms of direction and oversight of this effort, a body composed of the leaders of the major federal ocean agencies, such as a national ocean council, with the coastal states and territories serving in an advisory capacity, would be ideally positioned to assume responsibility for implementation.

A National Ocean Council

We have cited several times the urgent need for some sort of a policy and programmatic coordinating mechanism at the federal level in the field of ocean and coastal affairs. The need has grown acute as more and more single-purpose ocean legislation is enacted to address specific ocean and coastal problems with little regard for existing legislation and its goals.

The concept, of course, is not new. A Marine Science Council was established by legislation in 1966 and operated until 1971, when its funding terminated. Numerous subsets of ocean agency leaders have at times come together in interagency initiatives aimed at specific problems. The most recent probably was the interagency group brought together in 1995 by the Maritime Administration (at the urging of the White House) to review and streamline the regulatory processes connected with maintenance dredging of the nation's navigational channels. Perhaps most typical are bilateral get-togethers between two agencies to attack a particular issue. The cooperative work on non-point-source pollution by EPA and NOAA, which was mandated by statute (Section 6217 of 1990 legislation reauthorizing the Coastal Zone Management Act), is a good example.

The nature and complexity of the ocean and coastal policy issues that will have to be faced in the early years of the next century argue persuasively for a permanent ocean policy coordinating body. We mention three issues here to make our point.

1. Restoring the abundance of the nation's marine fishery resources.

2. Restoring and maintaining the quality of our coastal and estuarine waters.

3. Maintaining the utility and attractiveness of the nation's recreational beaches.

Success in each of these areas requires the coordinated, concerted effort by a number of federal agencies and other groups. Solving the fisheries problem will require that the Regional Fishery Management Councils, the National Marine Fisheries Service, state coastal management agencies, the EPA and its 319 program, NOAA and its 6217 program, and, probably, the marine protected-area programs of the state and federal governments all work together over an extended period of time toward an agreed outcome. Certainly, the prospects for success will be higher if such an effort is carried on within the framework of (and with the support of) an adequately empowered national ocean council.

Restoring coastal and estuarine waters to an acceptable level of swimmability and fishability will take the same kind of cooperative, sustained effort that the fisheries problem requires. The EPA and NOAA and their state counterparts will have to work closely with the Department of

Agriculture and its Natural Resources Conservation Service as well as with the local conservation districts to effectively address this problem. Without a significantly higher level of cooperation and political will than has been evidenced before, harmful algal blooms, *Pfiesteria*, and other dangerous organisms and pollutants will continue to threaten and degrade our coastal waters. There is no evidence to suggest that this higher level of sustained effort could come from the usual interagency committee, but it might be achievable through the oversight of a permanent federal council that had the support of the executive office.

Maintaining the utility of the nation's recreational beaches in the face of both accelerating sea-level rise and increased erosion, with federal support of beach replenishment projects declining, will also require a sustained effort by a number of federal agencies (Army Corps of Engineers, NOAA, FEMA, EPA, etc.) working very closely with their state counterparts and local beach cities, towns, and communities. The major economic impact of coastal tourism and recreation demands that this problem be addressed effectively and at a high level.

Among the functions to be performed by a national ocean council are the following:

- Consistent with existing legislation, develop national ocean policy positions as needed to fill gaps, deal with new and emerging issues, and resolve conflicts.
- Provide broad national guidance on overall goals for the nation's oceans and coasts in consultation with resource users, the coastal states, and other relevant user groups.
- Review and assess the progress of individual agency programs in achieving national ocean goals.
- Create and oversee the work of federal agency working groups formed to address specific ocean and coastal problems requiring concerted, high-level, and sustained attention.
- Serve as a liaison between the coastal states and territories and the federal government, especially in encouraging and overseeing state efforts to create ocean plans for the oceans adjacent to their coastal waters.
- Review the budget submissions of ocean agencies to ensure their consistency with agreed national ocean policy goals and objectives.

Ideally, a national ocean council should be a part of the executive office structure in a manner similar to the Council for Environmental Quality (CEQ), but this is not likely to occur given the desire of most chief executives to keep the executive office small and limited to a few overarching policy areas (national security, economic policy, domestic policy). However, some of the prestige and clout of the higher level might be gained if the new council were created by a presidential executive order, which enumerated specific duties and responsibilities such as those suggested above. Empowering the council to conduct annual reviews of the ocean budgets of the constituent agencies would also give it a role that would be taken seriously. Also, if the council had authority to "tax" its member agencies up to some small percentage of their annual budgets (say one-tenth of 1 percent) to support council-initiated programs, it would likely command respect and attention. Requiring that such a council report directly to the president would also enhance its standing and increase the likelihood that its recommendations would have impact.

It would be desirable that a national ocean council created by an executive order would eventually be made permanent by legislative action. The council would have greater durability and its actions would likely have greater impact if it were seen as a creature of both the executive branch and the Congress.

Partnerships with the Coastal States and Territories

The national ocean council concept outlined above will be difficult to achieve, in part, because of the rivalries and "turf" concerns that characterize the relations between "sister" agencies, especially when they are competing for the same resources and new missions. Different but equally thorny problems exist when considering the means and modalities for improving relationships between levels of government—national, state, and local. Each level of government fully understands the prerogatives associated with that governmental level and anything that threatens those prerogatives is taken as a serious threat.

Superimposed on these relationships is a strong tendency for creation of a "pecking order" that places the federal government and its agencies on top, local governments and their communities on the bottom, and state governments in between. This order flies in the face of several realities, not the least of which is the enumeration of powers provision of the Constitution, which makes it clear that the federal government is not superior to the states and localities in all areas of policy. The federal government has certain enumerated powers and can supercede state actions in those areas and in other areas where Congress has declared a national interest and has created a federal presence. The protection of endangered species, including marine mammals, would fall in this category. But when considering the oceans and their conservation and development, the coastal states and their coastal communities have legitimate and defensible interests in state waters (e.g., within their 3-mile boundaries), in the adjacent 9-mile belt of U.S. territorial sea (the "orphan ocean"), and in significant parts of the EEZ beyond the outer boundary of the territorial sea (at 12 miles) (Eichenberg 1992). The state of Oregon, for example, has presented convincing arguments that its economic and environmental interests extend up to 85 miles offshore (well into federal waters), depending on the location of fisheries and other resources of direct and immediate interests to the state's citizens (Oregon 1995). They define this zone, which generally coincides with the location of the continental shelf off its shores, as the "Oregon stewardship zone."

Of course, the federal government also has very legitimate interests in the coastal waters of the nation. It must guarantee national security and must ensure that the United States is fulfilling its obligations as a member of the community of nations. Protecting the commerce between states is another enumerated federal responsibility. And in federal waters, currently beyond 3 miles, federal agencies oversee the operation of numerous federal programs generally authorized by specific statutes. Oil and gas leasing, enforcement of fisheries laws, and supervising the Marine Sanctuaries Program are three of the many such programs.

We suggest the following set of principles that should undergird a strengthened federal-state partnership as well as a set of specific actions that should be taken.

Proposed Principles

1. Except for those relatively rare instances of clear national security concern, the federal government and the states and their coastal local governments have equally legitimate interests in the conservation and sustainable use of the U.S. ocean.

2. The 3-mile line (where formal state jurisdiction ends and formal federal jurisdiction begins) should not be seen as a hard-and-fast boundary delimiting the separation of state/local and federal interests in the ocean.

3. In conflicts that may occur, other things being equal, within the full territorial sea (0–12 miles), demonstrated state and local interest should prevail over latent national interests.

4. Coastal states and territories with significant interests in ocean resources and uses adjacent to their coastal zones should be incorporated into any EEZ strategic planning and management process as full partners with the federal government.

Proposed Specific Actions

1. Encourage coastal states to develop territorial sea plans as a part of their CZM programs.

2. Encourage states to develop plans for the 3- to 12-mile zone of territorial sea as an extension of their CZM programs.

3. As one of its early actions, a national ocean council (or similar body) should, in cooperation with interested coastal states and territories, develop the terms of reference of a new and enhanced federal-state partnership in ocean and coastal affairs.

4. As an early item of business, the new partnership should develop the parameters for implementation of a coastal energy impact assistance program. The situation is significantly different than in 1976 when the original coastal energy impact program was adopted. In the Gulf of Mexico, for example, most of the new oil and gas development is in very deep water, raising a different set of adjacent state concerns involving environmental impacts to the marine environment, pipeline and other transportation issues, and the abandonment of shutdown platforms. In other areas, better integration of development plans on federal leases with plans for developing leases in state waters is needed. Improved coordination between federal and state oil and gas development is essential if coastal zone impacts are to be properly addressed.

Some Experiences from Other Nations

As the United States considers strategies for improving its national ocean policy and the associated institutional arrangements, it will be useful to monitor and exchange information with other nations that are undergoing similar processes. As noted in Chapter 6, many nations have, in recent years, begun to adopt national and regional mechanisms for ocean and coastal policy coordination and have undertaken new planning efforts for their EEZs and other waters under national jurisdiction. Here is a brief highlight of the experiences of Australia, Canada, and Korea.

Australia's Oceans Policy

A significant national ocean policy initiative was announced by Australia in December 1998 with considerable public notice. Released by the prime minister's office, the document, "Australia's Oceans Policy," contains a detailed blueprint for the conservation and wise use of Australia's oceans. Senator Robert Hill, minister for the environment and heritage and leader of the government in the Senate, said that "the release of Australia's Oceans Policy in the International Year of the Ocean positions Australia as a world leader in implementing integrated oceans planning and management" (Australia 1998, p. 3).

The new plan contains new mandates for regional ocean planning, newly articulated principles to guide such planning, and new institutional arrangements. Australia has decided to develop regional marine plans based on identified large marine ecosystems that will ensure, as far as possible, the integration of planning and management across state and commonwealth waters. The new policy statement contains broad principles for ecologically sustainable ocean use that will be used in formulating regional marine plans. Four new institutions are called for: a National Oceans Ministerial Board (to oversee the regional marine planning process), a National Oceans Advisory Group (mainly made up of NGOs, to advise in the planning process), a National Oceans Office (to serve as a secretariat to the National Oceans Ministerial Board and the regional planning efforts, to be housed in Environment Australia), and Regional Marine Plan Steering Committees (to oversee the development of regional marine plans). The Australian government has committed $50 million over three years to support various aspects of this program.

With one of the world's largest and most biologically diverse exclusive economic zones and its oceans strategy, Australia has clearly moved to center stage on the global ocean policy scene.

Canada's Oceans Act

After several years of study, Canada passed its Oceans Act in December 1996. The act reiterates the boundaries of Canada's maritime zones and defines the jurisdictional division between the federal government and the provinces. It calls for the development of a national strategy for the management of estuarine, coastal, and marine ecosystems "in waters that form part of Canada or in which Canada has sovereign rights under international law" (Canada Oceans Act, Chapter C-31, 45 Elizabeth II, 1996). The act specifies that the national strategy will be based on the principles of sustainable development, integrated management of estuarine and coastal activities, and the precautionary approach. The minister of fisheries and oceans is called upon to "lead and facilitate the development and implementation of plans for the integrated management of all activities or measures in or affecting estuaries, coastal waters and marine waters. . . ." With regard to the implementation of these plans, the minister is given broad authority to coordinate with other ministers, boards, and government agencies to establish new advisory or management bodies and to recognize existing management bodies; and to establish marine environmental quality guidelines and objectives and criteria respecting estuaries, coastal waters, and marine waters.

While it is difficult to know precisely what direction the new Canadian oceans initiative will take, it is clear that Canada intends to develop and implement a more coherent national oceans policy. It is also likely that Canada's coastal management efforts, which have been largely community oriented to date, will begin to receive increased attention at the federal level.

Korea's Ministry of Maritime Affairs and Fisheries

As is well known, the Republic of Korea (herein Korea) witnessed very substantial economic development in the decades since about 1970. By the early 1990s, however, some of the environmental consequences of this development were becoming clear, especially in terms of degradation of its coastal waters (Hong and Lee 1995). Korea participated fully in the 1992 UNCED in Rio, establishing local Agenda 21 programs in the years that followed. In March of 1995, the Korean government formulated the "New Marine Policy Direction Toward the 21st Century." Included in the new policy statement were Law of the Sea issues, integrated coastal management, the development of ocean industries, and increasing public awareness of ocean-related culture (Hong and Lee 1995). Specifically, consistent with Chapter 17 of Agenda 21, the new policy called for Korea to enact a national coastal management law by 1997 and to prepare a management plan for the development and conservation of coastal resources by 1998.

In terms of institutional arrangements, in 1997 Korea reorganized its marine programs and created a new Ministry of Maritime Affairs and Fisheries and a new Korean Maritime Institute. The new ministry brings together most of Korea's programs in marine affairs, coastal management, fisheries, marine science, and port management. The declared intention of the reorganization is to permit Korea to develop and implement a more coherent oceans policy in the coming years. The newly created Korea Maritime Institute serves as the marine policy "think tank" for the nation, conducting a wide variety of policy analyses on marine issues.

Building a Coalition for Change

We have proposed a good number of changes to the existing system, many of them requiring legislative action. However, getting policy changes through the Congress and signed into law can be a notoriously difficult process usually requiring, among other things, the formation of a coalition of affected interests. In this section, we examine the kinds of coalitions that will likely be needed if the changes proposed here are to be realized, beginning with an assessment of the current political landscape in the ocean and coastal arena.

The Political Landscape in the Late 1990s: A Snapshot

The term "political" in this discussion refers not to the partisan politics of political parties but to the power exerted by various interests in the legislative process. Of particular interest is the extent to which various kinds of interest groups (e.g., the Coastal States Organization, the oil and gas companies, fish processors, environmental organizations) are active on an issue or set of issues. Active in this sense relates to the amount of time, energy, and/or money being expended to influence the outcome of legislation of one type or another. Rough estimates of the level of activity and the position of interest groups on specific issues can be obtained not only by monitoring news releases, newsletters, specialized publications to their membership, content and frequency of testimony to congressional committees, but also in discussion with committee staffers.

The power to influence legislation shifts over time. Changes in public perception can play a role. For example, fishers and the processing industry in the past have had power in Congress disproportionate to their numbers, but that influence may diminish as they bear (rightly or not) some of the blame for recent fishery declines. The political power of a particular interest at a given time depends on a number of factors but perhaps the most important is the centrality of the issue in question. If the issue is seen as a matter of burning importance to a particular interest group—one of their core issues—the views of the interest group will naturally be given more serious consideration, especially if the group has demonstrated technical competence in the area and has established a record for credibility. Other factors affecting the influence that interest groups can wield on particular issues depend on their relationship to the congressional subcommittee or full committee involved in the legislation. An interest group that has a record of being helpful to the committee and its key staffers by providing sound information in a timely fashion, by taking "reasonable" positions, and by attempting to work out problems with groups having interests different than its own, will receive more attention than a group with a reputation for contrariness.

A nonexhaustive snapshot of the current political landscape in the ocean and coastal field is shown in Table 7.2. The fifteen "issue area focal points" represent most of the principal issues on which upcoming ocean policy legislative debates will be centered. While there can be a considerable degree of overlapping interests on particular issues, each of these focal points represents a relatively free-standing, identifiable issue around which political power gathers when the issue or some aspect of it surfaces in the legislative process. The table also shows some of the major interest groups that make up the focal point, the current goals of the groups, and (if known) the positions of the groups on the 1998 Ocean Act legislation. The reader should note that this table represents a simplified view of reality, since a number of the major interests (e.g., Center for Marine Conservation, Natural Resources Defense Council, SeaWeb) are active in several issue areas.

The position and power of some subset of interest groups such as these will need to be taken into account as the legislative prospects for particular kinds of ocean policy change are assessed. Legislation proposing to change the balance of power between coastal state governments and their local governments, for example, clearly will get the attention of local governments, the coastal states, and almost certainly, members of the environmental community. Similarly, legislation that relates to the power of states to regulate the use of private property in the coastal zone will surely get close scrutiny by the property rights groups, among others.

Of course, Table 7.2 is only a snapshot of the situation as reflected in the positions of a number of the politically active interest groups. Clearly, of most importance in predicting the fate of a piece of legislation

TABLE 7.2 U.S. National Ocean Policy in the Late 1990s: The Political Landscape.

MAJOR ISSUE AREA FOCAL POINTS		
CATEGORY OF FOCAL POINT	NAMES OF SELECTED INTEREST GROUPS	SELECTED ISSUES OF CURRENT CONCERN
1. Marine and coastal environment, wildlife	Center for Marine Conservation (CMC)[1]	CMC is "committed to protecting ocean environments and conserving the global abundance and diversity of marine life." *Current Issues:* • Revitalize America's marine fisheries • Clean America's ocean waters • Invest in the future of America's oceans • Strengthen and expand marine protected areas • Save America's coral reefs • Protect endangered marine wildlife • Explore America's marine wildlife and ocean waters • Promote ocean stewardship and education "CMC enthusiastically supports the Oceans Act, the creation of a National Ocean Council, and the establishment of a Commission on Ocean Policy, and believes that through a new Oceans Act, United States would take needed action to fulfill the ocean stewardship responsibilities for the EEZ."
	American Oceans Campaign (AOC)[2]	AOC has taken a "leading role in protecting and restoring the vitality of the U.S. waters. It is educating the American public and policy makers about the need for establishing strong public policy to protect U.S. marine resources." *Current Issues:* • Lead the national effort to defend the Clean Water Act from an industry-written reauthorization bill (H.R. 961) • Concern that NMFS's fishery management guideline would not fulfill congressional mandate • Urge Congress to support increased funding for oceans • Launch public education campaign on the importance of fish habitat conservation "AOC supports passage of the Oceans Act, and firmly commits to the creation of a Commission on Ocean Policy, developing recommendations for a comprehensive national ocean policy."

	Natural Resources Defense Council (NRDC)[3]	NRDC's purposes are to protect the planet's wildlife and wild places and to ensure a safe and healthy environment for all living things. *Selective current marine-related issues:* • Restore ocean fisheries • Give North Atlantic swordfish a break • Stop beach pollution • Clean up our waters • Save the gray whale nursery • Protect marine wildlife
2. Wetlands and coastal water quality	Clean Water Network (CWN)[4]	A broad-based alliance of national, regional, and local groups have come together as the Clean Water Network to protect and strengthen the federal Clean Water Act. *CWN has developed an agenda for clean water based on three key issues:* • Prevent pollution and achieve zero discharge • Protect critical ecosystems • Enforce the law: Protect the people
	The Wildlife Society (TWS)[5]	TWS brings the combined expertise of its membership to bear on wildlife policy issues. *Selective issue priorities:* • Clean Water Act reauthorization • Ecosystem management • Endangered Species Act reauthorization • Federal budgets for wildlife protection • National wildlife refuges management • Teaming with wildlife • Wildlife research priorities and funding
	Environmental Defense Fund (EDF)[6]	EDF is dedicated to protecting the environmental rights of all people, in the issue areas of clean air, clean water, healthy and nourishing food, and a flourishing ecosystem. *Key concern on marine issues:* • Prevention of ocean pollution • Depletion of fish stocks

continued

TABLE 7.2 *(continued)*

MAJOR ISSUE AREA FOCAL POINTS		
CATEGORY OF FOCAL POINT	**NAMES OF SELECTED INTEREST GROUPS**	**SELECTED ISSUES OF CURRENT CONCERN**
		• Problems of bycatch • Establishment of a network of Marine Protected Areas • Coastal water chemical contamination from fish farming
3. Shoreline and its management (Shoreline management and beach) renourishment	American Coastal Coalition (ACC)[7]	ACC is a national membership organization composed of governmental entities, business people, academics, coastal community residents, etc. It has been organized to serve as the voice of the nation's coastal communities. *Selective current issues:* • Support the Shore Protection Act of 1996 (Section 227 of the Water Resources Development Act of 1996) and urge implementation of the Act • Promote environmental policies that benefit coastal communities • Work with and support the House and Senate Coastal Caucuses "ACC supports the Oceans Act and an initiative that will promote interagency and intergovernmental cooperation in the formulation and implementation of coastal policies."
4. Fisheries (Commercial fisheries, processing, fisheries conservation)	National Fisheries Institute (NFI)[8]	NFI is "the national voice of the fish and seafood industry." It represents the interests of its diverse members who are producers, processors, wholesalers, distributors, brokers, importers, aquaculturists, restaurants, retailers, exporters, etc. • NFI releases a Seafood Top 10 species for 1997 • NFI calls for ratification of Sea Turtle Treaty • NFI expresses that it is time for foreign fishermen to match U.S. efforts in swordfish conservation • NFI calls for enactment of S. 1489 • NFI welcomes WTO decision on shrimp embargo "NFI supports the Oceans Act and the establishment of a Commission on Ocean Policy. NFI is confident that through public education and cooperation between all levels of the government, the U.S. can develop long-range national ocean policies, focusing on the oceans [EEZ] as a source of food."
	American Fisheries Society (AFS)[9]	AFS is dedicated to conserving North America's fisheries and aquatic ecosystems by promoting professional excellence in fisheries science, management, and education, in the issue areas of: • Biodiversity • Effects of surface mining on aquatic resources in North America • North American fisheries policy

		• Special fishing regulations for managing freshwater sport fisheries • Floodplain management • Human use of fish and other living aquatic resources
	National Audubon Society (NAS)[10]	Living Oceans Program is the marine conservation program of NAS, aimed at influencing federal governmental actions such as the Magnuson-Stevens Sustainable Fisheries Act and its regional fishery management plans. *Action Alert:* • Overfishing of giant ocean fishes such as sharks, swordfish, bluefin tuna, blue and white marlins • Decline of Atlantic bluefish • Severe depletion of sharks in the Atlantic and Gulf of Mexico
	Ocean Wildlife Campaign (OWC)[11]	OWC is a coalition of five conservation groups: National Audubon Society, National Coalition for Marine Conservation, Natural Resources Defense Council, New England Aquarium, Wildlife Conservation Society, and World Wildlife Fund's Endangered Seas Campaign. *OWC is fighting to:* • Stop overfishing • Change the way fish are caught • Establish and implement management and recovery plans
	Greenpeace[12]	With nearly 3 million members around the globe, Greenpeace purposes are to create a green and peaceful world. *Selective current issues on oceans:* • High-tech floating fish factories are devastating our seas • Greenpeace urges all governments to eliminate overfishing, to rebuild depleted fish populations, to adopt a precautionary approach to fisheries management, and to develop more selective fishing practices to minimize bycatch and waste • Greenpeace promoted congressional passage of a bill banning factory trawlers from East Coast fisheries • Greenpeace campaigns to contact senators and representatives to encourage their support of S. 1212, which provides for the closure of U.S. fisheries to any new factory trawlers and begins the phase-out of the entire U.S. factory trawler fleet

continued

TABLE 7.2 *(continued)*

MAJOR ISSUE AREA FOCAL POINTS		
CATEGORY OF FOCAL POINT	**NAMES OF SELECTED INTEREST GROUPS**	**SELECTED ISSUES OF CURRENT CONCERN**
5. Coastal Grants (State coastal management programs [grants, etc.])	Coastal States Organization (CSO)[13]	CSO represents the governors of the coastal states, territories, and commonwealths as an advocate for improved management of the nation's coasts, oceans, and Great Lakes. *Selective current issues on coastal grants:* • CZMA sec. 306 /309 state grants • CZMA sec. 309 coastal nonpoint grants • CZMA sec. 6217 coastal nonpoint pollution grants • CZMA sec. 315 NERRS grants • NERRS Construction Account • Coastal Services Center grants • CWA sec. 106 state water quality management grants • CWA sec. 319 non-point-source management grants • Clean Water Action Plan Implementation grants • Hazard mitigation grant program (Stafford Act's sec. 404) • Mitigation banking "CSO supports the Oceans Act and a presidentially appointed Commission on Ocean Policy, to review ocean and coastal activities, identify key issues, assess federal programs and funding priorities, and recommend a comprehensive national ocean and coastal conservation and management plan."
6. Offshore Oil and Gas Industry	National Ocean Industries Association (NOIA)	NOIA: Exploration and development of domestic offshore oil and natural gas resources.
	American Petroleum Institute (API)	API: A major national trade association representing the entire petroleum industry including exploration, production, transportation, refining, and marketing.
	Domestic Petroleum Council (DPC)	DPC: Natural gas and crude oil exploration, production, and trading.
	Independent Petroleum Association of America (IPAA)	IPAA: Crude oil and natural gas exploration drilling more than 85 percent of new U.S. wells.
	International Association of Drilling Contractors (IADC)	IADC: Contract-drilling, well-servicing, and producing and supplying of oilfield equipment.

	National Mid-Continent Oil and Gas Association (NMCOGA)	NMCOGA: A national trade group representing both major and independent oil and gas companies on domestic exploration and production issues. "It is premature to recommend the establishment of a Council before the Commission has even met and deliberated. "It is not clear that there is an ocean policy problem that warrants the establishment of a Commission on Ocean Policy. "Offshore oil and gas operations are already highly regulated to meet stringent environmental standards. Also, existing federal statutes governing our industry in the offshore and OCS are numerous, complex, and comprehensive."
7. Marine Mammal Protection	World Wildlife Fund (WWF)—USA[14]	WWF (USA) works to influence U.S. policy makers to support strong international action to protect endangered species on Earth. *Major concerns with marine mammals:* • The establishment of the 11-million-square-mile circumpolar Southern Ocean Whale Sanctuary • The creation of the Turtle Island Heritage Protected Area • Cracking down on illegal wildlife trade and ending pirate whaling
	Defenders of Wildlife (DOW)[15]	DOW's mission is to protect all native animals and plants in their natural communities. *Three primary issue areas:* • Species Conservation (i.e., endangered species recovery, predator protection and restoration, wild bird protection, marine mammal protection) • Wildlife Habitat Conservation • Policy Leadership (i.e., state and federal wildlife laws, biodiversity education, wildlife trade, international treaties) *Representative accomplishments on marine mammal protection:* • DOW drafted and helped win enactment of laws to protect dolphins from death in tuna nets • DOW won a U.N. ban on high-seas large-scale drift-net fishing to help save dolphins and other marine mammals • DOW strengthened Marine Mammal Protection Act in 1994 reauthorization • DOW helped create a 3,700-square-mile marine sanctuary for the vaquita, the world's rarest marine mammal

continued

TABLE 7.2 *(continued)*

MAJOR ISSUE AREA FOCAL POINTS		
CATEGORY OF FOCAL POINT	NAMES OF SELECTED INTEREST GROUPS	SELECTED ISSUES OF CURRENT CONCERN
	Friends of the Sea Otter (FSO)[16]	FSO is an advocacy group dedicated to actively pressuring state and federal agencies, not only to maintain the current protections for sea otters but also to increase and broaden these preservation efforts. *Selective current issues:* • Translocation of the southern sea otters to San Nicolas Island • Shellfisheries interactions with range expansion of the southern sea otter migration • Prevention of oil spills • The Federal Wildlife Service's draft of the Southern Sea Otter Recovery Plan • Sea otters interacting with fish traps, nets, shellfish traps, and kelp commercial harvesting • Rehabilitation and captive management programs
	The Sea Shepherd Conservation Society (SSCS)[17]	SSCS is involved with the investigation and documentation of violations of international laws, regulations, and treaties protecting marine wildlife species. *Major concern on marine mammal protection:* • SSCS is defending the great whales by enforcement of International Whaling Commission (IWC) regulations • SSCS opposes drift-netting activities • SSCS negotiated an end to the slaughter of dolphins near Japan's Iki Island • SSCS is leading to protect seals
8. Marine Transportation (Ports, waterways, marine transportation)	American Association of Port Authorities (AAPA)[18]	AAPA represents public port authorities in the United States, Canada, Latin America, and the Caribbean and promotes the common interests of the port communities. *Selective current issues:* • AAPA supports a multiyear reauthorization of the Intermodal Surface Transportation Efficiency Act (ISTEA) • AAPA supports enactment of a Water Resources Development Act • AAPA urges the Federal Government to implement a National Dredging Policy

		• AAPA supports legislation to extend the phaseout period for methyl bromide until a feasible alternative is available • AAPA advocates increased funding for NOAA's navigation programs, including programs for mapping and real-time tide and current systems "AAPA is pleased to support the Oceans Act and the creation of a new Commission on Ocean Policy, but it hopes the Commission includes at least one port representative, and maritime commerce as a priority in any recommendations. "AAPA emphasizes the importance of the EEZ that the United States uses in its trade lanes connecting world markets."
9. The Law of the Sea (Pro LOS Convention national security)	The Navy League[19]	The Navy League is a civilian organization dedicated to educating the American people, the media, and the executive and legislative branches of government about the need for the United States, a maritime nation, to maintain a strong maritime force. • The Navy League supports ratification of the LOS Convention
10. Ocean Exploration Discovery	SeaWeb[20] also very involved in fisheries	SeaWeb is a project designed to raise awareness of the world ocean and the life within it. *Selective current issues:* • Declines in swordfish, tuna • Trawling and long-lines • Shrimp and salmon farming • Algal blooms • Marine sanctuaries and marine zoning • Shark finning • Florida Bay as a microcosm of marine problems • Land-based toxic pollutants
11. Marine Recreation, Tourism (recreational fishing, boating, coastal tourism, resorts)	American Sportfishing Association (ASA)[21]	ASA is an industry trade association that is dedicated to serving the needs of the entire sport fishing community. *Selective current issues:* • ASA is leading an effort by the Fishable Waters Coalition to propose an amendment to the Clean Water Act

continued

TABLE 7.2 *(continued)*

MAJOR ISSUE AREA FOCAL POINTS		
CATEGORY OF FOCAL POINT	**NAMES OF SELECTED INTEREST GROUPS**	**SELECTED ISSUES OF CURRENT CONCERN**
		• The Sport Fishing and Boating Partnership Council (SFBPC) monitors the progress of federal agencies in improving federal sportfisheries activities • ASA strongly supports three important changes in the Magnuson-Stevens Sustainable Fisheries Act: 1. outlaws overfishing 2. mandates rebuilding of overfished fisheries, and 3. requires minimization of commercial bycatch "ASA supports the Oceans Act and creation of a Commission on Ocean Policy that can provide the basis for a coordinated national ocean policy. "ASA emphasizes the importance of the EEZ because of highly migratory species in the Atlantic Ocean."
	Professional Association of Diving Instructors (PADI): The Project AWARE Foundation[22]	The PADI Project AWARE Foundation is dedicated to protecting the fragile ocean environment and its diversity of life through public education, innovative projects, research, and advocacy. *Selective current issues:* • More effective use of the immense PADI network to aggressively address the most critical instances of the degradation of marine resources and coral reef ecosystems • Environmentally sound artificial fish habitat development • Promotion of coral reef mooring buoys • Beach and reef cleanup activities in more than 70 countries • Monitoring on legislative and agency regulatory actions: i.e., California Clean Coastal Waters and Rivers Bond Act (AB 1000), California Marine Life Management Act (AM 1241), U.S. Oceans Act (S. 1213) "The Project AWARE Foundation supports the Oceans Act (S. 1213). "The Project AWARE Foundation supports the establishment of a National Ocean Council and a Commission on Ocean Policy."
12. Local Governments	National Association of Counties[23]	NACO's membership totals nearly 1,800 counties, representing over 85 percent of the nation's population. It acts as a liaison with other levels of government, works to improve public understanding of counties, and serves as a national advocate for counties, dealing with such issues as the environment, sustainable communities, and volunteerism.

		Current issues: • community-based environmental protection • environmentally preferable purchasing • global warming • non-point-source water pollution • pollution prevention • source water protection • watershed management • wetland protection
	National League of Cities[24]	NLC represents more than 18,000 municipalities. It aims at influencing national policy and building understanding and support for cities and towns. *The five key objectives:* • Develop and advocate policies that strengthen and support cities • Strengthen the ability of city officials to serve their communities • Retain and expand membership by delivering innovative, effective, and quality services • Promote the image and enhance the stature and influence of NLC and the municipalities it represents • Provide an organizational structure that is flexible, efficient, and responsive to the needs of municipalities and State Municipal League
	U.S. Conferences of Mayors[25]	*Current issues:* • Arts, culture, and recreation • Community development and housing • Criminal and social justice • Energy and environment • Health and human services • International affairs • Jobs, education, and the workplace • Transportation and communications • Urban economic policy

continued

TABLE 7.2 *(continued)*

MAJOR ISSUE AREA FOCAL POINTS		
CATEGORY OF FOCAL POINT	NAMES OF SELECTED INTEREST GROUPS	SELECTED ISSUES OF CURRENT CONCERN
13. Maritime, Shipbuilding	Sailor's Union of the Pacific (SUP)[26]	SUP is a union of unlicensed sailors that work in the deck, engine, and steward's departments in U.S. flag vessels under contract of the Union. It has been involved since 1885 in political action to use its influence for effecting changes in the maritime law of the United States. *Current issues:* The SUP remains at the forefront on the West Coast in defending America's cabotage laws on which its members' livelihood are dependent—the Passenger Services Act and the Merchant Marine Act of 1920 (the Jones Act).
	International Organization of Masters, Mates, and Pilots (MM&P)[27]	MM&P is the International Marine Division of the International Longshoremen's Association, AFL-CIO. With 6,800 members living throughout inland and coastal states of the U.S. and outside the U.S., MM&P represents licensed deck officers, state pilots, and other marine personnel. *Current issues:* • Protest against California Ballot Initiative, Proposition 226, which would require that unions obtain annual written permission from each union member authorizing the use of their dues money for legislative activity, issues advocacy, or political activity • Defend the Jones Act • Support the Maritime Security Program: FY 99 Budget • Maritime tax reform—support tax equity policy between American mariners working aboard vessels in the foreign trades and other Americans working outside the United States • Double hull requirements under the Oil Pollution Act of 1990
	The Jones Act Reform Coalition (JARC)[28]	JARC represents 1,000,000 affiliated members in several economic sectors. Its purpose is to educate the Congress and public on the need for significant reform of the Jones Act and related laws. *Current issues:* • Reform the Jones Act of 1920 • Reform the 1886 Passenger Services Act • Reform the Longshore and Harbor Workers Act

14. Private Property Rights	Citizens for Private Property Rights (CPPR)[29]	CPPR's purpose is to conduct research and educate the public concerning: (1) civil and constitutional rights affecting private property; (2) public policy issues affecting private property rights; (3) the promotion and protection of constitutional and civil rights; (4) the governmental regulation and taxation of private property rights; (5) citizen involvement on governmental decision making; and (6) the prevention of governmental interference with private property rights. *Current issues:* The following governmental programs are opposed by the CPPR because of their usage to justify private land takings without just compensation as mandated by the Fifth Amendment of the U.S. Constitution • Endangered Species Act • Habitat Conservation Programs • The proposed National Wildlife Refuges • Community Conservation Program
	Defenders of Property Rights (DPR)[30]	DPR is a national public interest law foundation devoted exclusively to helping private property owners who have been deprived of their property by government actions. *Current issues:* DPR supports proposed laws to protect property rights. • The Tucker Act Shuffle Relief Act (H.R. 992) • The Private Property Rights Implementation Act (H.R. 1534) • The Railway Abandonment Clarification Act (H.R. 2438) • The Omnibus Patent Act of 1997 (S. 507) • The Property Owners' Access to Justice Act (S. 1204) • The Citizens' Access to Justice Act (S. 709) • The Private Property Rights Act (S. 709)
15. Coastal Agricultural Interests	American Farm Bureau Foundation (AFBF)[31]	AFBF represents more than 4.7 million members and is the voice of farmers and ranchers in local meetings, at state legislatures, and in the national capital.

continued

TABLE 7.2 *(continued)*

MAJOR ISSUE AREA FOCAL POINTS		
CATEGORY OF FOCAL POINT	NAMES OF SELECTED INTEREST GROUPS	SELECTED ISSUES OF CURRENT CONCERN
		Current issues: • Vote yes for a fast-track trade negotiating authority renewal • Vote yes for International Monetary Fund Funding—$18 billion • Vote to exempt agriculture from U.S. unilateral trade sanctions • EPA's *National Water Quality Inventories 1990–1994 report* (containing the message that agriculture, in general, and livestock specifically, are responsible for 60 to 70 percent of the pollution in U.S. surface waters) has a strong bias against agriculture • The criminalization of livestock production • *Pfiesteria*—no conclusive connection between nutrients and *Pfiesteria* problems, no conclusive evidence on human health impacts

Source: Table compiled by Dosoo Jang, Center for the Study of Marine Policy, with thanks from the Authors.

1. From Roger E. McManus. Testimony of Roger E. McManus, President of the Center for Marine Conservation Before the Subcommittee on Fisheries Conservation, Wildlife and Oceans of the House Resources Committee on H.R. 2547 and H.R. 3445, The Oceans Act of 1997/1998. March 19, 1998, and CMC web-site at http://www.cmc-ocean.org/
2. From David Younkman. Prepared Statement of David Younkman, Executive Director of American Oceans Campaign, Before the House Transportation and Infrastructure Committee Water Resources and Environment Subcommittee. August 6, 1998,Thursday, and the AOC web-site at http://www.americanoceans.org/
3. From the NRDC web-site at http://www.nrdc,organization/
4. From the CWN web-site at http://www.cwn.org/
5. From the TWS web-site at http://www.wildlife.org/
6. From the EDF web-site at http://www.edf.org/
7. From Howard Marlowe. Prepared Statement of Howard Marlowe, President of the American Coalition, Before the House Committee on Resources, Fisheries Conservation, Wildlife and Oceans Subcommittee, Subject—Activities Related to the International Year of the Oceans. October 30, 1997, and the ACC web-site at http://www.coastalcoalition.org/
8. From Richard E. Gutting Jr. Prepared Statement of Richard E. Gutting Jr., Executive Vice President of the National Fisheries Institute, Before the House Committee on Resources, Fisheries Conservation, Wildlife and Oceans Subcommittee. March 19, 1998, and the NFI web-site at http://www.nfi.org/
9. From the AFS web-site at http://www.fisheries.org/
10. From the NAS web-site at http://www.audubon.org/
11. From the OWC web-site at http://www.audubon.org/campaign/lo/ow/index.htm
12. From the Greenpeace web-site at http://www.greenpeaceusa.org/

13. From R. Kerry Kehoe Esq. Prepared Testimony of R. Kerry Kehoe Esq. The Coastal States Organization, Inc., Before the House Committee on Transportation and Infrastructure, Water Resources and Environment Subcommittee Subject—H.R. 2094, S. 1213, and H.R. 3445, ACSO Fax Alert. Sent by Chris Darnell, Policy Analyst at CSO, on September 21, 1998, and the CSO web-site at http://www.sso.org/cso/
14. From the WWF web-site at http://www.worldwildlife.org/
15. From the DOW web-site at http://www.defenders.org/
16. From the FSO web-site at http://www.seaotters.org/
17. From the SSCS web-site at http://www.seashepherd.org/org/
18. From Testimony March 19, 1998, The American Association of Port Authorities, House Resources Committee, Fisheries Conservation, Wildlife and Oceans Subcommittee, Ocean Bills. March 19, 1998, Thursday, and the AAPA web-site at http://www.aapa-ports.org/
19. From the Navy League web-site at http://www.navyleague.org/
20. From the SeaWeb web-site at http://www.seaweb.org/
21. From Michael Nussman. Prepared Statement of Michael Nussman, Vice-President of American Sportfishing Association, Before the House Transportation and Infrastructure Committee Water Resources and Environment Subcommittee. August 6, 1998, and the ASA web-site at http://www.asafishing.org/
22. From the PADI web-site at http://www.padi.com/Aware/
23. From the NACO at http://www.naco.org
24. From the NLC at http://www.nic.org
25. From the USCM at http://www.usmayors.org
26. From the SUP at http://www.sailors.org
27. From MM&P at http://www.bridgedeck.organization/mmp_htmlcode/mmp_about/
28. Source from JARC at http://www.jonesactreform.org
29. Source from CPPR at http://members.aol.com/proprts/cppr/
30. From DPR at http://www.defendersproprights.org
31. From AFBF at http://www.fb.com

is the predisposition of the members of Congress who comprise the subcommittees and committees that will consider the measure. Usually, the chairs of these bodies are also very instrumental in the outcome of legislation pending before their committees.

In addition, the executive branch of the federal government is not as passive a player in legislative matters as one might think. Ocean agencies will often have strong feelings on pending ocean policy changes and will find ways to make these feelings known to the relevant congressional committee. The administration's formal opinion on pending legislation (as approved by the Office of Management and Budget) is usually of considerable importance to a congressional committee. Unless there is outright warfare between a committee and the federal agency it oversees, it is usually difficult to get major changes in policy approved without the agency's agreement. Hence, part of the coalition building needed for ocean policy change will almost certainly involve getting at least some of the more influential ocean agencies or departments on board. Indeed, staff members from federal agencies often respond positively to requests from congressional committees for help in actually writing a new piece of legislation.

Finally, experience shows that a window of opportunity for policy change often opens when an urgent problem suddenly appears on the scene. Oil pollution legislation languished in the Congress for a decade but passed in only a few months after the dramatic *Exxon Valdez* spill occurred. Political scientist John Kingdon (1995) has suggested that the opening of this window of opportunity also often involves a political change such as the election of a new president or a change in parties controlling Congress. In the Kingdon model, policy entrepreneurs are urged to have their policy reforms tested and ready to go, pending the more or less simultaneous occurrence of a major problem and a political change.

Summary of Policy Changes Needed

Table 7.3 lists the three structural issues that we have highlighted earlier in this chapter as well as nine other narrower but also important problem areas drawn from the wider array of policy issues discussed in Chapter 5—the issue areas that are most in need of attention in the first five to ten years of the twenty-first century. We feel that a major effort needs to be mounted to build effective coalitions, including the appropriate federal and state agencies, to articulate the need for these changes

TABLE 7.3 A Summary of Needed Changes in National Ocean Policy.

MAJOR OVERARCHING ISSUES	IMPORTANT OCEAN ISSUES DEMANDING POLICY ATTENTION
1. The need for a national strategy for the sustainable development of the U.S. ocean—an EEZ plan. 2. The need for an effective policy and program coordination mechanism at the federal level—a national ocean council. 3. The need to strengthen and enhance the partnership between the federal government and the thirty-five coastal states and territories.	1. Restoring the abundance of America's fisheries. 2. Returning coastal and estuarine waters to a swimmable and fishable condition. 3. Making ports and waterways globally competitive. 4. Maintaining recreational beaches in the face of increasing erosion. 5. Revising coastal planning and emergency management programs, including flood insurance, to encourage a measured retreat from hazardous coastal areas. 6. Formulating a credible and workable plan for encouraging the development of marine aquaculture. 7. Reforming the federal offshore oil and gas programs in a way that is satisfactory to the affected interests. 8. Promoting the development of promising new ocean-related technology such as marine biotechnology. 9. Regaining U.S. leadership in ocean affairs at the international level by, among other things, accession to the Law of the Sea Convention and the Convention on Biological Diversity.

and to see them through the legislative and executive branch processes. Only when the reform legislation to deal with these issues is enacted and properly implemented, can we feel confident that U.S. national ocean policy is finally on the kind of firm footing needed to confront both the problems and the opportunities in the new century.

Concluding Remarks: Looking to the Future

The oceans that surround the United States hold inestimable wealth and value. This is a legacy that will accrue to the benefit of all Americans. And so far as we know, there are no irreversible calamities on the horizon.

Our plea is simply that the ocean deserves better of us. Quick to use its benefits, we are slower to recognize our duties of stewardship and careful management. The bureaucratic complexity that we have described here is probably no worse for the oceans than it is for other public policy areas. But that does not excuse us from the duty to improve it where we can.

A recent poll of the American people (Mellman Group 1996) revealed strongly held feelings about the ocean and its condition, suggesting that U.S. citizens would support aggressive efforts to restore and properly maintain this great natural resource.

The ocean that surrounds us also defines us. We are a continental nation that values the fact that broad seas separate us from our sometimes quarrelsome neighbors. Early on, the oceans, too, were the lifeblood of our nation.

But the ocean is far more than a defining feature of our national geography. It provides the Americans that it encloses with some of our most desirable food, our least expensive transportation, and, along its shorelines, wonderful places to recreate and restore the soul. But this list, too, is incomplete.

The ocean is one of the most vital links in the vast life support system that sustains all life on our planet. It is a complex set of interconnected parts, physical and biological, that operate as a whole producing the goods and services that we have come to take for granted. This ocean system must be planned and managed as the coherent, integrated system that it is. To fragment it to correspond to our pragmatic political and administrative subdivisions is to ignore this fundamental truth.

We must begin to set policy for the oceans in ways that respect this inherent integrity. Surely it is easier to achieve the effective harmonization of federal ocean programs than it is to part the seas along agency lines!

Among the richest and largest in the world and eagerly claimed at the first opportunity in 1983, our 200-mile zone still awaits the attention it clearly deserves. Serious plans for the conservation and sustainable use of this rich zone are urgently needed. This planning, best done at the regional scale, must include the adjacent coastal states as full partners in a fully transparent process. The twentieth anniversary of the U.S. EEZ declaration (year 2003) should see these plans in hand and responsible stewardship finally underway.

Last, we must comment on the appalling state of our fisheries. Perhaps more than anything else, this is a bellwether bottom line of the failure of U.S. ocean policy. Is it the fault of science or of the will of decision makers to take tough positions, or is it because we have been trying to manage fisheries in a one-dimensional way without regard to issues of habitat loss, pollution, and other factors? High priority must be given to restoring public confidence in our national capacity to achieve sensible policy goals. Putting marine fisheries on a sustainable footing would be a good first step.

The end of the twentieth century finds the United States the richest and single most powerful nation on earth. Surely, we can muster the resources and the political will to develop and put in place a better integrated approach to national ocean policy making in the twenty-first century. Americans of this generation and those to follow deserve no less.

APPENDIX

Chronology of Selected Major Events in U.S. Ocean Policy since 1945

1945 The Truman Proclamation. President Truman issues a proclamation declaring that the natural resources of the subsoil and seabed of the continental shelf were subject to the jurisdiction and control of the United States. A second proclamation, concerning fisheries, is also issued.

1945 *United States v. California* is argued before the Supreme Court in response to California leasing of near-shore tracts for oil development. The United States maintains that the territorial sea concept of the marginal sea did not arise until after the American Revolution; thus, no property rights existed for the states to succeed to at the time of independence. In 1947, the Court decides that the federal government retains "paramount rights" in coastal waters.

1947 An interstate compact creating the Pacific Marine Fisheries Commission with five participating states is approved by Congress. (Earlier, in 1942, the Congress had created another interstate compact, the Atlantic States Marine Fisheries Commission, with fifteen participating members.)

1949 The International Convention for the Northwest Atlantic Fisheries (ICNAF) is signed. This is the first multinational fisheries agreement signed by the United States. The Inter-American Tropical Tuna Commission (IATTC) is also established in 1949.

1949 An interstate compact approved by Congress creates the Gulf States Marine Fisheries Commission with five participating states.

1953 President Dwight D. Eisenhower (R) takes office on 20 January 1953. He serves two full terms and remains in office until 20 January 1961. During the first two years of Eisenhower's presidency, the government is unified (the president, the House majority and the Senate majority belong to the same party). For the remainder of Eisenhower's time in office, the House and Senate majorities are held by the Democrats.

1953 Enactment of the Submerged Lands Act grants states title to the natural resources (oil, gas, and all other minerals) located within 3 miles of their coastline (Texas and the Gulf coast of Florida are granted title to natural resources located within 3 marine leagues, or about 10 miles, off their coastlines). This legislation, in effect, reverses the 1947 Supreme Court ruling.

1953 Enactment of the Outer Continental Shelf Lands Act establishes federal jurisdiction over submerged lands on the outer continental shelf seaward of state boundaries.

1954 Enactment of the Saltonstall-Kennedy Act provides that 30 percent of the custom duties collected on imported fish be allocated to research and marketing to promote the use of domestic fish.

1958 The First United Nations Law of the Sea Conference is held in Geneva, Switzerland. Four conventions dealing with the high seas, the territorial sea and contiguous zone, fisheries, and the continental shelf were approved. No agreement was reached on the width of the territorial sea.

1959 *Oceanography 1960–1970* is published by the National Academy of Sciences. This report presents a blueprint for a major increase in federal support for ocean sciences.

1959 The Federal Council for Science and Technology (FCST) is established by an executive order from President Eisenhower. The council is created to enhance science and technology planning, to foster greater cooperation between federal agencies, and to advise the president regarding federal programs that have impacts upon multiple federal agencies.

1960 The Second United Nations Law of the Sea Conference is held in Geneva specifically to resolve the question of the width of the territorial sea. The Second Law of the Sea Conference also fails to reach an agreement on this issue.

1960 Congress approves a federal fishing vessel construction subsidy program that provides up to one-third (later one-half) of the cost of new fishing boats if they are to be used in fisheries threatened by foreign imports of fish.

1960 Passage of the Commercial Fisheries and Research Development Act authorizes a program of grants to the states (ultimately authorized at a level of $36 million per year) to study and improve commercial fisheries.

1960 The Interagency Committee on Oceanography (ICO) is created by the Federal Council for Science and Technology.

1961 President John F. Kennedy (D) takes office on 20 January 1961. He serves until his assassination on 22 November 1963. During President Kennedy's time in office, both the House and the Senate are in Democratic control.

1961 Passage of the first Anadromous Fish Conservation Act, authorizes a cooperative grants program with the states.

1963 President Lyndon B. Johnson (D) takes office following the assassination of President Kennedy on 22 November 1963. Johnson serves the remainder of Kennedy's term as well as one full term of his own and remains in office until 20 January 1969. Throughout Johnson's presidency, both the House and the Senate have Democratic majorities.

1964 Congress enacts legislation that provides for reimbursement of fines paid by owners of fishing boats seized by foreign governments for "unlawful" fishing within their extended jurisdiction zones.

1965 California enacts legislation making the (San Francisco) Bay Conservation and Development Commission (BCDC) permanent. BCDC becomes the nation's first coastal management program.

1966 The U.S. State Department officially supports a 3- to 12-mile-wide exclusive fishery zone for the United States and the enactment of legislation creating such a zone.

1966 Federal legislation creates the national sea grant college program, an ocean analogue to the successful land grant college program.

1966 The Marine Resources and Engineering Development Act creates the Marine Science Council and the Commission on Marine Science, Engineering, and Resources (COMSER), or the Stratton Commission.

1969 President Richard M. Nixon (R) takes office on 20 January 1969. He serves until his resignation from office on 9 August 1974, one and a half years into his second term. Throughout Nixon's presidency, both the House and the Senate are in Democratic control.

1969 *Our Nation and the Sea*, the influential report of the Stratton Commission, is released. It recommends, among other things, the creation of a national ocean and atmosphere agency and the adoption of a national coastal management program.

1970 The National Environmental Policy Act becomes law, establishing the environmental impact statement process for all federal actions affecting the environment and creating the Council on Environmental Quality in the office of the president.

1970 Reorganization of the federal government creates the National Oceanic and Atmospheric Administration and the Environmental Protection Agency.

1972 Enactment of the Coastal Zone Management Act (CZMA) creates a national coastal zone management program to be undertaken primarily by the coastal states with the cooperation and assistance of the federal government.

1972 Enactment of the amendments to the Federal Water Pollution Control Act (popularly known as the Clean Water Act) sets standards and timetables for improving the nation's coastal waters and regulates, among other things, the dredging and filling of the nation's wetlands.

1972 Enactment of the Marine Protection, Research, and Sanctuaries Act of 1972 establishes (1) a regulatory framework for ocean dumping in U.S. waters and (2) the National Marine Sanctuaries Program (a mechanism to designate and protect special ocean areas off the edge of the continental shelf).

1972 The Marine Mammal Protection Act establishes a moratorium on the take of marine mammals (with some exceptions) and a complete ban on the importation into the United States of marine mammals and marine mammal products.

1973 Enactment of the Endangered Species Act is designed to protect all threatened and endangered species, terrestrial and marine alike, and to protect habitats critical to their survival.

1974 President Gerald R. Ford (R) takes office following the resignation of President Nixon on 9 August 1974. He serves the remainder of Nixon's term and remains in office until 20 January 1977. During Ford's presidency, both the House and the Senate had Democratic majorities.

1974 Enactment of the Deepwater Port Act, entailing a process to license deepwater ports being considered for construction in U.S. coastal waters.

1976 The Fishery Conservation and Management Act (FCMA) establishes a 200-mile fishery zone for the United States and creates a series of regional councils to manage fisheries in this zone.

1977 President Jimmy Carter (D) takes office on 20 January 1977. He serves one full term and remains in office until 20 January 1981. During Carter's presidency, both the House and the Senate are in Democratic control.

1978 The Outer Continental Shelf Lands Act Amendments (OCSLAA) creates a complex new regulatory system for managing offshore oil and gas–related activities and their environmental impacts.

1980 Enactment of the Deep Seabed Hard Mineral Resources Act, creating a regulatory scheme to license U.S. mining companies seeking to explore and develop seabed hard mineral resources in international waters.

1980 Enactment of the Ocean Thermal Energy Conversion Act (OTEC), establishing a licensing system for OTEC facilities proposed to be located in the coastal zone or U.S. waters.

1980 Enactment of the American Fisheries Promotion Act, providing assistance for fisheries development and related activities.

1981 President Ronald Reagan (R) takes office on 20 January 1981. He serves two full terms and remains in office until 20 January 1989. During the first six years of Reagan's presidency, the president and the Senate majority belong to the Republican party while the House is in Democratic control. During Reagan's last two years in office, both the House and Senate majorities are Democratic.

1981 President Reagan calls for a major review of the emerging Law of the Sea (LOS) Convention.

1982 President Reagan announces that the United States will not sign the recently completed LOS Convention and, furthermore, will not participate in the work of the LOS Preparatory Committee.

1982 Enactment of the Coastal Barriers Resources Act, under which undeveloped barrier beaches and islands are to be formally designated and on which development will be actively discouraged through restrictions on federal financial assistance.

1983 By proclamation, President Reagan claims a 200-mile Exclusive Economic Zone for the United States consistent with the 1982 LOS Convention.

1984 In *Secretary of Interior v. California* (464 U.S. 312 [1984]), the Supreme Court holds that OCS lease sales are exempt from consistency review under section 307(C) (1) of the CZMA.

1987 Major amendments to the Clean Water Act create, among other things, a new National Estuary Program to improve the condition of U.S. estuaries of national importance and a new program aimed at controlling nonpoint sources of marine pollution.

1988 By proclamation, President Reagan extends the U.S. territorial sea from 3 miles to 12 miles in breadth, also consistent with the 1982 LOS Convention.

1989 President George Bush (R) takes office on 20 January 1989. He serves one full term and remains in office until 20 January 1993. Throughout Bush's presidency, both the House and the Senate are in Democratic control.

1989 The Ocean Dumping Ban Act prohibits the dumping of sewage sludge in U.S. waters after December 1991.

1990 Enactment of Coastal Zone Act Reauthorization Amendments. Section 6217 of the Amendments of 1990 requires the twenty-nine coastal states with federally approved coastal zone management plans in 1990 to develop and submit coastal nonpoint-source pollution control programs for approval by NOAA and the EPA. These amendments also restore the federal consistency provision. Under the amendments, federal offshore oil leasing activities are, once again, subject to the federal consistency provisions of the CZMA (thus, overturning *Secretary of Interior v. California*).

1990 Prompted by the *Exxon Valdez* tanker grounding in Alaskan waters, the Oil Pollution Act (OPA) is enacted. OPA amends section 311 of the Federal Water Pollution Control Act to clarify federal response authority, increase penalties for spills, establish U.S. Coast Guard response organizations, require tank vessel and facility response plans, and provide for contingency planning in designated areas.

1992 The United Nations Conference on Environment and Development (the Earth Summit) is held in Rio de Janeiro, Brazil.

1993 President Bill Clinton (D) takes office on 20 January 1993 and is reelected, in 1996, for a second term. During the first two years of Clinton's presidency, the House and the Senate are in Democratic control. From 1995 onward, both the House and the Senate have Republican majorities.

1993 The Convention on Biological Diversity enters into force. The Convention serves as the main international forum for addressing biodiversity. Its three objectives are conservation of biological diversity, sustainable use of its components, and a fair and equitable sharing of the benefits of genetic resources. The United States signs the Convention in 1993 but, as of 1999, does not ratify it.

1994 The Agreement Relating to the Implementation of Part XI of the United Nations Convention on the Law of the Sea of 10 December 1982 is adopted. The Agreement removed the obstacles that had led many industrialized nations to withhold ratification of the original Convention, opening the way for near universal acceptance of the Law of the Sea Convention.

1994 The 1982 Law of the Sea Convention enters into force. The Clinton administration signals its intention to seek Senate ratification of the treaty, but, as of 1999, the Senate does not ratify it.

1994 The 1994 Outer Continental Shelf Lands Act Amendments authorizes the secretary of the interior to negotiate agreements (rather than conduct a competitive lease sale) for sand, gravel, and shell resources for use in projects undertaken by federal, state, or local governments for shore protection and beach or coastal wetlands restoration, or for use in other types of construction projects that are wholly or partly funded, or authorized, by the federal government.

1994 The Framework Convention on Climate Change, which had been ratified by the United States in 1992, enters into force. The Convention's major objective is to achieve the stabilization of greenhouse gas emissions and to develop national adaptation strategies.

1995 Adoption of the Agreement for the Implementation of the Provisions of the United Nations Convention on the Law of the Sea of 10 December 1982, Related to the Conservation and Management of Straddling Fish Stocks and Highly Migratory Fish Stocks. The Agreement holds that conservation and management measures established for the high seas (i.e., beyond EEZ or national fishing boundaries), and those adopted for areas under national jurisdiction should be compatible; and that coastal states and states fishing in the high seas shall have a duty to cooperate in the management of straddling stocks and highly migratory species; and calls for the creation and/or strengthening of regional and subregional fishery management bodies to be governed in a transparent manner and with appropriate involvement by NGOs.

1995 The Global Programme of Action for the Protection of the Marine Environment from Land-Based Activities is adopted at an international conference held in Washington, D.C. The Programme seeks to prevent the degradation of the marine environment from land-based activities by providing guidance to parties to the agreement with respect to a list of nearly a dozen substances and/or land-based activities that can degrade the marine environment.

1996 The Sustainable Fisheries Act, amending the Fisheries Conservation and Management Act and renaming it the Magnuson-Stevens Fishery Conservation and Management Act, mandates the prevention of overfishing and ending overfishing of depressed stocks; rebuilding depleted stocks; reducing bycatch; designating and conserving essential fish habitat; reforming the approval process for fishery management plans; reducing conflict of interest on regional councils; and establishing user fees.

1997 Introduction of S.1213, the "Oceans Act of 1997," by Senator Hollings calls for the creation of a Commission on Ocean Policy as well as a National Ocean Council. S.1213 is passed in the Senate on 13 November 1997.

1998 H.R. 3445, introduced by Representative Saxton and calling for the establishment of a Commission on Ocean Policy, is passed by the House. However, differences between the Senate and the House versions of the legislation are not resolved before the adjournment of Congress.

Notes

CHAPTER 1

1. The Supreme Court gave Texas and Florida (gulf coast only) jurisdiction out to 3 marine leagues (about 10 statute miles) based on the jurisdictions existing when the states entered the Union (see *United States v. Florida et al*, 363 U.S. 121 [1960] and *United States v. Louisiana et al*, 363 U.S. 1 [1960]).

2. This discussion is adapted, in part, from Knecht and Cicin-Sain 1983.

3. We first noted the need to move from a "first generation" single-purpose system of ocean governance to a multi-purpose "second generation" system in Cicin-Sain and Knecht 1985. Wilder 1998 also emphasizes this theme.

4. Information on the Australian Oceans Policy may be found on the Internet at http://www.environment.gov.au/marine/oceans/index.htm

5. On the controversies surrounding U.S. maritime boundaries, see, for example, Charney 1994, 1987; Briscoe 1992; Forman, Jarman, and Van Dyke 1992; Van Dyke, Morgan, and Gurish 1988; Collins and Rogolf 1986; Broder and Van Dyke 1982; Feldman and Colson 1981.

6. On the challenges involved in ocean exploration, see, for example, Earle 1996; YOTO 1998 L.

7. On the development, current status, and future challenges of the ocean sciences in the United States, see, for example, NRC 1992b.

8. On natural science–based marine education, see YOTO 1998; regarding social sciences-based marine education, although this field has experienced significant expansion in the past twenty years, there is no readily available source documenting this field. For an international perspective of growth in education and training in integrated coastal management, see Cicin-Sain et al. 1999.

CHAPTER 2

1. On the same day, President Truman also issued another proclamation on fisheries, establishing that the United States "regards it as proper to establish conservation zones in those areas of the high seas contiguous to the coasts of the United States . . . [and] to establish explicitly bounded conservation zones in which fishing activities shall be subject to the regulation and control of the United States . . . " (U.S. Presidential Proclamation 2668 of 28 September 1945). This proclamation merely declared the right of the United States to establish such a zone; a fishery conservation zone was not actually established until 1976 with the passage of the Fishery Conservation and Management Act (Mangone 1988).

2. For a good history of the evolution of the Law of the Sea conference, see Hollick 1981; for an excellent account of the dynamics of its negotiation, see Friedheim 1993; for a concise account of the elements of the 1982 Law of the Sea agreement, see Jagota 1985.

3. On the controversies surrounding the U.S. position on the Law of the Sea, see Oxman, Caron, and Buderi 1983.

4. This section relies, in part, on Mangone (1988), who offers an excellent and detailed account of the role of the oceans in the history of the United States.

5. This section relies, in part, on Cicin-Sain and Knecht 1987.

6. The term "submerged lands" refers to "lands beneath navigable waters," which partly means "all lands within the boundaries of each of the respective States which are covered by nontidal waters that were navigable under the laws of the United States . . . up to the ordinary high water mark as heretofore or hereafter modified by accretion, erosion, and reliction" (Submerged Lands Act of 1953, sec. 1301).

7. This discussion is adapted, in part, from Knecht, Cicin-Sain, and Archer 1988.

CHAPTER 3

1. The concept of the oceans as the "common heritage of mankind" was proposed by Dr. Arvid Pardo, Malta's ambassador to the United Nations, in a now-famous speech to the United Nations in 1967.

2. Reasons why the United States did not agree to the convention in 1982 are detailed in Oxman, Caron, and Buderi 1983.

3. This section relies, in part, on the personal experience of author Robert W. Knecht, who directed the implementation of the Coastal Zone Management Act.

4. The Marine Science, Engineering and Resources Council had been created by the Marine Resources and Engineering Development Act of 1966 (*MREDA* 1966, secs. 1101 et seq.).

5. Parts of this discussion are adapted from Cicin-Sain 1982a and 1982b.

6. Residents of the Pribilof Islands (located in the Bering Sea north of the Aleutian chain) conduct an annual fur seal hunt. The hunt has been the major economic activity of the islanders since the original settlers of the islands (transplanted Eskimos and Aleuts) were transported to the Pribilofs nearly 150 years ago by Russian fur traders to harvest the fur seals. It should be noted that these seal populations are not threatened or endangered and that the hunt is regulated both by U.S. law (the Fur Seal Act of 1966) and by international agreement (the North Pacific Fur Seal Convention). See Orbach 1982b.

7. The term "take" means to harass, hunt, capture, or kill, or attempt to harass, hunt, capture, or kill any marine mammal (*MMPA* 1972, sec. 1362 [12]).

8. "Maximum sustainable yield" may be defined as the largest annual harvest that can be taken from a given species while allowing the harvested species to replenish itself.

9. "Optimum carrying capacity" was originally defined in the act as "the ability of a given habitat to support the optimum sustainable population of a

species or population stock in a healthy state without diminishing the ability of the habitat to continue that function." "Optimum sustainable population" means, "with respect to any population stock, the number of animals which will result in the maximum productivity of the population or the species, keeping in mind the optimum carrying capacity of the habitat and the health of the ecosystem of which they form a constituent element" (*MMPA* 1972, sec. 1362 [8]).

10. The analysis of congressional enactment of the MMPA relies on the following sources: (1) congressional documents, that is, U.S. Congress, House, Subcommittee on Fisheries and Wildlife Conservation, 1971; U.S. Congress, Senate, Subcommittee on Oceans and Atmosphere, 1972; U.S. Congress, House, House Report, 1972; U.S. Congress, Conference Report, 1972; and (2) personal interviews conducted in June 1981 by Laura L. Manning (Sea Grant trainee, University of California, Santa Barbara, and research assistant to Professor Biliana Cicin-Sain) in Washington, D.C., on the enactment of the MMPA and the Act. See Baysinger 1981, Garrett 1981, Grandy 1981, Lenzini 1981, Parsons 1981, Poser 1981, and Potter 1981.

11. For an explanation of the concept of animals as a "moral resource," see Partridge 1982.

12. The act was originally enacted as the Fishery Conservation and Management Act of 1976 (Public Law 94-265, 94th Congress, sec. 104, 90 Stat. 1, codified at 16 U.S.C. secs. 1801–1882 [1976]). It was signed into law by President Ford on April 13, 1976, and was initially enforced on March 1, 1977. In 1980, the act was renamed the Magnuson Fishery Conservation and Management Act in honor of Senator Warren G. Magnuson (D-Washington), a key figure in the passage of the act (L. 95-561, Title II, sec. 238 (b), 94 Stat. 3300, December 22, 1980).

13. Scientific testimony received by the Senate Commerce Committee reported that the following list of stocks of direct interest to fishers were depleted or threatened: haddock, herring, menhaden, and yellowtail flounder in the Atlantic; mackerel, sablefish, and shrimp in the Pacific; and halibut (in both the Atlantic and the Pacific). Three other species of lesser interest to U.S. fishers—Alaska pollock, yellowfin sole, and hake (all in the Pacific)—were also severely damaged (U.S. Congress 1976a, p. 670).

14. Anadromous species are "species of fish which spawn in fresh or estuarine waters of the United States and which migrate to ocean waters" (*MFCMA* 1976, sec. 3 [1]).

15. Highly migratory species are defined as species of tuna that, in the course of their life cycle, spawn and migrate over great distances in waters of the ocean (*MFCMA* 1976, sec. 3 [14]).

16. The fishery conservation zone is defined as a "zone contiguous to their territorial sea of the United States. The inner boundary of the fishery conservation zone is a line coterminus with the seaward boundary of each of the coastal States, and the outer boundary of such zone is a line drawn in such a manner that each point on it is 200 nautical miles from the baseline from which the territorial sea is measured (*MFCMA* 1976, sec. 1811).

17. The New England Fishery Management Council, composed of seventeen voting members, includes Maine, New Hampshire, Massachusetts, Rhode Island, and Connecticut. The Mid-Atlantic Council, composed of nineteen voting members, consists of New York, New Jersey, Delaware, Pennsylvania, Maryland, and Virginia. The South Atlantic Council, composed of thirteen voting members, consists of North Carolina, South Carolina, Georgia, and Florida. The Caribbean Council, composed of seven voting members, includes the Virgin Islands and the Commonwealth of Puerto Rico. The Gulf Council, composed of seventeen voting members, consists of Texas, Louisiana, Mississippi, Alabama, and Florida. The Pacific Council, composed of thirteen voting members, consists of California, Oregon, Washington, and Idaho. The North Pacific Council, composed of eleven voting members, includes Alaska, Washington, and Oregon. The Western Pacific Council, composed of eleven voting members, consists of Hawaii, American Samoa, and Guam (*MFCMA* 1976, sec. 1852).

18. The act described the qualifications of public members and procedures for their appointment as follows: "The members required to be appointed by the Secretary shall be appointed by the Secretary from a list of qualified individuals submitted by the Governor of each applicable constituent State . . . As used in this subparagraph, (i) the term "list of qualified individuals" shall include the names (including pertinent biographical data) of not less than three such individuals for each applicable vacancy, and (ii) the term "qualified individual" means an individual who is knowledgeable or experienced with regard to the management, conservation, or recreational or commercial harvest, of the fishery resources of the geographical area concerned" (*MFCMA* 1976, sec. 1852 [b]).

19. The Pacific Council has an additional nonvoting member, a person appointed by the Governor of Alaska (*MFCMA* 1976, sec. 1852 [c] [2]).

20. A discussion of the differences between the Magnuson Act and the Law of the Sea prescriptions on extended jurisdiction may be found in Vargas 1979.

21. As part of a larger study on the implementation of the Magnuson Act organized by Professors Biliana Cicin-Sain, John E. Moore, and Alan J. Wyner (University of California, Santa Barbara), a number of personal interviews with congressional staff involved in the passage of the 200-mile limit were carried by Lauren Holland (Sea Grant trainee for the project) in July 1977 in Washington, D.C. See Everett 1977, Leggett 1977, Norlin 1977, Robinson 1977, Schaeffer 1977, Sloan 1977, Utz 1977, and Weddig 1977.

22. "Coastal fishers" denotes those fishers working closer to shore, in contrast to long-distance or high-seas fishers, such as those involved in the tuna and shrimp fisheries.

23. Author Cicin-Sain observed during the first hearing implementing the act in Astoria, Oregon, in 1977 the fisher's disbelief at the extent of the new governmental intrusion on their activities—a sentiment epitomized by a memorable quote, "[F]irst we had the Russians, now we have the bureaucrats."

24. As noted by Murphy and Belsky (1980), "Representatives of the Department of the Interior adopted a common theme: OCS leasing procedures were sound; OCS regulatory enforcement was adequate; objections were one-sided and unbalanced. In short, the expressed attitude was that the need to increase

domestic sources of energy outweighed criticisms as to present practices, possible risks, and inadequate law" (p. 302).

25. The role of Secretary Andrus in the enactment of the OCSLAA is explicitly acknowledged in the legislative history of the act. On taking office, he urged quick passage of the amendments to the 1953 act and indicated that he would stress environmental concerns, require "diligence" from OCS operators, and ensure that coastal state officials had more input into Interior's decisions on OCS development (U.S. Congress, House 1977, pp. 112–113). But President Carter set his administration's new course with respect to OCS development and supported the amendments to the 1953 act repeatedly during his campaign and his first year in office (U.S. Congress, House 1977, pp. 87–88, pp.111–116).

26. As noted by the Congress, "The basic purpose of [the OCSLAA] is to promote the swift, orderly and efficient exploitation of our almost untapped domestic oil and gas resources in the Outer Continental Shelf" (U.S. Congress, House, 1977, p. 53).

27. Interior's anticipation and accommodation to the impending policy changes wrought by the OCSLAA have not been sufficiently appreciated. The revised OCS leasing, exploration, and development process put into effect by the agency during the early and mid-1970s, before the enactment of the OCSLAA, made substantial changes and improvements in this process. In fact, almost all elements of the 1978 amendments were established through rule making before passage of the OCSLAA. See, for example, U.S. Congress, House 1977 at pp. 62–63 and the very helpful chronology of OCS events during the period 1969–1977 at pp. 74–89. However, both Interior's OCS rules and the later 1978 amendments preserved the extraordinary degree of discretion lodged in the agency to manage OCS development, despite the perception of many to the contrary at the time of the amendments.

CHAPTER 4

1. The National Estuary Program (NEP) was established in 1987 by amendments to the Clean Water Act to identify, restore, and protect nationally significant estuaries of the United States. Unlike traditional regulatory approaches to environmental protection, the NEP targets a broad range of issues and engages local communities in the process. The program focuses not just on improving water quality in an estuary, but on maintaining the integrity of the whole system —its chemical, physical, and biological properties, as well as its economic, recreational, and aesthetic values. Stakeholders work together to identify problems in the estuary, develop specific actions to address those problems, and create and implement a formal management plan to restore and protect the estuary.

2. The Coastal Barrier Resources Act of 1982 restricts federal subsidies that encourage or facilitate development of undeveloped coastal barriers along U.S. coasts. The act establishes certain coastal areas to be protected by prohibiting the expenditure of federal funds for new or expanded facilities or programs to serve such areas.

3. The Ocean Dumping Act was passed in 1972 as Title I of the Marine Protection, Research, and Sanctuaries Act (P.L. 92-532). The act provides a framework for managing ocean dumping activities and for conducting research related to such activities. The law bans ocean dumping of radiological, chemical, and biological warfare agents and high-level radioactive waste. Amendments in 1988 extended this ban to sewage sludge, industrial waste, and medical wastes.

4. The Oil Pollution Act (OPA) of 1990 establishes liability for damages resulting from oil pollution and establishes a fund for the payment of compensation for such damages, as well as for other purposes. The act streamlined and strengthened the ability of the Environmental Protection Agency (EPA) to prevent and respond to catastrophic oil spills. A trust fund financed by a tax on oil is available to clean up spills when the responsible party is incapable or unwilling to do so.

5. Charges in the four-year investigation originally involved an Arkansas land deal and an "about to become defunct" savings and loan company; it later was expanded to include several White House problems and, eventually, the case of Monica Lewinsky.

6. The endangered snail darter proved to be the first test in the application of the protective provision of the 1973 Endangered Species Act. It pitted a little-known species—the snail darter—against huge economic interests involved in the Tellico Dam, a multistate hydroelectric project in the southeastern United States (see *TVA v. Hill*, 437 U.S. 153 [1978]). Similarly, under the Marine Mammal Protection Act, offshore oil operations in the Arctic were slowed down to assess and address the effects of noise emanating from offshore oil operations on the endangered bowhead whale.

7. See *American Petroleum Institute (API) v. Knecht*, 456 F. Supp. 889, 922 (C.D. Cal. 1978), *aff'd*, 609 F. 2d 1306 (9th Cir. 1979).

8. See U.S.C. Section 1536 (g)(2). (Supp. v 181) as amended by the 1982 Amendments (Section 4 [a][5][c]).

9. The other nations were Turkey, Venezuela, and Israel.

10. For the president's statement of 10 March, 1983, on the establishment of the zone, see the *Weekly Compilation of Presidential Documents*, vol. 19, p. 383.

11. On formal passage of the legislation, Dr. White asked Robert Knecht, who had prepared NOAA's implementation plan, to leave his post as deputy director of NOAA's Environmental Research Laboratories (Boulder, Colorado) and come to Washington to lead NOAA's implementation effort of the CZMA.

12. The 1990 amendments are contained in the Omnibus Budget Reconciliation Act of 1990, L. No. 101–508, sessions 6202–6217, 104 Stat. 1388.

13. Following the mid-1980s, however, there is a concerted effort to put more representatives from recreational and environmental interests on the councils.

14. Category I fisheries are ones that have frequent incidental taking of marine mammals (MMPA Amendments of 1988, section 114[b][1][A]).

15. These estimates could be unreliable due to the lack of full observer coverage (Twiss 1998).

CHAPTER 5

1. The twelve papers are: (1) The U.S. Marine Transportation System; (2) The Oceans and National Security; (3) Ensuring the Sustainability of Ocean Living Resources; (4) Ocean Energy and Minerals: Resources for the Future; (5) Perspectives on Marine Environmental Quality Today; (6) Coastal Tourism and Recreation; (7) Impacts of Global Climate Changes—With Emphasis on U.S. Coastal Areas; (8) Mitigating the Impacts of Coastal Hazards; (9) Opportunities and Challenges for Marine Science, Technology, and Research; (10) A Survey of International Agreements; (11) Marine Education U.S.A.: An Overview; and (12) The Legendary Ocean: The Unexplored Frontier. They are available from the Office of the Senior Scientist, NOAA, Department of Commerce, Washington, DC 20230.

2. The discussion in this section relies on the report of the Marine Board, National Research Council on opportunities and constraints in the development of marine aquaculture (NRC 1992c); co-author Cicin-Sain, a member of the committee, contributed to the discussions of policy contained in the report.

3. The elements of an appropriate policy framework for offshore aquaculture were discussed at a 1996 conference and reported in a special issue of the *Ocean and Coastal Law Journal* in 1997 (see e.g., Rieser 1997; Hopkins, Goldburg, and Marston 1997).

4. Cicin-Sain and Knecht were lead authors of the YOTO report on tourism.

CHAPTER 6

1. In this chapter, we are emphasizing the U.S. role in recent international agreements (largely since the Earth Summit in 1992). We are thus not covering, due to space and resource limitations, other important international agreements related to oceans and coasts to which the United States is a party, such as those mentioned at the outset of the chapter.

2. Many thanks are due to Nigel Bradly for his research work and for preparing initial drafts of sections of this chapter.

CHAPTER 7

1. "Majority of Americans Say That Oceans Should Be Priority Over Space Exploration." The poll was conducted by the Mellman Group for SeaWeb (URL reference: http://www.seaweb.org/Mell.html), a nonpartisan, multimedia educational initiative on the ocean and a project of The Pew Charitable Trusts. The poll's release came on the eve of the 1998 United Nations–designated International Year of the Ocean, focusing global attention on the plight of the world's ocean.

References

Abel, Robert B. 1981. The History of United States Ocean Policy. In *Making Ocean Policy: The Politics of Government Organization and Management*, eds. Francis W. Hoole, Robert L. Friedheim, and Timothy M. Hennessey, 3–48. Boulder, Colorado: Westview.

Adler, Robert W., Jessica C. Landman, and Diane M. Cameron. 1993. *The Clean Water Act 20 Years Later.* Washington, D.C.: Island Press.

Albright, M. K. 1998. *Earth Day 1998: Global Problems and Global Solutions.* Address by Secretary of State Madeline K. Albright, April 21, National Museum of Natural History, Washington, D.C.

Alker, Susan C. 1996. The Marine Mammal Protection Act: Refocusing the Approach to Conservation. *UCLA Law Review* 44:527–577.

American Law Institute. 1971. *A Model Land Development Code, Tentative Draft No. 3.* Washington, D.C.

Anderson, Lee G. 1995. A Commentary on the Views of Environmental Groups on Access Control in Fisheries. *Ocean and Coastal Management* 28:1–3: 165–188.

Anderson, John W. 1998. *Looking Toward Buenos Aires: After Bonn, Climate Negotiations Get More Difficult.* Weathervane Feature Essay, Resources for the Future, Washington, D.C. Internet reference: http://www.weathervane.rff.org/features/feature042.html

Archer, Jack H. 1989. Resolving Intergovernmental Conflicts in Marine Resource Management: The U.S. Experience. *Ocean and Shoreline Management* 12(3): 253–271.

———. 1991. Evolution of Major 1990 CZMA Amendments: Restoring Federal Consistency and Protecting Coastal Water Quality. *Territorial Sea Journal* 1(2): 191–222.

———. 1999. Note on Oceans Legislation. *Ocean and Coastal Policy Network News* 1 (2): 8–9.

Archer, Jack H., and Joan Bondareff. 1988. Implementation of the Federal Consistency Doctrine: Lawful and Constitutional. *Harvard Environmental Law Review* 12(1):115–156.

Archer, Jack H., Donald L. Connors, Kenneth Laurence, Sarah C. Columbia, and Robert Bowen. 1994. *The Public Trust Doctrine and the Management of America's Coasts.* Amherst: University of Massachusetts Press.

Archer, Jack H., and M. Casey Jarman. 1992. Sovereign Rights and Responsibilities: Applying Public Trust Principles to the Management of EEZ Space and Resources. *Ocean and Coastal Management* 17(3&4):251–270.

Archer, Jack H., and Robert W. Knecht. 1987. The U.S. National Coastal Zone Management Program: Problems and Opportunities in the Next Phase *Coastal Management* 15:103–120.

Armitage, Sarah. 1984. Federal "Consistency" Under the Coastal Zone Management Act: A Promise Broken by *Secretary of the Interior v. California. Environmental Law* 15:153–180.

Aron, William. 1988. The Commons Revisited: Thoughts on Marine Mammal Management. *Coastal Management* 16: 99–110.

Australia (Commonwealth of Australia). 1998. *Australia's Oceans Policy.* Oceans Policy. Marine Group. Environment Australia.

Bailey, Robert J. 1994. Notes from the Field: Implications of Oregon's Ocean Program. In *Moving Ahead on Ocean Governance,* eds. Biliana Cicin-Sain and Lori L. Denno, pp. 36–38. Newark: Center for the Study of Marine Policy, Graduate College of Marine Studies, University of Delaware.

Baldwin, Pamela L., and Malcolm F. Baldwin. 1975. *Onshore Planning for Offshore Oil: Lessons from Scotland.* Washington, D.C.: Conservation Foundation.

Barker, Rodney. 1997. *And the Waters Turned to Blood: The Ultimate Biological Threat.* New York: Simon and Schuster.

Baysinger, Earl. 1981. Office of Endangered Species, Fish and Wildlife Service, Department of Interior. Interview by Laura L. Manning, Sea Grant Trainee, University of California at Santa Barbara, on enactment of Marine Mammal Protection Act and Endangered Species Act, 18 and 24 June, Washington, D.C.

Bean, Michael J. 1983. *The Evolution of National Wildlife Law.* New York: Praeger.

Bollens, J. C., and H. J. Schmandt. 1965. *The Metropolis.* New York: Harper & Row.

Bolze, Dorene A. 1990. Outer Continental Shelf Oil and Gas Development in the Alaskan Arctic. *Natural Resources Journal* 30 (1): 17–64.

Born, S. M., and A. H. Miller. 1988. Assessing Networked Coastal Management Programs. *Coastal Management* 16: 229–243.

Bowen, R. E. 1981. The Major United States Federal Government Marine Organization Proposals. In *Making Ocean Policy,* eds. F. W. Hoole, R. L. Friedheim and T. M. Hennessey, 56. Boulder, Colorado: Westview.

Branson, Jim H. Douglas M. Larson, and Ronald W. Miller. 1986. *U.S. Fisheries*

Management: Process without Purpose? Paper presented at the American Fisheries Society Annual Meeting, Ithaca, New York, August 1986.

Brennan, William J. 1997. To Be or Not to Be Involved: Aquaculture Management Options for the New England Fishery Management Council. *Ocean and Coastal Law Journal* 2 (2): 261–271.

Brewer, Garry D., and Peter deLeon. 1983. *The Foundations of Policy Analysis.* Homewood, Illinois: The Dorsey Press.

Briscoe, John. 1992. The Effect of President Reagan's 12-mile Territorial Sea Proclamation on the Boundaries and Extraterritorial Powers of the Coastal States. *Territorial Sea Journal* 2:225.

Broder, Sherry, and Jon M. Van Dyke. 1982. Ocean Boundaries in the South Pacific. *University of Hawaii Law Review* 4:1.

Brooks, Douglas L. 1984. *America Looks to the Sea: Ocean Use and the National Interest.* Boston: Jones and Bartlett.

Brown, Naomi. 1999. "The New York–New Jersey Harbor Dredging Issue: An Appraisal of the Process and Early Efforts at Meeting Dredged Material while Protecting the Environment." Master's Thesis, Marine Policy Program, University of Delaware.

Buck, Eugene H. 1995. *Summaries of Major Laws Implemented by the National Marine Fisheries Service. Congressional Research Service, Report to Congress, March 24, 1995. 95–460 ENR.* Internet reference: http://www.cnie.org/nle/leg-11.html

Carey, John J. and Richard B. Mieremet. 1992. Reducing Vulnerability to Sea Level Rise: International Initiatives. *Ocean and Coastal Management Journal* 18:161–177.

Caron, David, and Harry Scheiber, eds. 1993. *Issues in Ocean Governance.* Newark: Ocean Governance Study Group, Center for the Study of Marine Policy, University of Delaware.

Carson, Rachel. 1962. *Silent Spring.* Boston: Houghton Mifflin.

Center for Marine Conservation and World Wildlife Federation. 1994. *Limiting Access to Marine Fisheries: Keeping the Focus on Conservation.* Center for Marine Conservation and World Wildlife Fund.

Center for Marine Conservation (CMC). 1998. *An Agenda for the Oceans. 5. Strengthen and Expand Marine Protected Areas.* Internet reference: http://www.cmc-ocean.org/agenda/5.html

Center for Oceans Law and Policy and the Center for National Security Law. 1997. Capitol Hill forum: *Is the 1982 Law of the Sea Convention Ripe for Ratification?* University of Virginia School of Law, November 19.

Chandler, A. D. 1988. The National Marine Fisheries Service. *Audubon Wildlife Report 1988/89*, 3–98.

Charney, Jonathan I. 1987. The Delimitation of Ocean Boundaries. *Ocean Development and International Law* 18:497.

———. 1994. Progress in International Maritime Boundary Delimitation Law. *American Journal of International Law* 88:227.

CIAWG (Coastal Impact Assistance Working Group). 1997. *Coastal Impact Assistance.* A report from the Coastal Impact Assistance Working Group to the Outer Continental Shelf Policy Committee.

Cicin-Sain, Biliana. 1980. Evaluative Criteria for Making Limited Entry Decisions: An Overview. In *Limited Entry as a Fishery Management Tool,* Bruce Rettig and Jay J. C. Ginter, eds. 230–250. Seattle: University of Washington Press.

———. 1982a. Managing the Ocean Commons: U.S. Marine Programs in the Seventies and Eighties. *Marine Technology Society Journal* 16(4):6–18.

———. 1982b. Introduction: Exploring Conflicts Between Marine Mammals and Fisheries. In *Social Science Perspectives on Managing Conflicts Between Marine Mammals and Fisheries,* eds. Biliana Cicin-Sain, Phyllis M. Grifman, and John B. Richards, 1–15. Santa Barbara: Marine Policy Program, University of California.

———. 1982c. Sea Otters and Shellfish Fisheries in California: The Management Framework. In *Social Science Perspectives on Managing Conflicts Between Marine Mammals and Fisheries,* eds. Biliana Cicin-Sain, Phyllis M. Grifman, and John B. Richards, 195–232. Santa Barbara: Marine Policy Program, University of California.

———. 1986. Offshore Oil Development in California: Challenges to Governments and to the Public Interest. *Public Affairs Reports.* Berkeley: Institute of Governmental Studies, University of California.

———. 1990. California and Ocean Management: Problems and Opportunities. *Coastal Management* 18:311-335.

———. 1992. Multiple Use Conflicts and Their Resolution: Toward a Comprehensive Research Agenda. In *Ocean Management in Global Change,* ed. Paolo Fabbri, 280–307. New York: Elsevier Applied Science.

———. 1993. Sustainable Development and Integrated Coastal Management. *Ocean & Coastal Management Journal* 21: 11–43.

———. 1994. "Essay: A National Ocean Governance Strategy for the United States is Needed Now," *Coastal Management* 22: 171–176.

———. 1995. National and Regional Perspectives on Ocean Governance. Paper presented at the Pacific Coast Ocean Management Workshop, Portland, Oregon, September 9–11, 1995.

———. 1996. Earth Summit Implementation: Progress Since Rio. *Marine Policy* 20(2):123–143.

Cicin-Sain, Biliana, and Lori L. Denno, eds. 1994. *Moving Ahead on Ocean Governance.* Newark: Ocean Governance Study Group, Center for the Study of Marine Policy, University of Delaware.

Cicin-Sain, Biliana, Marc J. Hershman, Richard Hildreth, and John Isaacs. 1990. *Ocean Management Capacity in the Pacific Coast Region: State and Regional Perspectives.* Newport, Oregon: National Coastal Resources Research and Development Institute.

Cicin-Sain, Biliana, and Robert W. Knecht. 1985. The Problem of Governance of U.S. Ocean Resources and the New Exclusive Economic Zone. *Ocean Development and International Law* 15(3–4): 289–320.

———. 1987. Federalism Under Stress: The Case of Offshore Oil and California. In *Perspectives on Federalism,* ed. Harry Scheiber, 149–176. Berkeley: University of California, Institute for Governmental Studies.

———. 1992. Research Agenda on Ocean Governance. In *Ocean Governance: A New Vision,* ed. Biliana Cicin-Sain, 9–16. Newark: Ocean Governance Study Group, Center for the Study of Marine Policy, University of Delaware.

———. 1993. Implications of the Earth Summit for Ocean and Coastal Governance. *Ocean Development and International Law* 24:323–353.

———. 1998. *Integrated Coastal and Ocean Management: Concepts and Practices.* Washington, D.C.: Island Press.

Cicin-Sain, Biliana, Robert W. Knecht, Lori Denno Bouman, and Gregory W. Fisk. 1996. Emerging Policy Issues in the Development of Marine Biotechnology. In *Ocean Yearbook,* eds. Elizabeth Mann Borgese and Joe Morgan, 179–206. Chicago: University of Chicago Press.

Cicin-Sain, Biliana, Robert W. Knecht, and Nigel J. Bradly, in preparation. Perspectives on Ecosystem Management: Lessons from Six Cases in the Mid-Atlantic U.S.

Cicin-Sain, Biliana, Robert W. Knecht, and Gregory W. Fisk. 1995. Growth in Capacity for Integrated Coastal Management Since UNCED: An International Perspective. *Ocean & Coastal Management,* Special issue on Earth Summit Implementation: *Progress Achieved on Oceans and Coasts,* ed. Biliana Cicin-Sain and Robert W. Knecht. Vol. 29(1–3):93–123.

Cicin-Sain, Biliana, Robert W. Knecht, and Nancy Foster, eds. 1999. *Trends and Future Challenges for U.S. National Ocean and Coastal Policy.* Proceedings from a workshop organized by the National Ocean Service, NOAA; Center for the Study of Marine Policy, University of Delaware; and the Ocean Governance Study Group.

Cicin-Sain, Biliana, Robert W. Knecht, Adalberto Vallega, and Ampai Harakunarak. Education and Training in Integrated Coastal Management: Lessons from the International Arena. *Ocean & Coastal Management*, forthcoming 1999.

Cicin-Sain, Biliana, and Katherine A. Leccese, eds. 1995. *Implications of Entry into Force of the Law of the Sea Convention for U.S. Ocean Governance.* Newark: Ocean Governance Study Group, Center for the Study of Marine Policy, University of Delaware.

Cicin-Sain, Biliana, and Michael K. Orbach. 1986. Mutual Mysteries: Washington/Regional Interactions in the Implementation of Fisheries Management Policy. *Policy Studies Review* 6(2, November):348–357.

Cicin-Sain, Biliana, and Art Tiddens. 1989. Private and Public Approaches to Solving Oil/Fishing Conflicts Offshore California. *Ocean and Shoreline Management* 12(3):233–253.

Clark, Walter F., and Steven E. Whitesell. 1994. *North Carolina's Ocean Stewardship Area: A Management Study.* Raleigh: North Carolina Department of Environment, Health and Natural Resources, Division of Coastal Management.

Clean Water Action Plan. 1998. U.S. Government Printing Office. Superintendent of Documents, Mail Stop: SSOP, Washington D.C. 20402-9328. ISBN 0-16-049536-9.

Clean Water Network. 1993. Briefing Papers on the Clean Water Act Reauthorization. Washington, D.C. March, 1993.

Coastal Ocean Policy Roundtable. 1992. *The 1992 Coastal States Report: A Pilot Study of the U.S. Coastal Zone and its Resources.* Newark: University of Delaware.

Coastal States Organization (CSO). 1998. *Periodical Legislative Update,* July. Washington, D.C.

Cobb, W. Roger, and Charles D. Elder. 1975. *Participation in American Politics: The Dynamics of Agenda Building.* Boston: Allyn and Bacon.

Coggins, George Cameron. 1975. Legal Protection of Marine Mammals: An Overview of Innovative Resource Conservation Legislation. *Environmental Law* 16:1–59.

Collins, Edward Jr., and Martin A. Rogolf. 1986. The Gulf of Maine Case and the Future of Ocean Boundary Delimitation. *Maine Law Review* 38:1.

Commission of Geosciences, Environment, and Resources. 1990. *Assessment of the U.S. Ocean Continental Shelf Environmental Studies Program.* Washington, D.C.: National Academy Press.

Committee for Humane Legislation Inc. v. Richardson. 414 f supp. 297 (d.d.c), AFFD, 540 F. 2D 1141 (D.C. Cir. 1976).

Congressional Research Service. 1985. *Interjurisdictional Fisheries Management: Issues and Options.* Washington, D.C.: The Library of Congress.

Costanza, R.R., d'Arge, R. S. de Groot, R. Farber, S. Grasso, M. Hannon, B. Naeem, S. Limburg, K. Paruelo, J. O'Neill, R.V. Raskin, R. Sutton, and P.M. van den Belt. 1997. The Value of the World's Ecosystem Services and Natural Capital. *Nature,* London, May 15.

Coy, Peter, Gary McWilliams, and John Rossant. 1997. The New Economics of Oil. *Business Week.* 3 November: 140–144.

Crutchfield, J A. 1961. An Economic Evaluation of Alternative Methods of Fishery Regulation. *Journal of Law and Economics* 4:131–143.

Cullitan, Thomas J., J. J. McDonough III, D. G. Remer, and D. M. Lott. 1990. *Building Along America's Coasts: 20 Years of Building Permits, 1970–1989.* Silver Spring, Maryland: NOAA, Strategic Environmental Assessment Division.

Davidson, Bruce B. 1995. Implementing the United Nations Convention on the Law of the Sea: The Department of Defense View. *The Georgetown International Environmental Law Review* 7:659–666.

de Fountabert, A. Charlotte, David R. Downes, and Tundi S. Argady. 1996. *Biodiversity in the Seas. Implementing the Convention on Biological Diversity in Marine and Coastal Habitats.* IUCN Environmental Policy and Law Paper No. 32. Gland and Cambridge: IUCN.

De Voe, M. R., and A. S. Mount. 1989. An Analysis of Ten State Aquaculture Leasing Systems: Issues and Strategies. *Journal of Shellfish Research* 8(1):223–239.

Dernbach, John. 1997. U.S. Adherence to Its Agenda 21 Commitments: A Five-Year Review. Unpublished paper, Widener University Seminar on Law and Sustainability, Spring Semester.

Domestic Policy Council. 1986. *The Status of Federalism in America.*

Donaghue, M. forthcoming. Conservation Services Levies in New Zealand—Recovery of the Environmental Costs of Fishing Activities. *Ocean & Coastal Management.*

Downs, Anthony. 1972. Up and Down with Ecology: The "Issue-Attention Cycle." *Public Interest* 28(summer):38–50.

Earle, Sylvia. 1996. *Sea Change: A Message of the Oceans.* London: Constable.

Eichenberg, Tim. 1992. State Jurisdiction Under the Coastal Zone Management Act After Extension of the U.S. Territorial Sea. *Territorial Sea Journal* 2(1):118–151.

Eichenberg, Tim, and Jack Archer. 1987. The Federal Consistency Doctrine: Coastal Zone Management and "New Federalism." *Ecology Law Quarterly* 14:9–68.

Eliopoulous, P. 1982. Coastal Zone Management, Program at Crossroads. *Environmental Reporter* 13(20):1–24.

Elliot, L. 1998. *Global Politics of the Environment.* London: Macmillan Press Limited.

Engler, Robert. 1961. *The Politics of Oil: A Study of Private Power and Democratic Directions.* New York: Macmillan.

ENN (Environmental News Network). 1998. Clinton Extends Ban on Offshore Oil Drilling. *ENN Daily News,* 19 June.

Environmental Health Center. 1998. *Coastal Challenges: A Guide to Coastal and Marine Issues.* Washington, D.C.: The National Safety Council's Environmental Health Center.

Everett, Ned. 1977. Staff member, Merchant Marine and Fisheries Committee, U.S. House of Representatives. Interview by Lauren Holland, Sea Grant Trainee, University of California at Santa Barbara, on enactment of Magnuson Act, July, Washington, D.C.

Feldman, Mark, and David Colson. 1981. The Maritime Boundaries of the United States. *American Journal of International Law* 75:729.

Finch, R. 1985. Fishery Management Under the Magnuson Act. *Marine Policy* 9(3)176–179.

Fitzgerald, Edward A. 1985. *Secretary of the Interior v. California*: Should Continental Shelf Lease Sales Be Subject to Consistency Review? *Boston College Environmental Affairs Law Review* 12:425–471.

Food and Agricultural Organization (FAO). 1995. *Code of Conduct for Responsible Fisheries.* Adopted by the 28th Session of FAO Conference on 31 October 1995.

———. 1996. *State of the World Fishery and Aquaculture.* (FAO Fisheries Circular.) Food and Agriculture Organization. United Nations.

Forman, David M., M. Casey Jarman, and Jon M. Van Dyke. 1992. Filling In a Jurisdictional Void: The New U.S. Territorial Sea. *Territorial Sea Journal* 2:1.

Foster, N.M. and J.H. Archer. 1988. The National Marine Sanctuary Program: Policy, Education, and Research. *Oceanus* 31(1, Spring):5–17.

Friedheim, Robert L. 1979. The Political, Economic, and Legal Ocean. In *Managing Ocean Resources: A Primer,* ed. Robert L. Friedheim, 26–42. Boulder, Colorado: Westview.

———. 1993. *Negotiating the New Ocean Regime* Columbia: University of South Carolina Press.

Galdorisi, George, and Kevin Vienna. 1997. *Beyond the Law of the Sea: New Directions for U. S. Ocean Policy.* Westport, Connecticut: Praeger.

GAO (General Accounting Office). 1985. By the Comptroller General Report to the Chairman, Subcommittee on Oversight and Investigations Committee on Energy and Commerce House of Representatives of the United States *Early Assessment of Interior's Area-Wide Program for Leasing Offshore Lands.* GAO/RCED-85–66. 15 July.

Garrett, Tom. 1981. International Whaling Commissioner. Interview by Laura L. Manning, Sea Grant Trainee, University of California at Santa Barbara, on enactment of Marine Mammal Protection Act and Endangered Species Act, 25 June, Washington, D.C.

Gay, Joel. 1996. Magnuson-Stevens Law Bodes Change. *Pacific Fishing,* December, 26–28.

Goodman, Peter S. 1999. Maryland Plans to Shift Pollution Liability. *Washington Post,* 19 March 1999, B1.

Gordon, William G. 1986. Critical Assessment of the Current Fisheries Management System. In *Rethinking Fisheries Management,* eds., Jon G. Sutinen, and Lynne Carter Hanson, 9–12 Proceedings from the Tenth Annual Conference held June 1–4, 1986, Center for Ocean Management Studies, University of Rhode Island, Kingston.

Gramling, Robert. 1996. *Oil on the Edge: Offshore Development, Conflict, Gridlock.* Albany: State University of New York Press.

Grandy, John. 1981. Executive Vice President, Defenders of Wildlife. Interview by Laura L. Manning, Sea Grant Trainee, University of California at Santa Barbara, on enactment of Marine Mammal Protection Act and Endangered Species Act, 25 June, Washington, D.C.

Greenberg, Eldon V. C. and Michael E. Shapiro. 1982. Federalism in the Fishery Conservation Zone: A New Role for the States in an Era of Federal Regulatory Reform. *Southern California Law Review* 55:641–690.

Greenstein, Fred I. 1978. Change and Continuity in the Modern Presidency. In *The New American Political System,* eds. Samuel H. Beer and Anthony S. King, 45–85. Washington, D.C.: American Enterprise Institute for Public Policy Research.

Grotius, Hugo. 1633. *Hugh Grotius De mari libero. Et P. Merula De Maribus.* Lugd. Batavorum: Ex Officina Elzeviriana.

Gulf of Maine Council on the Marine Environment. 1991. *The Gulf of Maine Action Plan 1991–2000.* July.

Gutting, Richard E. Jr. 1986. An Assessment of the Current System from a Commercial Perspective. In *Rethinking Fisheries Management,* eds. Jon G. Sutinen and Lynn Carter Hanson, 21–45. Proceedings from the Tenth Annual Conference held June 1–4, 1986, Center for Ocean Management Studies, University of Rhode Island, Kingston.

Hanna, Susan. 1998. Personal communication with Biliana Cicin-Sain, December.

Hardin, Garrett. 1968. The Tragedy of the Commons. *Science.* 162:1243–1248.

Hawaii Ocean and Marine Resources Council. 1991. *Hawaii Ocean Resources Management Plan.*

Heath, Milton S. Jr. 1971. Description of Illustrative State Programs of Estuarine Conservation. In *Coastal Zone Resource Management,* eds. James C. Hite and James M. Stepp, 154–165. New York: Praeger.

Heclo, Hugh. 1978. Issue Networks and the Executive Establishment. In *The New American Political System,* eds. Samuel H. Beer and Anthony S. King, 87–124. Washington, D.C.: The American Enterprise Institute for Public Policy Research.

Heinz Center. 1998. *Our Ocean Future: Themes and Issues Concerning the Nation's Stake in the Oceans Developed for Discussion During 1998, The Year of the Ocean.* Washington, D.C.: The H. John Heinz III Center for Science, Economics and the Environment.

Hershman, Marc J. 1987. The Coastal Decision-Making Framework as a Model for Ocean Management. In *Proceedings of the National Conference on the States and the Extended Territorial Sea: Coastal Law and Policies, December 9–11, 1985,* eds. Lauriston R. King, and Amy Broussard, 92–100 College Station: Texas A&M University.

———. 1996. Ocean Management Policy Development in Subnational Units of Government: Examples from the United States. *Ocean & Coastal Management* 31:25–40.

Hershman, Mark J., James W. Good, Tina Bernd-Cohen, Robert F. Goodwin, Virginia Lee, and Pam Pogue. 1999. The Effectiveness of Coastal Zone Management in the United States. *Coastal Management* 27 (2-3):113–138.

Hildreth, Richard G. 1989. Marine Use Conflicts Arising from Development of Seabed Hydrocarbons and Minerals: Some Approaches from the United States West Coast. *Ocean and Shoreline Management* 12(3):271–284.

Hildreth, Richard G., and Ralph W. Johnson. 1985. CZM in California, Oregon, and Washington. *Natural Resources Journal* 25(1):103–165.

Hinrichsen, Don. 1998. *Coastal Waters of the World: Trends, Threats, and Strategies.* Washington, D.C.: Island Press.

Hoagland, P., and Timothy K. Eichenberg. 1988. The Channel Islands National Marine Sanctuary. *Oceanus* 31(1, Spring 1988):66–75.

Hoffman, Robert J. 1989. The Marine Mammal Protection Act: A First of Its Kind Anywhere. *Oceanus* 32(1):21–28.

Hollick, Ann I. 1981. *U.S. Foreign Policy and the Law of the Sea.* Princeton, New Jersey: Princeton University Press.

Hong, Seoung-Yong, and Jihyun Lee. 1995. National Level Implementation of Chapter 17: The Korean Example. *Ocean & Coastal Management,* Special issue on Earth Summit Implementation: *Progress Achieved on Oceans and Coasts,* eds. Biliana Cicin-Sain and Robert W. Knecht. Vol. 29(1–3):231–249.

Hopkins, D. Douglas, Rebeca J. Goldburg, and Andrea Marston. 1997. An Environmental Critique of Government Regulations and Policies for Open Ocean Aquaculture. *Ocean and Coastal Law Journal* 2(2):235–261.

Houston, James R. 1995. The Economic Value of Beaches. *CERCular,* Coastal Engineering Research Center, Vol. CERC-95–4, December.

———. 1996. International Tourism and U.S. Beaches. *Shore and Beach* 64(2):3–4.

Husing, Onno. 1984. A Matter of Consistency: A Congressional Perspective of Oil and Gas Development on the OCS. *Coastal Zone Management Journal* 12:301–319.

Imperial, Mark T., and Timothy M. Hennessey. 1996. An Ecosystem-Based Approach to Managing Estuaries: An Assessment of the National Estuary Program. *Coastal Management* 24:115–139.

Independent World Commission on the Oceans (IWCO). 1998. *The Ocean . . . Our Future: The Report of the Independent World Commission on the Oceans.* Cambridge, England: Cambridge University Press.

Iudicello, Suzanne; Scott Burns, and Andrea Oliver. 1996. Putting Conservation into the Fishery Conservation and Management Act: The Public Interest in Magnuson Act Reauthorization. *Tulane Environmental Law Journal* 9:339–347.

Jacobson, Jon; Daniel Conner, and Robert Tozer. 1985. *Federal Fisheries Management.* Eugene: Ocean and Coastal Law Center, University of Oregon Law School.

Jagota, S. P. 1985. The United Nations Convention on Law of the Sea. In *Ocean Yearbook 5,* eds. Elisabeth Mann Borgese, Norton Ginsburg, and Joseph R. Morgan, 10–28. Chicago: University of Chicago Press.

Jones, Charles O. 1984. *An Introduction to the Study of Public Policy.* Third edition. Monterey, California: Brooks/Cole.

Juda, Lawrence. 1996. *International Law and Ocean Use Management: The Evolution of Ocean Governance.* London and New York: Routledge.

Juda, Lawrence, and R. H. Burroughs. 1990. The Prospects for Comprehensive Ocean Management. *Marine Policy* (January): 23–35.

Kagan, Robert A. 1994. Dredging Oakland Harbor: Implications for Ocean Governance. *Ocean and Coastal Management* 23:49–63.

Kalo, Joseph, Richard G. Hildreth, Alison Rieser, Donna R. Christie, and Jon L. Jacobson. 1994. *Coastal and Ocean Law, Second edition.* Houston, Texas: John Marshall Publishing.

Kelly, Paul. 1998a. Testimony of Paul L. Kelly, Rowan Companies, Inc., before the House Subcommittee on Fisheries Conservation, Wildlife, and Oceans on the Oceans Act Legislative Proposals H.R. 2547, S. 1213, H.R. 3445. 19 March.

———. 1998b. Extension of the Offshore Oil Moratorium: An Unwise Step. *Ocean Coastal Policy Network News.* 1(1):10–11.

Kildow, Judith. 1999. National Ocean Economic Study Begins. *Ocean and Coastal Policy Network News* 1(2):1–2.

King, Lauriston, and S. G. Olson, 1988. Coastal State Capacity for Marine Resources Management. *Coastal Management* 16(4):305–318.

Kingdon, John W. 1995. *Agendas, Alternatives, and Public Policies,* Second edition. New York: HarperCollins College Publishers.

Kitsos, Thomas R. 1981. U.S. Ocean Policy and the Uncertainty of Implementation in the 80s: A Legislative Perspective. *Marine Technology Society Journal* 15(3):3–11.

———. 1994a. The Clinton Administration, the 103rd Congress, and Environmental Policy: Strange Things Are Happening. *Moving Ahead on Ocean Governance.* Biliana Cicin-Sain and Katherine Lecesse, eds. Newark: The Ocean Governance Study Group, Center for the Study of Marine Policy, University of Delaware.

———. 1994b. Troubled Waters: A Half Dozen Reasons Why the Federal Offshore Oil and Gas Program Is Failing: A Political Analysis. In *Moving Ahead on Ocean Governance.* Biliana Cicin-Sain and Katherine Lecesse, eds. Newark: The Ocean Governance Study Group, Center for the Study of Marine Policy, University of Delaware.

Klee, Gary A. 1999. *Toward Integrated Coastal and Marine Sanctuary Management.* Upper Saddle River, New Jersey: Prentice Hall.

Knecht, Robert W. 1979. Coastal Zone Management: The First Five Years and Beyond. *Coastal Zone Management Journal* 6:259–272.

———. 1986. The Exclusive Economic Zone: A New Opportunity in Federal-State Ocean Relations. In *Ocean Resources and U.S. Intergovernmental Relations in the 1980s,* ed. Maynard Silva, 263–275. Boulder, Colorado: Westview.

———. 1992. National Ocean Policy in the United States: Less Than the Sum of Its Parts. In *Ocean Management in Global Change,* ed. P. Fabbri, 184–204. New York: Elsevier Applied Science.

———, ed. 1993. *The National Estuarine Research Reserve System: Building a Valuable National Asset. An Assessment by the Review Panel on the National Estuarine Research Reserve System.* Newark: Center for the Study of Marine Policy, University of Delaware.

Knecht, Robert W., and Jack H. Archer. 1993. Integration in the U.S. Coastal Zone Management Program. *Ocean & Coastal Management* 21:183–200.

Knecht, Robert W., Biliana Cicin-Sain, and Jack H. Archer. 1988. National Ocean Policy: A Window of Opportunity. *Ocean Development and International Law* 19:113–142.

Knecht, Robert W., Biliana Cicin-Sain, and Gregory W. Fisk. 1996. Perceptions on the Performance of State Coastal Zone Management Programs in the United States: National Perspectives. *Coastal Management* 24:141–163.

———. 1997. Perceptions on the Performance of State Coastal Zone Management Programs in the United States: Regional and State Comparisons. *Coastal Management* 25:325–343.

Knecht, Robert W., Biliana Cicin-Sain, and Nancy Foster. 1998. *The Stratton Roundtable: Looking Back, Looking Forward: Lessons from the 1969 Commission on Marine Science, Engineering, and Resources: Proceedings, May 1, Washington, D.C.* Washington, D.C.: National Oceanic and Atmospheric Administration.

Knecht, Robert W., Biliana Cicin-Sain, and Dosoo Jang. In preparation. *Policy Issues in the Development of Marine Biotechnology.*

Knecht, Robert W., Biliana Cicin-Sain, and Willett Kempton. In preparation. *Perceptions of Threats from Sea Level Rise Among State Coastal Managers.*

Koester, G. Thomas. 1990. State-Federal Jurisdictional Conflict in the U.S. 120-mile Territorial Sea: An Opportunity to End the Seaweed Rebellion. *Coastal Management* 18:195–211.

Kopp, Raymond J., Richard D. Morgenstern, and Michael A. Toman. 1998a. *The Kyoto Protocol: Unresolved Issues.* Weathervane Feature Essay, Resources for the Future, Washington, D.C. Internet reference: http://www.weathervane.rff.org/features/feature026.html

———. 1998b. *The Kyoto Protocol: The Realities of Implementation.* Weathervane Feature Essay, Resources for the Future, Washington, D.C.

Korfmacher, Katrina S. 1998. Invisible Successes, Visible Failures: Paradoxes of Ecosystem Management in the Albemarle-Pamlico Estuarine Study. *Coastal Management* 26:191–211.

Kubasek, Nancy, M. Neil Browne, Melissa Young, and Wesley Hiers. 1995. Protecting Marine Mammals: Time for a New Approach. *Journal of Environmental Law* 13:1–30.

Kurtz, Randy. 1994. The Taking of Marine Mammals Under the Marine Mammal Protection Act: Domestic and International Implications. *Journal of Energy, Natural Resources, and Environmental Law* 14:395–414.

LaBelle, Robert. 1998. Environmental Forum Report. April. Held by Minerals Management Service, Department of the Interior.

Lanchberry, John. 1997. What to Expect from Kyoto? *Environment* 39(9, November): 5–11.

Leggett, Robert. 1977. Representative to U.S. Congress. Interview by Lauren Holland, Sea Grant Trainee, University of California at Santa Barbara, on enactment of Magnuson Act, July, Washington, D.C.

Leitzell, Terry. 1980. Director of the National Marine Fisheries Service, National Oceanic and Atmospheric Administration. Interview by Michael K. Orbach, co-researcher, conducted under the NOAA, National Sea Grant College Program, grant number NA80AA-D-00120, project number R/MA8, through the California Sea Grant College Program and the California State Resources Agency. Washington, D.C.

Lenzini, Paul. 1981. Legal Counsel for the International Association of Fish and Wildlife Commissioners and the American Federation of Wildlife Agencies. Interview by Laura L. Manning, Sea Grant Trainee, University of California at Santa Barbara, on enactment of Marine Mammal Protection Act and Endangered Species Act, 26 June, Washington, D.C.

Lester, Charles. 1996a. Reforming the Oil and Gas Program: Rediscovering the Public's Interests in the Outer Continental Shelf Lands. *Ocean and Coastal Management* 30(1):1–42.

———. 1996b. Cumulative Impact Management in the Coastal Zone: Regulation and Planning in the Monterey Bay Region of California. Paper presented at the 1996 symposium of the Ocean Governance Study Group, 21–23 July, 1996. Boston.

Lineberry, Robert L. 1980. *Government in America: People, Politics, and International Law.* Los Angeles: University of California Press.

Lowi, Theodore J. 1969. *The End of Liberalism: The Second Republic of the United States.* New York: Norton.

Lund, Thomas A. 1980. *American Wildlife Law*. Berkeley: University of California Press.

Mair, Lucy, P. 1962. *Primitive Government.* Baltimore: Penguin Books.

Malone, James L., 1990. Testimony before the Senate Foreign Relations Committee on May 1, 1990 (available on the CIS Congressional Universe website: web.lexis-nexis.com/congcomp/docum)

Mangone, Gerard J. 1988. *Marine Policy for America,* Second edition, revised and expanded. New York: Taylor & Francis.

Manning, Laura L. 1990. *The Dispute Processing Model of Public Policy Evolution: The Case of Endangered Species Policy Changes from 1973 to 1983.* New York: Garland.

———. 1989. Marine Mammals and Fisheries Conflicts: A Philosophical Dispute. *Ocean and Shoreline Management* 12(3):217–233.

Marine Conservation News. 1995. Gray Whale Removed from Endangered Species List. Autumn, 5.

Marine Mammal Commission (MMC). 1992. *Annual Report to Congress, 1991* (31 January 1992). Washington, D.C.

———. 1996. *Annual Report to Congress, 1995* (29 February 1996). Washington, D.C.

Marine Technology Society and the Institute of Electrical and Electronics Engineers Council on Oceanic Engineering. 1984. *Exclusive Economic Zone Papers.* Exclusive Economic Zone Symposium, September. Reprinted from OCEAN '84 Conference proceedings by NOAA Ocean Assessments Division, Rockville, Maryland 20852.

Massachusetts Coastal Zone Management. 1995. *Aquaculture White Paper* and *Aquaculture Strategic Plan,* 1995. (MCZM web page: http://www.magnet.state.ma.us/czm/aquatoc.htm)

Mazmanian, Daniel A., and Paul A. Sabatier, eds. 1981. *Effective Policy Implementation.* Lexington, Massachusetts: D.C. Heath.

McCay, Bonnie J. 1997. Personal communication with Biliana Cicin-Sain. November.

McCay, Bonnie J., and James M. Acheson, eds. 1987. *The Question of the Commons.* Tucson, Arizona: University of Arizona Press.

McCay, Bonnie J., C. F. Creed, A. C. Finlayson, R. Apostle and K. Mikalsen. 1995. Individual Transferable Quotas (ITQs) in Canadian and U.S. Fisheries. *Ocean and Coastal Management* 28(1–3):85–117.

McEvoy, Arthur F. 1983. Law, Public Policy, and Industrialization in California Fisheries, 1900–1925. *Business History Review* 57(Winter):494–521.

McKernan, Donald L. 1976. The New National Fisheries Law: Solution or Problem? *Washington Public Policy Notes* (Summer). Seattle: Institute of Government Research, University of Washington.

McWilliams, Gary. 1997. The New Economics of Oil. 1997. *Business Week*, 3 November 147–148.

M'Gonigle, R. Michael, and Mark W. Zacher. 1979. *Pollution, Politics, and International Law.* Los Angeles: University of California Press.

Mellman Group. July 1996. *Results of a Public Opinion Survey on the Oceans.* Washington D.C.: The Mellman Group.

Mieremet, Richard B. 1995. The International Coral Reef Initiative: A Seed from the Earth Summit Tree Which Now Bears Fruit. *Ocean & Coastal Management,* Special issue on Earth Summit Implementation: *Progress Achieved on Oceans and Coasts,* ed. Biliana Cicin-Sain and Robert W. Knecht. Vol. 29 (1–3):303–328.

Miles, Edward L. 1992. Future Challenges in Ocean Management: Towards Integrated National Ocean Policy. In *Ocean Management in Global Change.* ed. Paolo Fabbri, 595–620. London: Elsevier Applied Science.

Miller, Daniel S. 1984. Offshore Federalism: Evolving Federal-State Relations in Offshore Oil and Gas Development. *Ecology Law Quarterly* 11(3):401–450.

Miloy, John. 1983. *Creating the College of the Sea.* College Station: Texas A&M Sea Grant Program.

Mitchell, James K. 1982. Coastal Zone Management: A Comparative Analysis of National Programs. In *Ocean Yearbook 3,* eds. Elisabeth Mann Borgese and Norton Ginsburg, 258–319. Chicago: The University of Chicago Press.

Morrow, Lance. 1987. A Change in the Weather. *Time,* 30 March 129(13):28–37.

Murphy, John M., and Martin H. Belsky. 1980. OCS Development: A New Law and a New Beginning. *Coastal Zone Management* 7(2–4):297–337.

Nash, A. E. Keir; Dean E. Mann, and Phil G. Olsen. 1972. *Oil Pollution and the Public Interest: A Study of the Santa Barbara Oil Spills.* Berkeley: Institute of Government Studies, University of California.

National Academy of Public Administration. 1995. *Setting Priorities, Getting Results: A New Direction for EPA.* A National Academy of Public Administration Report to Congress.

National Fisherman. 1995. Killing Seattle Sea Lions Approved as Last Resort. March, 9.

Natural Resources Defense Council (NRDC). 1997. *Hook, Line, and Sinking: The Crisis in Marine Fisheries.* New York: NRDC Publications.

Nelson, Barbara J. 1978. Setting the Public Agenda: The Case of Child Abuse. In *The Policy Cycle,* eds. Judith V. May and Aaron B. Wildavsky, 17–41. Beverly Hills: Sage Publications.

New York Times. 1999. Norway Whale Hunters Battle Activists and the Marketplace. August 5, p. A11.

Nicholls, Robert J. 1995. Coastal Megacities and Climate Change. *Geo Journal* 37 (3): 369–379.

Nielsen, Larry A. 1976. The Evolution of Fisheries Management Philosophy. *Marine Fisheries Review* 38(12, December):15–23.

NOIA (National Ocean Industries Association). 1998. Exploring New Frontiers NOIA. A catalogue printed by *Offshore Magazine.*

———.1997. *Offshore Magazine,* Special Issue, May.

Norlin, Richard. 1977. Staff member, Maritime and Education Subcommittee, Merchant Marine and Fisheries Committee, U.S. House of Representatives. Interview by Lauren Holland, Sea Grant Trainee, University of California at Santa Barbara, on enactment of Magnuson Act, July, Washington, D.C.

Norse, Elliot A., ed. 1993. *Global Marine Biological Diversity: A Strategy for Building Conservation in Decision Making.* Washington, D.C.: Island Press.

NRC (National Research Council). 1992a. *Working Together in the EEZ.* Final Report of the Committee on Exclusive Economic Zone Information Needs. Marine Board, Commission on Engineering and Technical Systems. Washington, D.C.: National Academy Press.

———. 1992b. *Oceanography in the Next Decade: Building New Partnerships.* Washington, D.C.: National Academy Press.

———. 1992c. *Marine Aquaculture: Opportunities for Growth.* Ocean Studies Board, Marine Board. Washington, D.C.: National Academy Press.

———. 1994. *Low-Frequency Sound and Marine Mammals: Current Knowledge and Research Needs.* Ocean Studies Board. Washington, D.C.: National Academy Press.

———. 1995. *Clean Ships, Clean Ports, Clean Oceans: Controlling Garbage and Plastic Wastes at Sea.* Marine Board. Washington, D.C.: National Academy Press.

———. 1996. *Stemming the Tide: Controlling Introductions of Nonindigenous Species by Ships' Ballast Water.* Marine Board. Washington, D.C.: National Academy Press.

———. 1997. *Striking a Balance: Improving Stewardship of Marine Areas.* Marine Board. Washington, D.C.: National Academy Press.

———. 1998. *Sharing the Fish: Toward a National Policy on Individual Fishing Quotas.* Ocean Studies Board. Washington, D.C.: National Academy Press.

Odell, Rice. 1972. *Saving of San Francisco Bay: A Report on Citizen Action and Regional Planning.* Washington, D.C.: The Conservation Foundation.

Olsen, Stephen, and Virginia Lee. 1991. A Management Plan for a Coastal Ecosystem: Rhode Island's Salt Pond Region. In *Case Studies of Coastal Management: Experience from the United States,* ed. Brian Needham, 57–69. Narragansett: Coastal Resources Center, University of Rhode Island.

Orbach, Michael. 1982a. Otters, Marine Mammals, and Man: The Human Dimension. In *Social Science Perspectives on Managing Conflicts Between Marine Mammals and Fisheries,* eds. Biliana Cicin-Sain, Phyllis M. Grifman, and John B. Richards, 76–98. Santa Barbara: Marine Policy Program, University of California.

———. 1982b. U.S. Ocean Policy and the Ocean Ethos. *Marine Technology Society Journal* 16(4):41–48.

———. 1989. Of Mackerel and Menhaden: A Public Policy Perspective on Fishery Conflict. *Ocean and Shoreline Management* 12(3):199–217.

Oregon Ocean Resource Management Program. 1991. *Oregon's Ocean Resources Management Plan.* The Oregon Ocean Resources Management Task Force, Portland, Oregon.

Oregon Coastal Zone Management Program. 1995. *Creating the Mosaic: An Ocean Management Conference for the Pacific Coast of the United States.* Northwestern School of Law of Lewis & Clark College. September 9–11.

Oxman, Bernard H., David D. Caron, and Charles L. O. Buderi, eds. 1983. *Law of the Sea: U.S. Policy Dilemma.* San Francisco: Institute for Contemporary Studies.

Palley, Marian L. 1994. From Iron Triangles to Issue Networks in American Politics. In *Comparative Study of Policy Making Structures,* Proceedings of the International Conference of the Korean Association for Public Administration, October.

Pardo, Arvid. 1979. Law of the Sea Conference: What Went Wrong. In *Managing Ocean Resources: A Primer,* ed. Robert L. Friedheim, 137–148. Boulder, Colorado: Westview Press.

Parsons, Richard. 1981. Chief of Federal Wildlife Permit Office, Fish and Wildlife Service. Interview by Laura L. Manning, Sea Grant Trainee, University of California at Santa Barbara, on enactment of Marine Mammal Protection Act and Endangered Species Act, 19 June, Washington, D.C.

Partridge, Ernest. 1982. Sea Otters as a Moral Resource: A Philosopher's View. In *Social Science Perspectives on Managing Conflicts Between Marine Mammals and Fisheries,* eds. Biliana Cicin-Sain, Phyllis M. Grifman, and John B. Richards, 47–75. Santa Barbara: Marine Policy Program, University of California.

Patterson, Samuel C. 1978. The Semi-Sovereign Congress. In *The New American*

Political System, eds. Samuel H. Beer and Anthony S. King, 125–177. Washington, D.C.: American Enterprise Institute for Public Policy Research.

Pedrick, John L. Jr. 1976. Land Use Control in the Coastal Zone: The Delaware Example. *Coastal Zone Management Journal* 2(4):345–368.

Pell, Claiborne. 1995. The United Nations Convention on the Law of the Sea: Prospects for Advice and Consent by the Senate. *The Georgetown International Environmental Law Review* 7:797–801.

Peterson, Russell W. 1999. *Rebel with a Conscience.* Newark: University of Delaware Press.

Petro, Richard. 1968. *In the Wake of the Torrey Canyon.* New York: McKay.

Pontecorvo, Giulio 1989. Contribution of the Ocean Sector to the United States Economy: Estimated Values for 1987: A Technical Note. *Marine Technology Society Journal* 23:2–7.

Pontecorvo, Giulio, Maurice Wilkinson, Ronald Anderson, and Michael Holdowsky. 1980. Contribution of the Ocean Sector to the United States Economy. *Science* 208:1000–1006.

Poser, Joseph. 1981. Representative for Fur Merchants Association. Interview by Laura L. Manning, Sea Grant Trainee, University of California at Santa Barbara, on enactment of Marine Mammal Protection Act and Endangered Species Act, 23 June, Washington, D.C.

Potter, Frank. 1981. Committee staff member for the House Energy and Communication Committee. Interview by Laura L. Manning, Sea Grant Trainee, University of California at Santa Barbara, on enactment of Marine Mammal Protection Act and Endangered Species Act, 19 June, Washington, D.C.

———. 1991. *National Marine Sanctuaries: Challenge and Opportunity.* A report to the National Oceanic and Atmospheric Administration. Washington, D.C.: National Oceanic and Atmospheric Administration.

Public Land Law Review Commission. 1970. *One-Third of the Nation's Land.* Washington, D.C.: U.S. Government Printing Office.

Quarterman Cynthia L., 1998. Testimony of Cynthia L. Quarterman, Director Minerals Managements Service. Department of the Interior before the Committee on Resources, Subcommittee on Energy and Mineral Resources. U.S. House of Representatives. May 14, 1998. United States Department of the Interior.

Ragsdale, L. 1996. *Vital Signs on the Presidency: Washington to Clinton.* Washington, D.C.: Congressional Quarterly.

Reilly, William K., et al., eds. 1973. *The Use of Land: A Citizen's Policy Guide to Urban Growth.* New York: Crowell.

Rieser, Alison. 1997. Defining the Federal Role in Offshore Aquaculture: Should It Feature Delegation to the States? *Ocean and Coastal Law Journal* 2(2): 209–235.

Robinson, Kip. 1977. Office of Congressional Affairs, National Oceanic and Atmospheric Administration. Interview by Lauren Holland, Sea Grant Trainee, University of California at Santa Barbara, on enactment of Magnuson Act, July, Washington, D.C.

———. 1981. Office of Congressional Affairs, National Oceanic and Atmospheric Administration. Interview by Laura L. Manning, Sea Grant Trainee, University of California at Santa Barbara, on enactment of Marine Mammal Protection Act and Endangered Species Act, 25 June, Washington, D.C.

Rogalski, William R. 1980. The Unique Federalism of the Regional Councils Under the Fishery Conservation and Management Act of 1976. *Boston College Environmental Affairs Law Review* 9:163–203.

Rogers, A. 1993. *The Earth Summit: A Global Reckoning.* Los Angeles: Global View Press.

Ruthgeber, David D. 1984. OCS Leasing Policy Prevails Over the California Coastal Commission. *Natural Resources Journal* 24:1133–1145.

Sabatier, Paul A., and Daniel A. Mazmanian. 1983. Policy implementation. In *Encyclopedia of Policy Studies,* ed. Stuart S. Nagel, 143–170. New York: Marcel Dekker.

Schaeffer, Richard. 1977. State/Federal Program, National Fisheries Service. Interview by Lauren Holland, Sea Grant Trainee, University of California at Santa Barbara, on enactment of Magnuson Act, July, Washington, D.C.

Scheiber, Harry N. 1998. The Stratton Commission: A Historical Perspective on Ocean Policy Studies in Ocean Governance, 1969 and 1998. In *The Stratton Commission: Looking Back, Looking Forward. Lessons from the 1969 Commission on Marine Science, Engineering, and Resources,* eds. Robert W. Knecht, Biliana Cicin-Sain, and Nancy Foster 31–38. Proceedings from a workshop organized by the Center for the Study of Marine Policy, University of Delaware; National Ocean Service, NOAA; and the Ocean Governance Study Group.

Scheiber, Harry N., and Chris Carr. 1992. Constitutionalism and the Territorial Sea: An Historical Study. *Territorial Sea Journal* 2(1):67–91.

Simmons, Malcolm M. 1984. *Accelerated OCS Leasing Program: Changes and Congressional Reaction.* Washington, D.C.: Congressional Research Service.

Sloan, Lucy. 1977. Executive Director, National Federation of Fishermen. Interview by Lauren Holland, Sea Grant Trainee, University of California at Santa Barbara, on enactment of Magnuson Act, July, Washington, D.C.

Soroos, Marvin S. 1997a. *The Endangered Atmosphere.* Columbia: University of South Carolina Press.

———. 1997b. The Evolution of Global Environmental Governance. In *Environmental Policy in the 1990s,* eds. Norman J. Vig and Michael E. Kraft, 278–299. Washington, D.C.: Congressional Quarterly.

Speer, Lisa, and Sarah Chasis. 1995. The Agreement on the Conservation and Management of Straddling and Highly Migratory Fish Stocks: An NGO Perspective. *Ocean & Coastal Management,* Special issue on Earth Summit Implementation: *Progress Achieved on Oceans and Coasts,* ed. Biliana Cicin-Sain and Robert W. Knecht. Vol. 29(1–3):71–77.

Stenberg, Carl. 1982. The States' Role in the New Federalism. In *The Newest Federalism: Framework for Coastal Issues,* ed. Thomas D. Galloway, 41–50. Wakefield, Rhode Island: Times Press.

Sutinen, Jon G., and Hanson, Lynne C., eds. 1986. *Rethinking Fisheries Management, Proceedings from the Tenth Annual Conference, June 1–4, 1986.* Kingston: Center for Ocean Management Studies, University of Rhode Island.

Szekeley, Alberto. 1979. *Latin America and the Development of the Law of the Sea.* Dobbs Ferry, New York: Oceana Publications.

Tarnas, David A. 1988. The U.S. National Marine Sanctuary Program: An Analysis of the Program's Implementation and Current Issues. *Coastal Management* 16:275–303.

Thorne-Miller, Boyce. 1999. *The Living Ocean: Understanding and Protecting Marine Biodiversity,* 2nd ed. Washington, D.C.: Island Press.

Tiddens, Arthur A. 1990. *Aquaculture in America: The Role of Science, Government, and the Entrepreneur.* Boulder, Colorado: Westview Press.

Twiss, John. 1998. Personal communication with authors Cicin-Sain and Knecht.

UNCED. 1992. Report of the United Nations Commission on Environment and Development. Rio de Janeiro, Brazil, 3–14 June.

Underdahl, Arild. 1980. Integrated Marine Policy: What? Why? How? *Marine Policy* 4(3):159–169.

U.S. Commission on Marine Science, Engineering, and Resources (U.S. COMSER). 1969. *Our Nation and the Sea: A Plan for National Action* (H. Doc. 91–42). Washington, D.C.: U.S. Government Printing Office. [91st Congress, 1st Session]

U.S. Congress. 1817. *State Papers and Public Documents of the United States from the Accession of George Washington to the Presidency,* Second edition, Volume One. Boston: Thomas B. Wait.

———. 1972. Marine Mammals: House Approves 5-year Ban on Killing. *Congressional Quarterly Weekly Report,* (18 March), 603–605.

———. Congressional Research Service. 1976a. *A Legislative History of the Fishery Conservation and Management Act of 1976.* Washington, D.C.: U.S. GPO. [Y4.C73/2:F53/13]

———. Congressional Research Service. 1976b. *Legislative History of the Coastal Zone Management Act of 1972, As Amended in 1974 and 1976 with a Section-by-Section Index.* Washington, D.C.: U.S. GPO. [Y4.C73/2:C63/7]

———. House. Committee on Merchant Marine and Fisheries. 1971. *Marine Mammals.* Hearing, 9, 13, 17, 23 September (Serial No. 92–10). Washington, D.C.: U.S. GPO. [92nd Congress, 1st Session]

———. House. Committee on Merchant Marine and Fisheries. 1975. *Marine Fisheries Conservation Act of 1975* (H. Rpt. 94–445). Washington, D.C.: U.S. GPO. [CIS/MF/4 Item 1008-A]

———. House. 1998. Global Warming Treaty. 105th Congress, 2nd Session. *Congressional Record* 144, no. 65 (20 May): H3494.

———. House. Outer Continental Shelf Committee. 1977. *Outer Continental Shelf Lands Act Amendments of 1978* (H. Rpt. 95–590). Washington, D.C.: U.S. GPO.

———. Senate. Committee on Interior and Insular Affairs. Subcommittee on Public Lands. 1967. *S. 2255 and H.R. 12121, 1967.* Hearing, 26 October, before the Public Land Law Review Commission. Washington, D.C.: U.S. GPO. [90th Congress, 1st Session]

———. Senate. 1969. Report of the Secretary of the Interior to the U.S. Congress. *U.S. Senate Document No. 91–58.* Washington, D.C.: U.S. GPO.

———. Senate. Committee on Commerce. Subcommittee on Oceans and Atmosphere. 1972. *Ocean Mammal Protection.* Hearing, 15, 16, 23 February, 7 March, and 11, 12, 13 May. Washington, D.C.: U.S. GPO. [92nd Congress, 2nd Session]

———. Senate. Committee on Commerce and National Policy Study. 1976. *Legislative History of the Coastal Zone Management Act of 1972, As Amended in 1974 and 1976 with a Section-by-Section Index; Part I, Legislative History of the Coastal Zone Management Act of 1972* (S. Res. 222). Washington, D.C.: U.S. GPO. [94th Congress, 2nd Session]

———. Senate. 1990. Law of the Sea. *Congressional Record* 136, no. 52: S5547.

———. Senate. 1992a. The UNCED Biodiversity Convention. Remarks. 102nd Congress, 2nd Session. *Congressional Record* 138, no. 77 (2 June): S7265.

———. Senate. 1992b. The Earth Summit. Extension of Remarks. 102nd Congress, 2nd Session. *Congressional Record* 138, no. 92 (24 June): E1966.

———. Senate. 1992c. The Environment. Remarks. 102nd Congress, 2nd Session. *Congressional Record* 138, no. 111 (31 July): S11048.

———. Senate. 1993a. Earth Day 1993: Time to Recommit to the Spirit of UNCED, Remarks. 103rd Congress, 1st Session. *Congressional Record* 139, no. 52 (22 April): S4795.

———. Senate. 1993b. Convention on Biological Diversity. Unanimous consent request and President's message. 103rd Congress, 1st session. *Congressional Record* 139, no. 162, part 2 (19 November): S16572.

———. Senate. 1995. Middle East Peace Facilitation Act and State Department Reorganization. Remarks. 104th Congress, 1st session. *Congressional Record* 141, no. 170 (31 October): S16401.

———. Senate. 1996. The 104th Congress. Remarks. *Congressional Record* 142, no. 141 (3 October): S12272.

U.S. Department of Commerce (U.S. DOC). National Oceanic and Atmospheric Administration. 1981a. *Calendar Year 1980: Report on the Implementation of the Magnuson Fishery Conservation and Management Act of 1976.* National Marine Fisheries Service. Washington, D.C.: U.S. GPO.

———. 1981b. Withdrawal of final rule. *Federal Register* 46:35, 253–257.

———. 1986. *NOAA Fishery Management Study.* Washington, D.C.

———. 1994. *1992–1993 Biennial Report to Congress on the Administration of the Coastal Zone Management Act. Volume 1—Executive Summary.* Washington, D.C.: U.S. GPO.

———. National Marine Fisheries Service. 1986. *An Evaluation of the Implementation of the Magnuson Fishery and Conservation Act. Final Report of the Council/NOAA Task Group.* NOAA, Washington, D.C. October 1986.

———. National Marine Fisheries Service. 1991. *Our Living Oceans Report on the Status of U.S. Living Marine Resources 1991.* Silver Spring, Maryland: National Marine Fisheries Service. NOAA Tech. Memo. NMFS-F/SPO.

———. National Marine Fisheries Service. 1992. *Our Living Oceans Report 1992.* Silver Spring, Maryland: National Marine Fisheries Services.NOAA Tech. Memo NMFS-F/SPO.

———. National Marine Fisheries Service. 1993. *Our Living Oceans Report 1993.* Silver Spring, Maryland: National Marine Fisheries Services.NOAA Tech. Memo NMFS-F/SPO.

———. National Marine Fisheries Service. 1994a. *Marine Mammal Protection Act of 1972 Annual Report,* 1 January 1992 to 31 December 1993. Washington, D.C.: U.S. GPO.

———. National Marine Fisheries Service. 1994b. The Marine Mammal Protection Act is Amended. *MMPA Bulletin* (September):1–2, 4. Silver Spring, Maryland: Office of Protected Resources.

———. National Marine Fisheries Service. 1994c. Marine Mammal Protection Act of 1972 Annual Report, 1 January 1992 to 31 December 1993. Washington, D.C.: U.S. GPO.

———. National Marine Fisheries Service. 1994d. *Report to Congress on Results of Feeding Wild Dolphins: 1988–1994.* Silver Spring, Maryland: Office of Protected Resources.

———. National Marine Fisheries Service. 1995a. *Marine Mammal Protection Act of 1972 Annual Report,* 1 January 1994 to 31 December 1995. Washington, D.C.: U.S. GPO.

———. National Marine Fisheries Service. 1995b. *Our Living Oceans Report on the Status of U.S. Living Marine Resources 1995.* Washington, D.C.:U.S. GPO. NOAA Tech Memo NMFS-F/SPO

———. National Marine Fisheries Service. 1995c. Flipper's Myth Proves Harmful. *MMPA Bulletin 6(3).* Silver Spring, Maryland: Office of Protected Resources.

———. National Marine Fisheries Service. 1997. Marine Mammals and Commercial Fisheries Take Reduction Team Fact Sheet. Silver Spring, Maryland: Office of Protected Resources.

———. National Marine Fisheries Service. 1998. NMFS Continues Campaign to Halt Feeding and Harrassment of Wild Dolphins. *MMPA Bulletin* 10(1). Silver Spring, Maryland: Office of Protected Resources.

———. National Ocean Service. 1985. *Draft Federal Consistency Study.* Silver Spring, Maryland: Office of Ocean and Coastal Resource Management.

———. National Ocean Service. 1987. *Final Evaluation of the California Coastal Zone Management Program.* Silver Spring, Maryland: Office of Ocean and Coastal Resource Management.

———. National Ocean Service. Strategic Assessment Branch, Office of Oceanography and Marine Assessment. 1991. *The 1990 National Shellfish Register of Classified Estuarine Waters.* Silver Spring, Maryland: National Ocean Service.

———. National Ocean Service. 1994. *1992–1993 Biennial Report to Congress on the Administration of the Coastal Zone Management Act, Volume II, April 1994.* Silver Spring, Maryland: Office of Ocean and Coastal Resource Management.

———. National Ocean Service. 1996. *Florida Keys National Marine Sanctuary, Volume I: The Management Plan.* Silver Spring, Maryland: Sanctuaries and Reserves Division, Office of Ocean and Coastal Resource Management.

———. National Ocean Service. 1997. *National Marine Sanctuaries: Accomplishments Report.* Silver Spring, Maryland: Sanctuaries and Reserves Division, Office of Ocean and Coastal Resource Management.

———. Office of the Administrator. 1982. *Administrator's Letter No. 37.* Silver Spring, Maryland. National Ocean Service.

U.S. DOC et al. 1998a. U.S. Department of Commerce, U.S. Department of the Navy, United States Coast Guard, U.S. Department of the Interior, U.S. Environmental Protection Agency, Federal Emergency Management Agency, National Aeronautics and Space Administration, and National Science Foundation. *National Ocean Conference, Ocean of Commerce, Ocean of Life.* 11–12 June, 1998. Naval Postgraduate School, Monterey, Califorina.

———. 1998. U.S. Department of Commerce, U.S. Department of Defense, U.S. Department of State, U.S. Department of the Interior, U.S. Department of Transportation and other agencies. The Convention on the Law of the Sea (information packet).

U.S. Department of Defense, U.S. Department of State, U.S. Department of the Interior, and U.S. Department of Transportation. May, 1998. *The Convention on the Law of the Sea: Joint Paper.* Washington, D.C.: U.S. GPO.

U.S. Department of Housing and Urban Development (U.S. DHUD). 1976. *Annual Report: Comprehensive Planning Assistance Grant (701) Program.* Washington, D.C.: U.S. GPO.

U.S. Department of the Interior (U.S. DOI). Fish and Wildlife Service. 1970. *National Estuary Study.* Washington, D.C.: U.S. GPO.

———. Minerals Management Service. 1986. *OCS National Compendium.* OCS Information Report MMS 86–0017 (May). Herndon, Virginia: OCS Information Program.

———. Minerals Management Service. 1987. *Federal Offshore Statistics: 1985.* OCS Information Report MMS 87–00008. Herndon, Virginia: OCS Information Program.

———. Minerals Management Service. 1991. *OCS National Compendium.* OCS Information Report MMS 91–0032 (June). Herndon, Virginia: OCS Information Program.

———. Minerals Management Service. 1993. Report of the OCS Policy Committee's Subcommittee on OCS Legislation. *The Outer Continental Shelf Oil and Gas Program—Moving beyond Conflict to Consensus.* October.

———. Minerals Management Service. 1995. *Federal Offshore Statistics: 1995.* OCS Information Report MMS 97–0007 (December). Herndon, Virginia: OCS Information Program.

U.S. Department of Transportation, Maritime Administration. 1994. *The Dredging Process in the United States: An Action Plan for Improvement.* Report to the Secretary of Transportation, The Interagency Working Group on the Dredging Process, Washington, D.C.

U.S. Environmental Protection Agency (U.S. EPA). 1975. *Interim Output Evaluation Handbook for Section 208 Areawide Waste Treatment Management Planning.* June. Washington, D.C.: U.S. EPA.

———. 1994. *Measuring Progress of Estuary Programs: A Manual.* EPA 842-B-94-008, November. Washington, D.C.: U.S. EPA, Office of Water.

———. 1999. *Action Plan for Beaches and Recreational Waters* (EPA/600/R-98/079, March 1999). Washington, D.C.: U.S. Environmental Protection Agency, Office of Research and Development and Office of Water (also on website: http://www.epa.gov/ORD/WebPubs/beaches/).

U.S. Executive Office of the President (U.S. EOP). 1983. President Ronald Reagan's Statement on the Establishment of the Exclusive Economic Zone, March 10, 1983. *Weekly Compilation of Presidential Documents* 19:383.

———. 1984. *Statement by Principal Deputy Press Secretary, 6 June.* Washington, D.C.: U.S. GPO.

———. President's Committee of Advisors on Science and Technology (PCAST). 1998. *Teaming with Life: Investing in Science to Understand and Use America's Living Capital.* Washington D.C.: Office of Science, Technology and Policy.

U.S. General Accounting Office (U.S. GAO). 1980. *Problems Continue in the Federal Management of the Coastal Zone Management Program.* CED-80–103, June 25. Washington, D.C.: U.S. GPO.

———. 1983. *Need to Improve Fishery Management Plan Process.* GAO/RECD-83-72. Washington, D.C.: U.S. GPO.

U.S. General Services Administration (U.S. GSA). 1979. Statement on Signing the Fishery Conservation and Management Act of 1976, April 13, 1976. *Public Papers of the Presidents of the United States: Gerald R. Ford, Book II,* 1116–1118. Washington, D.C.: U.S. GPO.

———. Letter of Transmittal from President Clinton to the U.S. Senate, 7 October 1994. 1995 *U.S. Department of State Dispatch Supplement* 6 (1 February).

———. Letter of Submittal from Secretary of State Christopher to President Clinton, 23 September 1994. 1995 *U.S. Department of State Dispatch Supplement* 6 (1, February).

U.S. Presidential Proclamation No. 2667 of 28 September 1945. 1957. *Policy of the United States With Respect to the Natural Resources of the Subsoil and Sea Bed of the Continental Shelf.* 3 *CFR,* 1943–1948 Comp., 67–68. Washington, D.C.: U.S. GPO.

U.S. Presidential Proclamation No. 5030 of 10 March 1983. *Exclusive Economic Zone of the United States*. 48 *F.R.* 10601, 3 *CFR*, 1983 Comp., 22–23. Washington, D.C.: U.S. GPO.

U.S. Presidential Proclamation No. 5928 of 27 December 1988. 1989. *Territorial Sea of the United States of America*. 3 *CFR*, 1988 Comp., 547. Washington, D.C.: U.S. GPO.

U.S. Public Land Law Review Commission. 1970. *One Third of the Nation's Land*. Washington, D.C.: U.S. GPO.

United States v. California, 67 *S.Ct.* 1658, 332 *U.S.* 19, 91 *L.Ed.* 1889 (1947).

United States v. Florida et al., 363 *U.S.* 121 (1960).

United States v. Louisiana, Texas, Miss., Al. & Fla., 80 *S.Ct.* 961, 363 *U.S.* 1, 4 *L.Ed.2d* 1025 (1960).

Utz, William. 1977. National Shrimp Congress. Interview by Lauren Holland, Sea Grant Trainee, University of California at Santa Barbara, on enactment of Magnuson Act, July, Washington, D.C.

Van de Kamp, John K., and John A. Saurenman. 1990. Outer continental shelf oil and gas leasing: What role for the states? *Harvard Environmental Law Review* 14(73):73–134.

Van Deman Magoffin, Ralph. 1916. T*he Freedom of the Seas: A Dissertation by Hugo Grotius, Translated with a Revision of the Latin Text of 1633*. New York: Oxford University Press.

Van Dyke, Jon M. 1992a. Substantive Principles for a Constitution for the U.S. Oceans. In *Ocean Governance: A New Vision*, ed. Biliana Cicin-Sain, 22–23. Newark: University of Delaware, Ocean Governance Study Group, Center for the Study of Marine Policy, Graduate College of Marine Studies.

———. 1992b. The Role of a Constitution for the U.S. Oceans. *Ocean & Coastal Management* 17(3–4):273–297.

Van Dyke, Jon M., and Jonathan Gurish. 1988. The Exclusive Economic Zone of Northwestern Hawaiian Islands: When Do Uninhabited Islands Generate an EEZ? *San Diego Law Review* 25:425.

Van Dyke, Jon M., Durwood Zaelke, and Grant Hewison, eds. 1993. *Freedom for the Seas in the 21st Century: Ocean Governance and Environmental Harmony*. Washington, D.C.: Island Press.

Vargas, Jorge A. 1979. *Mexico y la Zona de Pesca de Estados Unidos*. Mexico, D.F.: Universidad Nacional Autonoma de Mexico.

Vulovic, Rod. 1995. *World Trade, Container Ships and Ports: A Study in Commercial Symbiosis*. A presentation before the Marine Board National Research Council. Washington D.C.

Wagner, Thomas J. 1990. The Oil Pollution Act of 1990: An Analysis. *Journal of Maritime Law and Commerce* 21(4):569–587.

Warren, Robert N. 1966. Government in Metropolitan Regions: A Reappraisal of Fractionated Political Organization. Davis: Institute of Governmental Affairs, University of California.

Warrick, Joby. 1998a. Administration Signs Global Warming Pact. *Washington Post,* November 13, 1998, A26.

Warrick, Joby. 1998b. 160 Nations Endorse Pact on Global Warming Compliance; Accord Speeds Up Timetable for 'Action Plan' on Environment. *Washington Post,* November 15, 1998, A6.

Weber, Peter. 1993. *Abandoned Seas: Reversing the Decline of the Oceans.* Worldwatch Paper 116, November. Washington, D.C.

Weddig, Lee. 1977. Executive Director, National Fisheries Institute. Interview by Lauren Holland, Sea Grant Trainee, University of California at Santa Barbara, on enactment of Magnuson Act, July, Washington, D.C.

Wenk, Edward Jr. 1972. *The Politics of the Ocean.* Seattle: University of Washington Press.

Wenk, Edward. 1998. Creating the Stratton Commission—A Reprise. In *The Stratton Roundtable. Looking Back, Looking Forward: Lessons from the 1969 Commission on Marine Science, Engineering, and Resources,* eds. Robert W. Knecht, Biliana Cicic-Sain, and Nancy Foster, 13–23. Newark: National Oceanic and Atmospheric Administration, in partnership with the Center for the Study of Marine Policy, University of Delaware, and the Ocean Governance Study Group.

Whitaker, John C. 1976. *Striking a Balance: Environmental and Natural Resources Policy in the Nixon-Ford Years.* Washington, D.C.: American Enterprise Institute for Public Policy Research.

Whitehead, John C. 1998. Keeping Promises by Paying U.N. *Washington Times,* 9 October, A21.

Wilder, Robert J. 1993. Cooperative Governance, Environmental Policy, and Management of Offshore Oil and Gas in the United States. *Ocean Development and International Law* 24:41–62.

———. 1998. *Listening to the Sea: The Politics of Improving Environmental Protection.* Pittsburgh: University of Pittsburgh Press.

Wilson, Pete, and Douglas P. Wheeler. 1997. *California's Ocean Resources: An Agenda for the Future.* The Resources Agency of California. April.

Wise, Morton. 1991. *Federal Conservation & Management of Marine Fisheries in the United States.* Washington, D.C.: Center for Marine Conservation.

World Travel and Tourism Council (WTTC), World Tourism Organization, and Earth Council. No date. *Agenda 21 for the Travel and Tourism Industry: Towards Environmentally Sustainable Development.* World Travel and Tourism Council, London, United Kingdom; World Tourism Organization, Madrid, Spain; and Earth Council, San Jose, Costa Rica. 78 pp. (presented at the June 1997 United Nations General Assembly, The Earth Summit plus Five, New York City).

Year of the Ocean (YOTO) Discussion Papers. March 1998. Prepared by the U.S. federal agencies with ocean-related programs. Washington, D.C.

A. The U.S. Marine Transportation System

B. The Oceans and National Security

C. Ensuring the Sustainability of Ocean Living Resources

D. Ocean Energy and Minerals: Resources for the Future

E. Perspectives on Marine Environmental Quality Today

F. Coastal Tourism and Recreation

G. Impacts of Global Climate Changes—With Emphasis on U.S. Coastal Areas

H. Mitigating the Impacts of Coastal Hazards

I. Opportunities and Challenges for Marine Science, Technology, and Research

J. A Survey of International Agreements

K. Marine Education USA: An Overview

L. The Legendary Ocean: The Unexplored Frontier

Young, Nina M., and Suzanne Iudicello. 1997. Blueprint for Whale Conservation: Implementing the Marine Mammal Protection Act. *Ocean and Coastal Law Journal* 3(1&2):149–217.

Young, Nina M., William Robert Irvin, and Meredith L. McLean. 1997. The Flipper Phenomenon: Perspectives on the Panama Declaration and the "Dolphin Safe" Label. *Ocean and Coastal Law Journal* 3(1&2):57–115.

Young, Oran R. 1982. *Resource Regimes: Natural Resources and Social Institutions.* Berkeley: University of California Press.

Glossary

baseline Normally, the mean low-water line of the coast and the lines drawn across river mouths, the openings of bays, and in particular cases, along the outer points of complex coasts.

contiguous zone A zone adjacent to a nation's territorial sea, not extending beyond 24 nautical miles of the baseline from which the breadth of the territorial sea is measured (usually between 12 and 24 nautical miles offshore). In this zone, a coastal nation may exercise the control necessary to prevent infringement of its custom, fiscal, immigration, or sanitary laws.

continental shelf The seabed and subsoil of the submerged areas that extend beyond a coastal nation's territorial sea throughout the natural prolongation of its land territory and to the outer edge of the continental margin, or to a distance of 200 nautical miles from the baseline from which the breadth of the territorial sea is measured.

ecosystem management A holistic resource management approach that creates management units based on the spatial extent of a particular ecosystem, in contrast to the more traditional, single-species or single-use approach that tended to ignore the interactions that occur between uses and/or different species within an ecosystem.

enclosure movement A movement that began with the Truman Declarations in 1945, and has been characterized by national assertion of control over what had previously been viewed as common property ocean resources. The effect of the enclosure movement has been to reduce areas of ocean space available for nations to exploit at will, and has resulted in larger ocean areas managed by nation states.

exclusive economic zone (EEZ) The maritime zone beyond, and adjacent to, the territorial sea, but not exceeding 200 nautical miles from the baseline from which the territorial sea is measured. In the EEZ, the coastal nation has sovereign rights to explore and exploit, conserve, and manage the natural resources, whether living or nonliving, of the waters superadjacent to the seabed and of the seabed and its subsoil.

federal consistency provision The Coastal Zone Management Act 1972, section 307(c)(1), provides that each federal agency conducting or supporting activities in the coastal zone shall conduct or support those activities in a manner that is, to the maximum extent practicable, consistent with approved state coastal management programs.

fishery conservation zone A zone contiguous to the territorial sea designated for offshore fisheries management. The inner boundary of this zone is the seaward boundary of each of the coastal states, and the outer boundary is a line drawn in such a manner that each point on it is 200 nautical miles from the baseline from which the territorial sea is measured.

freedom of the seas The principle that waters beyond national jurisdiction are open to all and in which no nation can assert sovereignty. Nations are free to undertake normal activities, including navigation, fishing, and scientific research, as long as reasonable regard is given to the interests of other states.

governing international fishery agreements (GIFAs) Agreements regulating the fishing activities of foreign nations in U.S. waters. GIFAs require the foreign nations' agreement to acknowledge exclusive U.S. management authority, to abide by all fishery regulations, and to allow boarding and inspection for law enforcement purposes. Foreign vessels must have an annual permit and are required to pay for fish caught.

innocent passage The principle that ships of all states enjoy the right to pass through the territorial sea of another nation. Passage is regarded as "innocent" so long as it is not prejudicial to the peace, good order, or security of the coastal state. Military vessels must show their flag, and submarines must travel on the surface of the sea in territorial waters.

maximum sustainable yield A scientific determination of the level of harvest that can be taken consistently on a long-term basis without diminishing the stocks and that will ensure an inexhaustible and perpetually renewable resource.

optimum carrying capacity Under the Marine Mammal Protection Act, the ability of a habitat to support the optimum sustainable population of a species or population stock in a healthy state without diminishing the ability of the habitat to continue that function.

optimum yield The term "optimum," with respect to the yield from a fishery, means the amount of fish that will provide the greatest overall benefit, particularly with respect to food production and recreational opportunities. This takes into account the protection of marine ecosystems; is prescribed on the basis of the maximum sustainable yield from the fishery, as reduced by any relevant social, economic, or ecological factor; and in the case of an overfished fishery, provides for rebuilding to a level consistent with producing the maximum sustainable yield in such fishery.

outer continental shelf The submerged lands seaward of state boundaries and that appertain to the United States and are subject to its jurisdiction and control.

overcapitalization Investing more money in capital equipment or an industry than should occur given the actual economic worth of that sector. In the case of U.S. fisheries, government subsidies encouraged overcapitalization beginning as early as the 1960s, when the U.S. government began heavily subsidizing commercial fisheries in an effort to encourage the development of the industry.

regional fishery management councils Management agencies created under the Magnuson Act, responsible for fisheries planning and management in each of eight offshore regions in the United States.

submerged lands Lands under the surface of the sea from 0 to 3 statute miles offshore and under state, rather than federal, control.

territorial sea The area of sea adjacent to its land area over which the sovereignty of a coastal nation extends, and, in the case of an archipelagic nation, a belt of sea adjacent to its archipelagic waters. According to the 1982 Law of the Sea Convention, every nation has the right to establish the breadth of its territorial sea to a limit not exceeding 12 nautical miles, measured from baselines. The sovereignty of a coastal nation extends to the airspace over the territorial sea, and the territorial sea's water column, seabed, and subsoil.

watershed management This approach to management incorporates a holistic view of the environment. Management is undertaken on a spatial scale that incorporates an entire watershed in a single unit. This allows management of all factors that may influence a particular environmental issue under one program, rather than segmenting environmental issues by creating artificial administrative boundaries. This approach is even broader in scope than ecosystem management, as there will often be several ecosystems within a single watershed.

regional fishery management councils [illegible] Magnuson Act [illegible] responsible for fishery planning [illegible] each of the eight [illegible] regions of the United States.

submerged lands Lands under the [illegible]

territorial sea The area of sea adjacent to [illegible] coastal nation [illegible] According to the 1982 [illegible] Law of the Sea Convention, every nation has the right to establish the breadth of its territorial sea up to a limit not exceeding 12 nautical miles, measured from baselines [illegible] the sovereignty of a coastal nation extends to the airspace over the territorial sea, [illegible] and subsoil.

watershed management [illegible]

Acronyms

AID	Agency for International Development
AOC	American Oceans Campaign
API	American Petroleum Institute
AUV	autonomous underwater vehicle
ASMFC	Atlantic States Marine Fisheries Commission
BCDC	Bay Conservation and Development Commission
CEIP	Coastal Energy Impact Program
CEQ	Council for Environmental Quality
CITES	Convention on International Trade in Endangered Species
CMC	Center for Marine Conservation
COMSER	Commission on Marine Sciences, Engineering, and Resources
CSD	Commission on Sustainable Development
CSO	Coastal States Organization
CWA	Clean Water Act
CZM	Coastal Zone Management
CZMA	Coastal Zone Management Act
DOC	Department of Commerce
DOI	Department of the Interior
EEZ	exclusive economic zone
ECOSOC	United Nations Economic and Social Council
EFH	essential fish habitat
EPA	Environmental Protection Agency
ESA	Endangered Species Act
ESS-21	Environmentally Sound Ships of the 21st Century
ESSA	Environmental Sciences Services Administration
FAO	Food and Agriculture Organization of the United Nations
FCMA	Fishery Conservation and Management Act
FCZ	fishery conservation zone
FMP	fishery management plan
FWS	U.S. Fish and Wildlife Service

GAO	General Accounting Office
GDP	gross domestic product
GIFA	governing international fishery agreement
GSMFC	Gulf States Marine Fisheries Commission
HSUS	Humane Society of the United States
HUD	U.S. Department of Housing and Urban Development
IATTC	Inter-American Tropical Tuna Commission
ICCAT	International Convention for the Conservation of Atlantic Tuna
ICNAF	International Convention for the Northwest Atlantic Fisheries
ICRI	International Coral Reef Initiative
ICRW	International Convention for the Regulation of Whaling
IGY	international geophysical year
IS&R	intelligence, surveillance, and reconnaissance
ITQ	individual transferable quota
JSA	joint subcommittee on aquaculture
LORAN	long-range aid to navigation
LOS	Law of the Sea
LRA	list of recommended areas
MFCMA	Magnuson Fishery Conservation and Management Act
MMPA	Marine Mammal Protection Act
MMS	Minerals Management Service
MPA	marine protected area
MPRSA	Marine Protection, Research, and Sanctuaries Act
NEP	National Estuary Program
NERRS	National Estuarine Research Reserve System
NFF	National Federation of Fishermen
NFI	National Fisheries Institute
NMFS	National Marine Fisheries Service
NMSP	National Marine Sanctuaries Program
NOAA	National Oceanic and Atmospheric Administration
NOIA	National Ocean Industries Association
NOPS	National Ocean Policy Study
NRC	National Research Council
NRDC	Natural Resources Defense Council
NSF	National Science Foundation
NWF	National Wildlife Federation
OCC	optimum carrying capacity

OCS	outer continental shelf
OCSLA	Outer Continental Shelf Lands Act
OCSLAA	Outer Continental Shelf Lands Act Amendments
OECD	Organization for Economic Cooperation and Development
OMB	Office of Management and Budget
ONR	Office of Naval Research
OPEC	Organization of Petroleum Exporting Countries
OSP	optimum sustainable population
OTEC	ocean thermal energy conversion
OY	optimum yield
PBR	potential biological removal
PMFC	Pacific Marine Fisheries Commission
ROV	remotely operated vehicle
SEL	site evaluation list
SFA	Sustainable Fisheries Act
SLA	Submerged Lands Act
TALFF	total allowable level of foreign fishing
TEU	twenty-foot-equivalent unit
TRT	take reduction team
UNCED	United Nations Conference on the Environment and Development
UNCHE	United Nations Conference on the Human Environment
UNCLOS	United Nations Conference on the Law of the Sea
UNEP	United Nations Environment Programme
UNESCO	United Nations Educational, Scientific, and Cultural Organization
USDA	U.S. Department of Agriculture
WWF	World Wildlife Fund
ZMRG	zero mortality rate goal

Index

Ad Hoc Working Group of Legal and Technical Experts, 268
Advisory committees, 290
Agenda 21, 4, 10, 270–73
Agricultural interests, coastal, 321–22
Agriculture Department, U.S., 225, 227
Air pollution, 264–65
Alaska, 42, 84
Albemarle-Pamlico Estuarine Study, 182–84
Albright, Madeline, 270
Alexander VI (pope), 32
American Cetacean Society, 167
American Humane Association, 167
American Land Institute, 62, 63
American Law Institute, 62
American Oceans Campaign, 146
American Petroleum Institute, 90, 117
Anderson, Glen, 74
Andrus, Cecil D., 89, 170
Animal Protection Institute, 167
Antifouling agents in marine paints, 215
Apalachicola Bay, 192
Aquaculture, 206–7, 224–28, 248
Area-based approach, 287
Army Corps of Engineers, U.S., 22
Ashepoo River Basin, 194
Asia, 207
Asian Development Bank, 272
Atlantic States Marine Fisheries Commission in 1942, 43
Attitudinal mistakes and policy errors by implementing agencies, 174
Audubon Living Oceans Program, 147
Audubon Society, 62, 73
Australia, 219, 273, 306

Bay Conservation and Development Commission (BCDC), 47–48
Beach protection/restoration/nourishment, 205, 238–39, 242, 302, 312
Beaufort Sea, 188
Bering Sea, 42, 262
Biocides, 215
Biodiversity, coastal, 267–70
  see also Convention on Biological Diversity (CBD)
Biophilia (Wilson), 233
Bioprospecting, 207
Biotechnology, marine, 233–35
Breaux, Sen., 59
Bumblebee, 162
Bureau of Commercial Fisheries, 42, 51
Bush, George, 228, 259, 264, 269

California:
  Coastal Zone Management Act of 1972 (CZMA), 62, 116, 129
  development, controlling rampant, 47–48
  expanding scope of planning/policy making, 113
  fisheries, 41
  oil and gas development, 146
  Outer Continental Shelf Lands Act of 1953 (OCSLA), 172
  populations, coastal, 205
Canada:
  fisheries, 42–43, 218, 262
  integrated coastal management, 272–73, 293
  Oceans Act, 306–7
Carbon dioxide emissions, 265
Carter, Jimmy, 60, 89, 101, 110, 170, 188
Center for Marine Conservation, 147, 163, 167, 190
Channel Islands, 186, 188
Chesapeake Bay, 194
China, 206, 223
Christopher, Warren, 260
Citizens Committee on Natural Resources, 73
Civil War, 36
Clean Water Action Plan, 180
Clean Water Network, 240

Climate change, global, 205, 243–44, 263–67
Clinton, Bill:
  Clean Water Action Plan, 180
  climate change, 264–65
  Conference of Parties for the Climate Change, 266
  Convention on Biological Diversity, 269–70
  Law of the Sea Convention, 12, 260
  National Marine Sanctuaries Program, 189
  National Ocean Conference, 205, 210, 229
  oil and gas development, 172, 228
Coalition for change, building a, 308–10
Coastal Energy Impact Program (CEIP), 88, 108, 111, 171
Coastal power *vs.* sea power, 11
Coastal state interests, 201
  see also States' jurisdiction and ownership of offshore resources
Coastal States Organization (CSO), 63, 101
Coastal Zone Management Act of 1972 (CZMA):
  amendments, major, 119–24
  aquaculture, 226
  assessments of, 125–29
  attacks on, 108
  consistency doctrine, 112
  enactment, the politics of, 60, 62–65, 100
  expanding the focus, 125
  integrated coastal management, 294
  intergovernmental approach, 65–67
  land-use planning *vs.* coastal planning, 62–65
  non-point source pollution, 179
  origins, legislative, 61–62
  policy challenges in the new century, 245–46, 252
  pollution, water, 240
  programs, coastal zone management, 110, 118
  Reagan, Ronald, 173
  single-purpose ocean laws, 98
  states' jurisdiction and ownership of offshore resources, 65–67, 73–74, 112–14
  strengthening CZM programs of the coastal states, 245–46
  successes and shortcomings, 116–29, 199
  tensions in the law, 67–68
  territorial sea, 22
  see also Governance, ocean; Management practices regarding ocean/coastal resources
Cold War, end of the, 210, 211
Colonial America, 33
Columbus, Christopher, 204
Combahee River Basin, 194
Commerce Department, U.S., 51, 64, 71, 73, 76, 227
Commission on Marine Science, Engineering and Resources (COMSER), 1
  see also Stratton Commission
Commission on Sustainable Development (CSD), 273–74
Commissions, interstate, 43–44
Committee for Humane Legislation, 69, 75
Committee of Scientific Advisors, 71
Committee on Multiple Uses of the Coastal Zone, 61
Common property concept, 297–98
Competition and Outer Continental Shelf Lands Act Amendments of 1978 (OCSLAA), 87
Complexity of the ocean, 281–82
Comprehensive Conservation and Management Plan (CCMP), 181–84, 242
Conference of Parties for the Climate Change, 265–66
Conflicts among users of coastal resources/space:
  crosscutting problems, 4
  environmental impact statement, 23
  governance, toward a new system of national, 291
  Marine Mammal Protection Act of 1972 (MMPA), 150–52

oil and gas development, 39
Outer Continental Shelf Lands Act Amendments of 1978 (OCSLAA), 88–91
populations, coastal, 206
rise in, 3
see also Tensions in ocean policy
Congress, U.S.:
Coastal Zone Management Act of 1972 (CZMA), 63–64
Law of the Sea Convention, 261
Reagan years, 256
resurgence, congressional, 59–60
1990s, the late, 205, 309
see also Legislation
Consistency concept, 100, 108, 112, 124
Consortium on Ocean Research and Education (CORE), 204
Contract with America, 107
Convention on Biological Diversity (CBD), 10, 207, 233, 267–70, 272
Coral reefs, 205, 275–76
Cordell Bank, 187
Cotton, Norris, 63
Council for Environmental Quality (CEQ), 73, 303
Country Studies Program, 267
Court cases:
Committee for Humane Legislation Inc. v. Richardson, 152
dolphins and tuna, 152
lease sales, 173
oil and gas development, 39, 118, 124, 125
Secretary of Interior v. Watt, 125
Secretary of the Interior v. California in 1980, 112
Secretary of the Interior v. California in 1984, 124, 173
states' jurisdiction and ownership of offshore resources, 9, 20, 112
United States v. California in 1947, 20, 39
United States v. Florida in 1960, 39
United States v. Louisiana in 1960, 39
Cousteau, Jacques, 38
Crosscutting problems, assessment of, 4–6
Cuyahoga River, 176
Davidson, Bruce B., 211
Decision making, narrowly based and adversarial, 280–81
Defenders of Wildlife, 58, 72, 167
Defense Department, U.S., 21, 82
Defense of the nation, ocean's role in historical development of the, 35–36
Delaware, 48, 62
Delaware Bay, 195
Denmark, 273
Deregulation, 103–4, 124
Designing a strengthened ocean governance system, 25–26
Development, ocean resources:
controlling rampant development, 47–48
environmental protection *vs.,* 12–13, 202
expedited development, 86
Outer Continental Shelf Lands Act Amendments of 1978 (OCSLAA), 86
sustainable development, 4, 298–300
Dingell, John, 59, 74, 83
Displacement effects, 17, 18
Divided government, 105–7
Dolphins, 150–52, 161–63, 164
Douglas, Peter, 127
Dredging, 214–15
Drilling, deepwater, 230
Dumping, ocean, 214
Dutch East India Company, 32

Earth Summit, *see* Conference on Environment and Development in 1992 *under* United Nations
Economic activity of various ocean/coastal sectors, 208–10
Economy, globalization of the, 207
Edisto River Basin, 194
Educational system, 49–50
Efficiency and effectiveness in government operations, 295
Elkhorn Slough, 192
Enactment, the politics of, 60, 62–65, 100–102

Enclosure movement, 33–34
Endangered species, 22
Energy Department, U.S., 265
Energy issues, 22, 107–9
Environmental concern/protection, 12–13, 58–59, 202
Environmental groups and fisheries management, 146–47
 see also individual group
Environmental impact statement (EIS), 23, 86
Environmental Policy Institute, 90
Environmental Protection Agency (EPA), 21, 92, 182
Environmental Sciences Services Administration (ESSA), 51
Essential fish habitat (EFH), 148
Estuaries, 191–98, 324
 see also National Estuary Program
Ethics, code of stewardship, 288
European Blue Flag program, 239
European Union, 262
Everett, Ned, 81
Exclusive Economic Zone (EEZ), *see* 200-mile fishery limit
Exotic species, 215
Expedited development, 86
Extraterritoriality, 66
Exxon Valdez oil spill, 324

Fagatele Bay, 187
FAO, *see* Food and Agriculture Organization
Federal Coordinating Council on Science, Engineering, and Technology, 225
Federal Emergency Management Agency (FEMA), 243
Federalism, New, 111, 201
Federal jurisdiction:
 aquaculture, 224–25, 227–28
 Coastal Zone Management Act of 1972 (CZMA), 65–67
 domination, recent, 20–23
 federal waters, 21
 fisheries, 42, 44–45, 78
 hazards, coastal, 243
 historical background, 20
 jurisdictional split among levels of government, 12, 201, 279–80, 304–5
 Magnuson Fishery Conservation and Management Act of 1976, 81
 Marine Mammal Protection Act of 1972 (MMPA), 96, 201
 1970s, 96–97
 ocean resources development *vs.* environmental protection, 12–13
 oil and gas development, 39–40
 optimum yield, 78–79
 tourism, 239
Fish and Wildlife Service, U.S., 48, 71, 94, 191
Fisheries:
 appalling state of, 328
 Coastal Zone Management Act of 1972 (CZMA), 113
 commissions, interstate, 43–44
 declines in the abundance of fish stocks, 203, 206
 environmental groups, 146–47
 FAO Code of Responsible Fishing, 216
 federal jurisdiction, 42, 44–45
 Fishery Conservation and Management Act (FCMA), 113, 129–42
 foreign fishing, threats from, 44–45
 future, looking to the, 328
 integrated coastal management, 301
 international negotiations, 42–43
 Magnuson Fishery Conservation and Management Act of 1976, 142–46
 management practices regarding ocean/coastal resources, 40–46
 National Ocean Conference, 205
 1960s, fisheries at the end of the, 45–46
 overcapitalization, 220
 policy challenges in the new century, 216–20, 247
 political landscape of the late 1990s, 312–13
 privatization of, 218–20

states' jurisdiction and ownership of offshore resources, 41–42
straddling and highly migratory fish stocks, 262–63
subsidies for impacted fishing communities, 218
successes and shortcomings in managing, 199
Sustainable Fisheries Act of 1966, 147–49
200-mile fishery limit, 77–83
Fishing as an American right, 46–47
Fixed ocean space, 18
Florida, 129, 146, 151, 172, 187
Flower Garden Banks, 187
Food and Agriculture Organization (FAO), 216, 272
Food source, the ocean as a, 36
Ford, Gerald R., 60, 84, 101, 110
Forest Service, U.S., 191
Fragmented programs/policies, 5, 7, 22, 76, 278
Framework Convention on Climate Change, 10, 263–67, 272
Freedom of the seas doctrine, 31–33
Friends of Animals, 75
Friends of the Earth, 58, 62, 72, 101
Friends of the Sea Otter, 167
Fund for Animals, 72
Fur industry, 73
Fur Seal Treaty, 74
Future, looking to the, 327–28
see also Policy challenges in the new century

Gas development, *see* Oil and gas development
General Accounting Office (GAO), 124, 143
General Agreement on Tariffs and Trade (GATT), 162
Genetic biodiversity, 268
Geological Survey, U.S., 21
Georges Bank, 172, 188
Geosteering, 230
Gilmore, Jim, 148
Gingrich, Newt, 107
Global Conference on Sustainable Development of Small Island Developing States, 272
Global Coral Reef Monitoring Network, 275
Globalization of the economy, 207
Global Programme of Action (GPA), 274–75
Global warming, *see* Climate change, global
Gordon, William E., 144–45
Gore, Al, 180, 205, 264
Governance, ocean:
attitudinal mistakes and policy errors, 174
chronology of selected major events since 1945, 329–36
coalition for change, building a, 308–10
conflict resolution, building better capacity for, 291
definition, 14
designing a strengthened system, 25–26
divided government, 105–7
efficiency and effectiveness, 295
evolution of, 16
forms of, 16–17
fragmented programs/policies, 5, 7, 22, 76, 278
functions of government, 14, 15
governing international fishery agreements, 80
governmental attention, expanded, 49–51
guidance, national, 289–90
improvements, needed, 287–95
integrated coastal management, 271–72, 285–86, 291–95, 301–5
intergovernmental aspects of, 5, 19–25, 65–67
iron triangles, 48–49
local governments, 318–19
multiple-use, 284–87
need for, 14–15
1970s, 95–96
principles, guiding, 287–89
private *vs.* government role in resource development, 13, 202, 218–20
spatial use, 17–18

Stratton Commission, 50–51
structural improvements, 287
see also Federal jurisdiction; Management practices regarding ocean/coastal resources; Policy challenges in the new century; Problems with the existing system, basic; States' jurisdiction and ownership of offshore resources; U.S. international leadership
Grand Bay, 195
Grants, coastal, 314
see also Sea Grant Program
Gray's Reef, 186, 188
Great Barrier Reef Marine Park Authority, 275
Great Bay, 194, 195
Great Britain, 42
Great Society programs, 60
Greenhouse gas emissions, 264–65
Greenpeace, 147, 167
Grotius, Hugo, 31
Groundfish, 203
Groundings, vessel, 215
Guana River, 195
Guidance, governance and national, 289–90
Gulf Maine region, 293
Gulf of Mexico, 168, 293
Gulf of the Farallones, 188, 189
Gulf Oil, 49
Gulf States Marine Fisheries Commission in 1949, 43

Habitat, critical, 95
Habitat, essential fish, 148
Hagel, Chuck, 266
Harmonization, 292
Harrington, Alice, 75
Hawaii, 113, 129, 187
Hazards, coastal, 238, 242–45, 252
Heinz Center for Science, Economics and the Environment, 8, 204, 283
Helms, Jesse, 259, 261, 270
Historical thread of national ocean policy:
defense of the nation, 35–36
enclosure movement, 33–34
food source, the ocean as a, 36
freedom of the seas doctrine, 31–33
future, looking to the, 327–28
governmental attention, expanded, 49–51
historical development of the U.S., role of the ocean in the, 35–38
inward, turning, 36–37
major periods, 9–11
management practices, evolution of, 38–48
overview, 20
transportation and commerce, 35
World War II and the postwar prosperity, 37–38
see also specific time period
Hollings, Ernest, 59, 75, 116
Hudson River, 193
Humane Society of the United States, 72, 167
Humphrey, Hubert, 50
Hunting, 46–47
Hurricanes, 243

Illinois, 125
India, 206, 223
Individual fishing quotas (IFQs), 219–20
Individual transferable quotas (ITQs), 219–20
Informal Negotiating Committee (INC), 268
Information, the need for more, 208–10
Integrated coastal management (ICM), 271–72, 275, 285–86, 291–95, 301–5
Intellectual property rights, 36, 269
Interagency policy integration, 291–93
Inter-American Tropical Tuna Commission in 1949, 43
Interest groups, 309, 310–23
Intergovernmental aspects of ocean governance, 5, 19–25, 65–67
see also Federal jurisdiction; States' jurisdiction and ownership of offshore resources
Intergovernmental integration, 293–95

Intergovernmental Oceanographic Commission, 6, 272
Intergovernmental Panel on Climate Change (IPCC), 264, 267
Interior, U.S. Department of the, 21
  Coastal Zone Management Act of 1972 (CZMA), 63, 64, 116, 173
  estuaries, 61
  federal and state control, tensions between, 201
  fragmented programs/policies, 76
  Marine Mammal Protection Act of 1972 (MMPA), 71, 73
International Association of Fish and Game Commissioners, 73
International Association of Game, Fish and Conservation, 73
International Convention for the Northwest Atlantic Fisheries (ICNAF), 43
International Coral Reef Initiative (ICRI), 272, 275–76
International Decade of Ocean Exploration, 50
International Fund for Animal Welfare, 72
International Geophysical Year (IGY), 49, 50
Internationalism *vs.* unilateralism, 11–12, 201
International Ocean Institute, 272
International relations, 4, 104–5
International Wildlife Coalition, 167
International Year of the Ocean (1988), 204
Inward, turning, 36–37, 110–12
Iron triangles, 48–49, 107
Issue networks, 107
Iudicello, Suzanne, 147

Jakarta mandate, 270, 272
Japan, 42
Jefferson, Thomas, 33
Jobos Bay, 193
Johnson, Lyndon, 110
Joint Chiefs of Staff, 82
Joint Subcommittee on Aquaculture (JSA), 225
Jurisdictional split among levels of government, 12, 201, 279–80, 304–5

Kelly, Paul L., 229
Kennedy, Sen. Edward, 81
Key Largo, 186
Kingdom, John, 55
Kitsos, Thomas R., 174
Korea, 307–8
Kyoto protocol, 265–66

Land-use planning, 62–65, 244
Latin America, 34, 82, 162, 207
Law of the Sea Convention:
  Clinton, Bill, 12, 260
  domestic concerns and international obligations, balance between, 25
  enclosure movement, 34
  energy issues, 108–9
  focusing the world's attention to problems/opportunities, 58
  integrated coastal management, 291
  Magnuson Fishery Conservation and Management Act of 1976, 80
  mining regime, deep seabed, 10, 105, 109
  political landscape of the late 1990s, 317
  Reagan, Ronald, 111, 114, 259
  sovereign rights, 297
  200-mile fishery limit, 3
  U.S. international leadership, 207, 255, 256, 259–61
Leadership, 6
  see also U.S. international leadership
Leasing of offshore areas, 39–40, 171–73, 226
Leggett, Congressman, 81
Legislation:
  American Fisheries Promotion Act of 1980, 134–35
  Anadromous Fish Conservation Act of 1961, 45
  Appropriations Act of 1915, 42
  Clean Water Act of 1972, 92–93, 177–78, 180–81, 199–200, 240
  Coastal Barrier Resources Act of 1982, 104
  Commercial Fisheries and Research Development Act of 1960, 45

Deep Seabed Hard Mineral Resources Act of 1980, 111
Dolphin Protection Consumer Information Act of 1990, 162
Endangered Species Act of 1973 (ESA), 68–69, 94–95
Fishery Conservation and Management Act (FCMA), 113, 129–42
International Dolphin Conservation Act of 1992, 162
Lacey Act, 224
Marine Research, Protection, and Sanctuaries Act of 1972, 93–94, 104
National Aquaculture Act, 225
National Aquaculture Improvement Act of 1985, 225
National Aquaculture Policy, Planning, and Development Act of 1998, 225
National Environmental Policy Act, 23
Ocean Dumping Act of 1972, 104
Oceans Act of 1996 (Canada), 306–7
Oceans Act of 1998, 229
Oil Pollution Act of 1990, 104
Outer Banks Protection Act, 172
Outer Continental Shelf Lands Act of 1953 (OCSLA), 9, 13, 20, 39–40, 97, 168–76
Outer Continental Shelf Lands Act Amendments of 1978 (OCSLAA), 84–91, 98, 100, 102, 108, 112, 168–76, 201, 229–32
Public Law 70-454, 61
Public Law 89-753, 61
Saltonstall-Kennedy Act of 1954, 44
Submerged Lands Act of 1953, 9, 20, 39
Sustainable Fisheries Act of 1966, 140–42, 147–49, 216–17, 235
Water Pollution Control Act of 1948, 92
see also Coastal Zone Management Act of 1972 (CZMA); Magnuson Fishery Conservation and Management Act of 1976; Marine Mammal Protection Act of 1972 (MMPA)

Lennon, Alton, 63
List of recommended areas (LRA), 188
Local governments, 318–19
London Convention of 1972, 255
Looe Key, 186
Louisiana, 168

Magnuson, Warren, 59, 63, 81, 116
Magnuson Fishery Conservation and Management Act of 1976:
decision making process, 130
federal jurisdiction, 20
Marine Mammal Protection Act of 1972 (MMPA), 81
privatization of fisheries, 220
successes/problems and continuing issues, 142–46
200-mile fishery limit, 12, 77–83
Mammals, protecting marine, 22, 248, 315–16
see also Marine Mammal Protection Act of 1972 (MMPA)
Management practices regarding ocean/coastal resources:
appropriate, absence of, 5
aquaculture, 226–27
Coastal Zone Management Act of 1972, 60–68
development, controlling rampant, 47–48
evolution of, 38–48
fisheries resources, 40–46
hazards, natural, 242–45
hunting and fishing as an American right, 46–47
intergovernmental conflict, growing, 39
issues involved, 18–19
leasing of offshore areas, 39–40
living marine resources, 19
Magnuson Fishery Conservation and Management Act of 1976, 79–80
1970s, 60–68, 99–100
oil and gas development, 38–40
pollution, water, 240–42
successes and shortcomings, 116–29
tourism, 236–40
water quality, 19

see also Coastal Zone Management Act of 1972 (CZMA); Federal jurisdiction; Governance, ocean; Policy challenges in the new century; Problems with the existing system, basic; States' jurisdiction and ownership of offshore resources
Manatees, 151
Mannina, George, 81
Mare Liberum (Grotius), 31
Marine Board, 283
Marine industries, 5
see also Fisheries; *specific industry*
Marine Mammal Commission (MMC), 71, 167
Marine Mammal Protection Act of 1972 (MMPA), 68–69
amendments to, 138–39, 152–61, 221–23
conflicts with other resources and uses, 150–52
enactment, the politics of, 100
federal role, enhancement of the, 96
implementation, issues for, 76–77, 163–68
internecine fight, 72–76
key features, 70–72
Magnuson Fishery Conservation and Management Act of 1976, 81
policy challenges in the new century, 220–23
prohibitive and narrowly focused policy, 149
protection movement, three factors influencing rise of the, 69
regional approach to ocean resources and space utilization, 235
state and federal control, tensions between, 201
successes and shortcomings, 198
tuna-dolphin issue, 161–63
Marine Science, Engineering and Resources Council, 61
Marine Science Council, 1, 21, 50–51, 301
Maritime state interests, 201
Maryland, 172
Massachusetts, 48, 128
Matanzas River, 195
Megacities, 205
Mexico, 162
Michigan, 62, 116
Military, the, 36, 205, 210–13
Minerals Management Service, 21, 175
Mining, seabed, 10, 34, 105, 109, 260
Minnesota, 62, 116, 125
Mississippi, 129
Mobil, 49
Monitor (ship), 186, 188
Monterey Bay, 187, 190
Montreal Protocol of 1987, 255
Moore, John N., 261
Mosher, Charles, 63
Mullica River, 195
Multilateral approaches to global problems, 277, 284–87
Murphy, Congressman, 59, 81

Nader, Ralph, 177
Narragansett Bay, 193
National Academy of Sciences, 59
National Aquaculture Development Plan of 1984, 225
National Association of Counties, 90
National Audubon Society, 167
National Coalition for Marine Conservation, 147
National Dialogues, 8
National Estuarine Pollution Study, 61
National Estuarine Research Reserve Program, 191–98, 232
National Estuarine Research Reserve System: Building a Valuable National Asset, 196–97
National Estuary Program, 104, 107, 128, 180–84, 201, 242
National Estuary Study, 61
National Federation of Fishermen (NFF), 81, 90, 101
National Fisheries Institute (NFI), 83
National Fisherman, 167
National Flood Insurance Program (NFIP), 243, 244
National Governors Conference, 62, 63, 90

National League of Cities, 90
National Marine Fisheries Service (NMFS), 20, 48, 71, 94, 129, 131, 143, 301
National Marine Sanctuaries Program, 184–91, 229, 232, 293
National Ocean Conference, 205, 210, 229
National Ocean Council, 300–303
National Oceanic and Atmospheric Administration (NOAA):
- Agenda 21, 273
- aquaculture, 225
- climate change, 267
- Coastal Zone Management Act of 1972 (CZMA), 116, 173
- fisheries, 143
- fragmented programs/policies, 76
- hazards, coastal, 243
- integrated coastal management, 275
- leasing, 173
- National Dialogues, 8
- National Marine Sanctuaries Program, 185–87
- Nixon, Richard, 51
- non-point source pollution, 301
- Office of Ocean and Coastal Resource Management, 128
- pollution, water, 240, 301–2
- programs existing in the, 20–21
- Stratton Commission, 3, 64
- tourism, 238

National Ocean Industries Association, 90
National Ocean Partnership Program, 205
National Ocean Policy Commission, 205, 229
National Ocean Service, 20
National Park Service, 191
National Pollutant Discharge Elimination System (NPDES), 177
National Research Council, 7, 172, 175, 215, 219, 235, 283
National Science Foundation, 51
National Weather Service (NWS), 243
National Wildlife Refuge System, 191
Native Americans, 74
Natural Resources, Department of, 76
Natural Resources Defense Council, 62, 90, 146
Naval power, 36, 210–13
New England, 82, 84, 145–46, 218
New Federalism, 111, 201
New Zealand, 219
1960s, fisheries at the end of the, 45–46
1966 to 1969, ocean polices between, 51
1970s:
- Clean Water Act of 1972, 92–93
- enactment, the politics of, 100–102
- Endangered Species Act of 1973 (ESA), 94–95
- fisheries, 77–83
- laws in the, characteristics of ocean, 95–100
- mammals, protecting marine, 68–77
- managing the nation's coasts, 60–68, 99–100
- Marine Research, Protection, and Sanctuaries Act of 1972, 93–94
- national and international context of early, 57–60
- oil and gas development, 84–91
- policy initiation/formation, 53–56

1970s to the 1990s:
- fisheries management, 129–49
- mammal protection, marine, 147–68
- National Estuarine Research Reserve Program, 191–98
- National Marine Sanctuaries Program, 184–91
- oil and gas development, 168–76
- political and policy context of the period, 103–16
- pollution, water, 176–84
- successes and shortcomings, 116–29, 198–202

1990s, the late:
- aquaculture, 224–28
- challenges to be faced in the new century, 205–8
- context of the, changed, 203–5
- fisheries and marine mammals, 216–23

management, issues in marine coastal, 235–47
oil and gas development, 228–32
policy changes in the new century, summary of major, 246–54
political landscape in, snapshot of, 308–25
regional approach to ocean resources and space utilization, 235
security, maritime, 210–13
transportation and port infrastructure, 213–16
wealth of the nation, overall ocean/coastal, 208–10

Nixon, Richard, 110
Coastal Zone Management Act of 1972 (CZMA), 63–65, 116
foreign affairs, focus on, 60
indifferent to ocean affairs, 101
Marine Mammal Protection Act of 1972 (MMPA), 75
National Oceanic and Atmospheric Administration (NOAA), 51
oil and gas development, 84

Nongovernmental organizations (NGOs), 272
Non-point-source pollution, 178–80, 206, 240, 241, 301
Norlin, Rich, 81
North American Wildlife Foundation, 73
North Carolina, 113, 129, 146, 193
North *vs.* South, 211
Northwind Undersea Institute, 167

Ocean Bowl, 205
Ocean Expo in 1998, 204
Ocean Governance Study Group (OGSG), 7, 283
Oceanic Society, 167
Ocean Our Future, 204
Ocean Principals, 291
Ocean thermal energy conversion (OTEC), 109, 111
Office of Management and Budget (OMB), 22, 110, 143
Office of Ocean and Coastal Resource Management (OCRM), 128–29
Office of the Commissioner of Fish and Fisheries, 42
Ohio, 125
Oil and gas development:
challenges to be met in the new century, 207
Coastal Zone Management Act of 1972 (CZMA), 117–18, 124–25
embargoes of 1973 and 1974, 84, 107–9
historical thread of national ocean policy, 38–40
mammal protection, marine, 151
moratoria, outer continental shelf, 228–29
1980s, 146
Oil Pollution Act of 1990, 104
policy challenges in the new century, 229–32, 249
political landscape of the late 1990s, 314–15
revenue sharing and coastal impact assistance, 231–32
spills, 12, 40, 58, 84, 93, 215–16, 324
stalled policies, 168–76
subgovernment networks, 49
successes and shortcomings in managing, 199
see also Outer Continental *listings under* Legislation

Old Woman Creek, 192
Olympic Coast, 187
O'Neill, Tip, 88
Optimum sustainable population (OSP), 71, 76, 152
Optimum yield (OY), 78–79, 144–45, 148, 217
Oregon, 62, 113, 116, 129, 146
Otters, 77, 150–51, 165, 166
Our Living Oceans, 145
Our Nation and the Sea, 1, 50, 61
Our Ocean Future: Themes and Issues Concerning the Nation's Stake in the Oceans, 204, 216
Outer continental shelf (OCS), 228–29
see also Outer Continental Shelf *listings under* Legislation

Overcapitalization, 220

Pacific Coast states, 293
Pacific islands region, 293
Pacific Marine Fisheries Commission in 1947, 43
Packard Foundation, 147
Padilla Bay, 193
Paints used on vessels, 215
Panama Declaration in 1995, 163
Pell, Sen., 59
Perlis, Steve, 81
Pew Foundation, 147
Pfiesteria piscicida, 180, 203, 302
Planning, coastal and/or land-use, 62–65, 87
Point sources of pollution, 177–78, 198
Polar bears, 167
Policy challenges in the new century:
  aquaculture, 206–7, 226–28
  Australia, 306
  Canada, 306–7
  Coastal Zone Management Act of 1972 (CZMA), 245–46, 252
  dredging, 214–15
  fisheries, 216–20, 247
  hazards, coastal, 244–45
  information, the need for more, 208–10
  Korea, 307–8
  Marine Mammal Protection Act of 1972 (MMPA), 221–23
  National Ocean Council, 300–303
  oil and gas development, 229–32
  operationalizing suggestions for improvement, 296–308
  overview of, 205–8
  partnerships with the coastal states and territories, 303–5
  pollution, water, 241–42
  populations, coastal, 205–6
  port capacity, 213–14
  problems with the existing system, basic, 282–83
  protected areas, marine, 233–35
  security, maritime, 212–13
  summary of major, 246–54, 324, 325
  sustainable development, 299–300
  tourism, 239–40
  200-mile fishery limit, 296–99
  window of opportunity for policy change, 324
  see also Governance, ocean; Management practices regarding ocean/coastal resources; Tensions in ocean policy
Policy initiation/formation, 53–56
Policy Makers' Summary of the Scientific Assessment of Climate Change, 264
Political landscape in the late 1990s, snapshot of, 308–25
Pollution, air, 264–65
Pollution, water:
  Clean Water Action Plan, 180
  Clean Water Act of 1972, 92–93, 180–81
  Clean Water Network, 240
  Coastal Zone Management Act of 1972 (CZMA), 240
  Cuyahoga River, 176
  estuaries, 61, 181–84
  exotic species introduction, 215
  Global Programme of Action, 274–75
  integrated coastal management, 301–2
  management of water quality, 19
  navy, U.S., 212–13
  non-point source pollution, 178–80, 206, 240, 241, 301
  point sources of pollution, 177–78, 198
  policy challenges in the new century, 241–42
  political landscape of the late 1990s, 311
  successes and shortcomings in controlling, 177–78, 198–99
  tourism, 238
Populations in coastal areas, 3, 205–6
Porpoises, 69, 167
Port facilities, 108, 206, 213–14, 216, 247
Portugal, 32
Potter, Frank, 74
Precautionary principle, 262–63
Presidency, weakened, 60

President's Council for Sustainable Development, 273, 290
President's Council on Environmental Quality, 177
Principles, governance and guiding, 287–89
Priorities, setting, 98–99
Private property rights, 321
Private *vs.* government role in resource development, 13, 202, 218–20
Problems with the existing system, basic:
complexity of the ocean, 281–82
decision making, narrowly based and adversarial, 280–81
institutional and policy-related problems, 282–83
jurisdictional split among levels of government, 279–80
sector-by-sector approach, 280
short-term thinking, 281
structural, 279
Programme of Action for the Protection of the Marine Environment from Land-Based Activities, 272
Property rights, 36, 269, 321
Protected areas, marine, 232–35, 249, 274–75
see also National Estuarine Research Reserve Program; National Marine Sanctuaries Program
Public Land Law Review Commission, 62
Puerto Rico, 197

Quarterman, Cynthia, 175

Ray, Carlton, 74
Reagan, Ronald:
deregulation, 103–4, 124
development, ocean resources, 109
Exclusive Economic Zone, 10, 11
inward, turning, 110–12
Law of the Sea Convention, 111, 114, 259
leadership in international ocean affairs, 256
National Marine Sanctuaries Program, 189
12-mile territorial sea, 115–16
200-mile fishery limit, 10, 11, 114–15, 296
Recreational activities and marine mammals, 151
see also Tourism
Reef Check Program, 275
Regional approach to ocean resources and space utilization, 217–18, 235, 293
Regional Fishery Management Councils, 301
Renewable energy projects, 108
Retreat from hazardous areas of the coastal zone, 244–45
Revolutionary War, 36
Rhode Island, 48, 62, 116, 172
Rookery Bay, 192
Russia, 262

Safe operations and Outer Continental Shelf Lands Act Amendments of 1978 (OCSLAA), 87
Salmon, 43
San Francisco, 47–48, 60–61
Santa Barbara oil spill, 12, 40, 58, 84, 93
Sapelo Island, 192
Sarbanes, Sen., 270
Sardines, 41
Science and policy, the joining of, 5, 25, 49–50
Science Council, 290
Scientific experiments conflicting with protection of mammals, 151–52
Scotland, 84
Scripps Institution of Oceanography, 151
Seabed Committee, 255
Sea denial missions, 211
Sea Grant Program, 21, 50, 51, 110, 171
Sea-level rise, 5
Sea lions, 151, 165

Seals, 42–43, 74, 76, 163, 165
Seaports, 205
Sea power *vs.* coastal power, 11
Seaweb, 147
Sector-by-sector approach to ocean resources management, 97–98, 280
Security, maritime, 210–13, 247
Sharing the Fish: Toward a National Policy on IFQs, 219
Shellfishing, 241–42
Shipbuilding, maritime, 320
Shoreline management, 312
Short-term thinking, 281
Shrimp, 82
Sierra Club, 62, 90, 167
Silent Spring (Carson), 38, 58
Single-purpose ocean laws, 5, 16, 97–98, 101–2, 105, 287
Snail darter, 108
Soares, Mario, 204
Society for Animal Protective Legislation, 59, 72, 101, 167
South Slough, 192
South *vs.* North, 211
Sovereign rights, 297, 298
Soviet Union, the former, 42, 44, 57–58, 210
Spain, 32, 262
Spatial use, management of ocean, 17–18, 22
Special districts, 16–17
Sputnik, 49–50, 107
Standard Oil of California/New Jersey, 49
Starkist, 162
State Department, U.S., 73, 82, 291
States' jurisdiction and ownership of offshore resources:
  Agenda 21, 271
  aquaculture, 225–28
  Coastal Zone Management Act of 1972 (CZMA), 65–67, 73–74, 112–14
  commissions, interstate, 43–44
  decision making powers decided by the courts, 112
  development, controlling rampant, 47–48
  fisheries, 41–44
  growing capacity and action, 112–14
  historical background, 20
  jurisdictional split among levels of government, 12, 201, 279–80, 304–5
  Magnuson Fishery Conservation and Management Act of 1976, 81
  Marine Mammal Protection Act of 1972 (MMPA), 201
  oil and gas development, 39
  Outer Continental Shelf Lands Act Amendments of 1978 (OCSLAA), 87
  reemerging participant, 23–24, 103–4
  Secretary of the Interior v. California in 1980, 112
  state waters, 21
  Supreme Court, 9
  tourism, 239
  Watt, James, 171–72
Stellwagen Bank, 187
Stephens, Ted, 59, 75, 81
Stewardship ethics, code of, 288
Storm frequency, 5
Straddling and highly migratory fish stocks, 216, 262–63
Strategic stock, 161
Stratton, Julius, 50
Stratton Commission:
  Coastal Zone Management Act of 1972 (CZMA), 61, 63
  National Oceanic and Atmospheric Administration (NOAA), 3, 64
  National Ocean Policy Commission, 205, 229
  Nixon/Ford/Carter responding to, 110
  Our Nation and the Sea, 1, 50
  themes emerging from, 50–51
Striking a Balance: Improving Stewardship of Marine Areas, 235
Studds, Congressman, 59, 81
Subgovernment networks, 48–49
Subsidies for impacted fishing communities, 218
Supreme Court, 9
  see also Court cases

Sustainable development, 4, 298–300
Sweden, 272, 273

Taiwan, 206, 223
Talbott, Lee, 74
Tellico Dam, 108
Tension leg platforms (TLPs), 230
Tensions in ocean policy:
  Coastal Zone Management Act of 1972 (CZMA), 67–68
  environment *vs.* development, 202
  federal *vs.* state jurisdiction, 12, 201
  government *vs.* private role in resource development, 13
  internationalism *vs.* unilateralism, 11–12
  sea power *vs.* coastal power, 11
  South *vs.* North, 211
  see also Conflicts among users of coastal resources/space
Territorial sea, 3, 21, 22
Texaco, 49
Texas, 125, 168, 205
Thailand, 206, 223
3-mile territorial sea, 33
Tides, red and brown, 203
Tolomato River, 195
Total allowable catch (TAC), 219
Tourism, 206, 236–40, 250, 317–18
Trade, global, 206
Transient resource/spatial use, 17
Transportation Department, U. S., 21
Transportation industry, marine, 35, 206, 213–14, 247, 316–17
Treasury Department, U.S., 82
Truman Proclamation of 1945, 11, 12, 33–34, 40
Tuna, 73, 76, 82, 150, 152, 161–63, 164
12-mile territorial sea, 3, 115–16
200-mile fishery limit:
  definition, 21
  designing a strengthened ocean governance system, 25
  future, looking to the future, 328
  introduction, book, 1, 3
  Law of the Sea Convention, 34, 109
  Magnuson Fishery Conservation and Management Act of 1976, 12, 77–83
  map of, 2
  Marine Board, 283
  operationalizing suggestions for improvement in ocean policy, 296–99
  Reagan, Ronald, 10, 11, 114–15, 296

Unilateralism *vs.* internationalism, 11–12, 201
Union Oil, 40
United Nations:
  anti-United Nations sentiment, 256
  Commission on Sustainable Development, 273–74
  Conference on Environment and Development in 1992, 3, 10, 57, 105, 233, 255, 262
  Division of Ocean Affairs and Law of the Sea, 272
  Environment Programme (UNEP), 57, 264, 272, 274–75
University of Rhode Island, 273
Urban areas and non-point source pollution, 178
U.S. international leadership:
  Agenda 21, 270–73
  challenges to be met in the new century, 207
  Commission on Sustainable Development, 273–74
  conservative stance taken, 255
  Convention on Biological Diversity, 267–70
  Framework Convention on Climate Change, 263–67
  Global Programme of Action, 274–75
  guiding force from 1950s to 1979, 255
  International Coral Reef Initiative, 275–76
  Law of the Sea Convention, 207, 255, 256, 259–61
  participation in major recent international agreements, 257–58
  regaining, suggestions for, 276–78
  straddling and highly migratory fish stocks, 262–63

Van Camp Seafood, 162
Vasco de Gama, 204
Vessel groundings, 215
Vienna Protocol of 1985, 255

Walsh, Bud, 81
Waquoit Bay, 194
Washington, 62, 116, 146
Water quality, 19
  see also Pollution, water
Water Wasteland, 177
Watt, James, 170–72, 230–31
Wealth of the nation, overall ocean/coastal, 208–10
Weeks Bay, 194
Wells Bay, 193
Wenk, Edward, Jr., 50
Western Oil and Gas Association, 90
Wetlands, 128, 311
Whale Center, 167
Whales, 43, 69, 163–64
White, Robert M., 116
Wildlife Management Institute, 73
Wildlife Society, 73
Window of opportunity for policy change, 324
Winyah Bay, 195
Wisconsin, 62, 116
World Bank, 6, 272
World Climate Conference, 264
World Fairs dedicated to the sea, 204
World Federation for the Protection of Animals, 73
World War II, 37–38
World Wildlife Fund, 73, 147, 167

Year of the Ocean (YOTO) programs, 204
Yellowstone National Park, 190

Zero discharge, 92–93
Zero mortality rate goal (ZMRG), 221, 222